Lecture Notes in Artificial Intelligence   4399
Edited by J. G. Carbonell and J. Siekmann

Subseries of Lecture Notes in Computer Science

Tim Kovacs   Xavier Llorà
Keiki Takadama   Pier Luca Lanzi
Wolfgang Stolzmann   Stewart W. Wilson (Eds.)

Learning
Classifier Systems

International Workshops, IWLCS 2003-2005
Revised Selected Papers

 Springer

Series Editors

Jaime G. Carbonell, Carnegie Mellon University, Pittsburgh, PA, USA
Jörg Siekmann, University of Saarland, Saarbrücken, Germany

Volume Editors

Tim Kovacs
University of Bristol, Bristol, BS8 1UB, UK
E-mail: kovacs@cs.bris.ac.uk

Xavier Llorà
University of Illinois at Urbana-Champaign, Urbana, IL, 61801, USA
E-mail: xllora@illigal.ge.uiuc.edu

Keiki Takadama
Tokyo Institute of Technology, Tokyo, 152-8550, Japan
E-mail: keiki@hc.uec.ac.jp

Pier Luca Lanzi
Politecnico di Milano, 20133 Milano, Italy
E-mail: lanzi@elet.polimi.it

Wolfgang Stolzmann
Daimler Chrysler AG, 71059 Sindelfingen, Germany
E-mail: wolfgang.stolzmann@daimlerchrysler.com

Stewart W. Wilson
Prediction Dynamics, Concord, MA 01742, USA
E-mail: wilson@prediction-dynamics.com

Library of Congress Control Number: 2007923322

CR Subject Classification (1998): I.2, F.4.1, F.1.1, H.2.8

LNCS Sublibrary: SL 7 – Artificial Intelligence

ISSN   0302-9743
ISBN-10   3-540-71230-5 Springer Berlin Heidelberg New York
ISBN-13   978-3-540-71230-5 Springer Berlin Heidelberg New York

Springer is a part of Springer Science+Business Media

springer.com

© Springer-Verlag Berlin Heidelberg 2007

Typesetting: Camera-ready by author, data conversion by Scientific Publishing Services, Chennai, India
Printed on acid-free paper   SPIN: 12029266   06/3142   5 4 3 2 1 0

Preface

The work embodied in this volume was presented across three consecutive editions of the International Workshop on Learning Classifier Systems that took place in Chicago (2003), Seattle (2004), and Washington (2005). The Genetic and Evolutionary Computation Conference, the main ACM SIGEvo conference, hosted these three editions. The topics presented in this volume summarize the wide spectrum of interests of the Learning Classifier Systems (LCS) community. The topics range from theoretical analysis of mechanisms to practical consideration for successful application of such techniques to everyday data-mining tasks.

When we started editing this volume, we faced the choice of organizing the contents in a purely chronological fashion or as a sequence of related topics that help walk the reader across the different areas. In the end we decided to organize the contents by area, breaking the time-line a little. This is not a simple endeavor as we can organize the material using multiple criteria. The taxonomy below is our humble effort to provide a coherent grouping. Needless to say, some works may fall in more than one category. The four areas are as follows:

Knowledge representation. These chapters elaborate on the knowledge representations used in LCS. Knowledge representation is a key issue in any learning system and has implications for what it is possible to learn and what mechanisms should be used. Four chapters analyze different knowledge representations and the LCS methods used to manipulate them.

Mechanisms. This is by far the largest area of research. Nine chapters relate theoretical and empirical explorations of the mechanisms that define LCS on the following subjects: (1) bloat control for variable-length representations, (2) classifier manipulation techniques: classifier ensembles and post processing (3) error guidance and the exploration/exploitation dilemma, (4) internal-model driven multistep LCS, (5) effects of class imbalance, (6) bounding convergence criteria for reinforcement-based LCS, and (7) techniques for dealing with missing data.

New directions. This group of chapters focuses on LCS applied to new and unconventional problems. Two chapters present work on the usage of LCS as learning models for system composition where they are used to create complex strategies based on properly assembling basic capabilities. Two other chapters explore seminal work on LCS as function approximators, exploring different architectures and methods to efficiently achieve this goal. Another chapter describes a new way of using LCS for determining relevant variables for the predictive process, instead of only focusing on classification performance. The last chapter of this group explores formal relations between LCS and ant colony optimization for the traveling salesman problem, illustrating how LCS can also be used to solve such a class of problems.

Application-oriented research and tools. The last group of chapters describes applied research, mostly oriented to data-mining applications. Two chapters explore and analyze how to improve the performance and accuracy of LCS for data-mining tasks. Two other chapters explore a more practical path that involves the creations of tools for (1) assisting the process of knowledge discovery and its visualization for medical data, and (2) creating computer-aided design tools that can help designers to identify and explore application areas where LCS methods can provide an efficient solution.

As mentioned earlier, this volume is based on the 6th, 7th, and 8th editions of the International Workshop on Learning Classifier Systems and would not have been possible without all the authors who contributed to it via the workshop.

Jaume Bacardit	University of Nottingham, UK
Flavio Baronti	Università di Pisa, Italy
Ester Bernadó-Mansilla	Universitat Ramon Llull, Spain
Warren B. Bilker	University of Pennsylvania, USA
Lashon Booker	Mitre Corporation, USA
Larry Bull	University of West England, UK
Martin V. Butz	University of Würzburg, Germany
Grzegorz Dabrowski	IMG Information Managements, Poland
Lawrence Davis	NuTech Solutions, USA
Yang Gao	Nanjing University, China
Josep Maria Garrell i Guiu	Universitat Ramon Llull, Spain
Pierre Gérard	Université de Paris Nord, France
David E. Goldberg	University of Illinois at Urbana-Champaign, USA
Da-qian Gu	The University of Hong Kong, China
Ali Hamzeh	Iran University of Science and Technology, Iran
John H. Holmes	University of Pennsylvania, USA
Joshua Zhexue Huang	The University of Hong Kong, China
Osamu Katai	Kyoto University, Japan
Samuel Landau	Université de Paris-Sud, France
Pier Luca Lanzi	Politecnico di Milano, Italy
Xavier Llorà	University of Illinois at Urbana-Champaign, USA
Javier G. Marín-Blazquez	Universidad de Murcia, Spain
Toby O'Hara	University of West England, UK

Albert Orriols	Universitat Ramon Llull, Spain
Ian Parmee	University of West England, UK
Alessandro Passaro	Università di Pisa, Italy
Sebastien Picault	Université de Lille-I, France
Adel Rahmani	Iran University of Science and Technology, Iran
Hongqiang Rong	The University of Hong Kong, China
Jennifer A. Sager	University of New Mexico, USA
Kumara Sastry	University of Illinois at Urbana-Champaign, USA
Sonia Schulenburg	Edinburgh University, UK
Katsunori Shimohara	Kyoto University, Japan
Olivier Sigaud	Université Pierre et Marie Curie, France
Antonina Starita	Università di Pisa, Italy
Keiki Takadama	Tokyo Institute of Technology, Japan
Olgierd Unold	Wroclaw University of Technology, Poland
Atsushi Wada	ATR Human Information Science Laboratories, Japan
Stewart W. Wilson	Prediction Dynamics, USA
David Wyatt	University of West England, UK

We would also like to acknowledge the reviewers for the workshops and this volume.

IWLCS 2003 and 2004

Lashon Booker	Mitre Corporation, USA
Larry Bull	University of the West of England, UK
Martin V. Butz	University of Würzburg, Germany
Pierre Gerard	Université de Paris, France
John H. Holmes	University of Pennsylvania, USA
Tim Kovacs	University of Bristol, UK
Rick Riolo	University of Michigan, USA
Sonia Schulenburg	Edinburgh University, UK
Oliver Siguad	Université Pierre et Marie Curie, France
Robert E. Smith	University of the West of England, UK
Keiki Takadama	ATR International Institute, Japan
Pier Luca Lanzi	Politecnico di Milano, Italy
Wolfgang Stolzmann	Daimler Chrysler, Germany
Stewart W. Wilson	Prediction Dynamics, USA

IWLCS 2005 and Additional Reviewing for this Volume

Jaume Bacardit	University of Nottingham, UK
Alwyn Barry	University of Bristol, UK

Ester Bernadó-Mansilla	Universitat Ramon Llull, Spain
Andrea Bonarini	Politecnico di Milano, Italy
Lashon Booker	Mitre Corporation, USA
Will Browne	University of Reading, UK
Larry Bull	University of West England, UK
Martin V. Butz	University of Würzburg, Germany
Rob Egginton	University of Bristol, UK
Pierre Gérard	Université de Paris Nord, France
John H. Holmes	University of Pennsylvania, USA
Tim Kovacs	University of Bristol, UK
Xavier Llorà	University of Illinois at Urbana-Champaign, USA
Luis Miramontes Hercog	Instituto Tecnológico y de Estudios Superiores de Monterrey, Mexico
Sonia Schulenburg	Edinburgh University, UK
Olivier Sigaud	Université Pierre et Marie Curie, France
Wolfgang Stolzmann	Daimler Chrysler, Germany
Christopher Stone	University of West England, UK
Atsushi Wada	ATR Human Information Science Laboratories, Japan
Stewart W. Wilson	Prediction Dynamics, USA

January 2007

Tim Kovacs\
Xavier Llorà\
Keiki Takadama\
Pier Luca Lanzi\
Wolfgang Stolzmann\
Stewart W. Wilson

Organization

IWLCS Organizing Committee from 1999 to 2004

Pier Luca Lanzi, Wolfgang Stolzmann, Stewart W. Wilson

IWLCS Organizing Committee from 2005 to 2006

Tim Kovacs, Xavier Llorà, Keiki Takadama

IWLCS Advisory Committee from 2005 to 2006

Pier Luca Lanzi, Wolfgang Stolzmann, Stewart W. Wilson

Tim Kovacs
University of Bristol,

Bristol, BS8 1UB, UK
kovacs@cs.bris.ac.uk

Keiki Takadama
Tokyo Institute of Technology,
Tokyo, 152-8550, Japan
keiki@hc.uec.ac.jp

Wolfgang Stolzmann
Daimler Chrysler AG,
71059 Sindelfingen, Germany
wolfgang.stolzmann
@daimlerchrysler.com

Xavier Llorà
University of Illinois at
  Urbana-Champaign,
Urbana, IL 61801, USA
xllora@illigal.ge.uiuc.edu

Pier Luca Lanzi
Politecnico di Milano,
Milano, 20133, Italy
lanzi@elet.polimi.it

Stewart W. Wilson
Prediction Dynamics,
Concord, MA 01742, USA
wilson@prediction-dynamics.com

Table of Contents

Part I Knowledge Representation

Analyzing Parameter Sensitivity and Classifier Representations for
Real-Valued XCS 1
  *Atsushi Wada, Keiki Takadama, Katsunori Shimohara, and
  Osamu Katai*

Use of Learning Classifier System for Inferring Natural Language Grammar 17
  Olgierd Unold and Grzegorz Dabrowski

Backpropagation in Accuracy-Based Neural Learning Classifier
Systems 25
  Toby O'Hara and Larry Bull

Binary Rule Encoding Schemes: A Study Using the Compact
Classifier System.. 40
  Xavier Llorà, Kumara Sastry, and David E. Goldberg

Part II Mechanisms

Bloat Control and Generalization Pressure Using the Minimum
Description Length Principle for a Pittsburgh Approach Learning
Classifier System.. 59
  Jaume Bacardit and Josep Maria Garrell

Post-processing Clustering to Decrease Variability in XCS Induced
Rulesets 80
  Flavio Baronti, Alessandro Passaro, and Antonina Starita

LCSE: Learning Classifier System Ensemble for Incremental Medical
Instances 93
  Yang Gao, Joshua Zhexue Huang, Hongqiang Rong, and Da-qian Gu

Effect of Pure Error-Based Fitness in XCS 104
  Martin V. Butz, David E. Goldberg, and Pier Luca Lanzi

A Fuzzy System to Control Exploration Rate in XCS 115
  Ali Hamzeh and Adel Rahmani

Counter Example for Q-Bucket-Brigade Under Prediction Problem 128
  Atsushi Wada, Keiki Takadama, and Katsunori Shimohara

An Experimental Comparison Between ATNoSFERES and ACS 144
  Samuel Landau, Olivier Sigaud, Sébastien Picault, and Pierre Gérard

The Class Imbalance Problem in UCS Classifier System: A Preliminary
Study 161
    Albert Orriols-Puig and Ester Bernadó-Mansilla

Three Methods for Covering Missing Input Data in XCS 181
    John H. Holmes, Jennifer A. Sager, and Warren B. Bilker

Part III   New Directions

A Hyper-Heuristic Framework with XCS: Learning to Create Novel
Problem-Solving Algorithms Constructed from Simpler Algorithmic
Ingredients 193
    Javier G. Marín-Blázquez and Sonia Schulenburg

Adaptive Value Function Approximations in Classifier Systems 219
    Lashon B. Booker

Three Architectures for Continuous Action 239
    Stewart W. Wilson

A Formal Relationship Between Ant Colony Optimizers and Classifier
Systems 258
    Lawrence Davis

Detection of Sentinel Predictor-Class Associations with XCS:
A Sensitivity Analysis 270
    John H. Holmes

Part IV   Application-Oriented Research and Tools

Data Mining in Learning Classifier Systems: Comparing XCS with
GAssist 282
    Jaume Bacardit and Martin V. Butz

Improving the Performance of a Pittsburgh Learning Classifier System
Using a Default Rule 291
    Jaume Bacardit, David E. Goldberg, and Martin V. Butz

Using XCS to Describe Continuous-Valued Problem Spaces 308
    David Wyatt, Larry Bull, and Ian Parmee

The EpiXCS Workbench: A Tool for Experimentation and
Visualization 333
    John H. Holmes and Jennifer A. Sager

Author Index 345

Analyzing Parameter Sensitivity and Classifier Representations for Real-Valued XCS

Atsushi Wada[1,2], Keiki Takadama[1,3],
Katsunori Shimohara[1,2], and Osamu Katai[2]

[1] ATR Human Information Science Laboratories,
Hikaridai"Keihanna Science City" Kyoto 619-0288 Japan
{wada,katsu}@atr.co.jp
[2] Kyoto University, Graduate School of Informatics,
Yoshida-Honmachi, Sakyo-ku, Kyoto 606-8501 Japan
katai@i.kyoto-u.ac.jp
[3] Tokyo Institute of Technology,
Interdisciplinary Graduate School of Science and Engineering,
4259 Nagatsuta-cho, Midori-ku, Yokohama, Kanagawa 226-8502 Japan
keiki@dis.titech.ac.jp

Abstract. To evaluate a real-valued XCS classifier system, we present a validation of Wilson's XCSR from two points of view. These are: (1) sensitivity of real-valued XCS specific parameters on performance and (2) the design of classifier representation with classifier operators such as mutation and covering. We also propose model with another classifier representation (LU-Model) to compare it with a model with the original XCSR classifier representation (CS-Model.) We did comprehensive experiments by applying a 6-dimensional real-valued multiplexor problem to both models. This revealed the following: (1) there are critical threshold on covering operation parameter (r_0), which must be considered in setting parameters to avoid serious decreases in performance; and (2) the LU-Model has an advantage in smaller classifier population size within the same performance level over the CS-Model, which reveals the superiority of alternative classifier representation for real-valued XCS.

1 Introduction

XCS [6] is a learning classifier system which has the potential to evolve accurate, maximally general classifiers to cover the state space for each action [3,7]. XCS takes *bit string* inputs, the same as traditional learning classifier systems [2] (LCS). To facilitate XCS and broaden the range of applicable problem representation while keeping its generalization abilities, XCSR [8] was proposed by Wilson to deal with *real-valued* problems, and he found that XCSR could learn appropriately on the real-valued 6-multiplexor problem.

Although Wilson analyzed the potential of XCSR, its validity was insufficient in two respects. Firstly, the parameter settings used for the experiment seemed to be set ad hoc, especially for the two newly introduced parameters m_0 and

X. Llorà et al. (Eds.): IWLCS 2003-2005, LNAI 4399, pp. 1–16, 2007.
© Springer-Verlag Berlin Heidelberg 2007

r_0 that were used in the real-valued classifier operations of *mutation* and *covering*. Secondly, the reason he adopted proposed classifier representation is not discussed, despite the possibility of other classifier representations.

Therefore, what we focus in this paper are (1) an analysis of the settings of real-valued XCS specific parameters to evaluate the model; and (2) an analysis of classifier representation with classifier operators such as covering and mutation. To achieve the latter, we propose an opponent model that presents another real-valued classifier representation that was inspired by Wilson's other model XCSI to deal with integer-valued input [9]. Although the requirement of extending XCS to integer-valued input is basically similar to that of extending XCS to real-valued input, XCSI adopts a different design concept over classifier representation. This concept can easily be applied to design another real-valued classifier representation, which we propose and adopt in the opponent model. For convenience, we have called this opponent the LU-Model and the original model the CS-Model, names which originate from the attributes used in each classifier condition that will be described later.

The rest of the paper is organized as follows. Section 2 describes both the CS and LU-Models by revealing the part extended from the XCS to achieve real-valued input. Section 3 describes the real-valued 6-multiplexor problem. Section 4 presents some simulation experiments that were done by applying both the CS and LU-Models to the real-valued 6-multiplexor problem. Section 5 has discussions based on the experimental results to validate real-valued XCS. Section 6 is the conclusion.

2   Extensions to XCS for Real-Valued Input

Both CS and LU-Models are based on XCS but differ in their classifier representation. This section presents the CS-Model, which adopts XCSR classifier representation and the LU-Model, where classifier representation is inspired by XCSI. It is done by describing their classifier representations in detail, which are the extended parts from XCS[1].

2.1   XCSR-Based Classifier Representation (CS-Model)

This section explains the CS-Model regarding its difference from XCS, which is equivalent to describing XCSR classifier representation with classifier operators such as covering, mutation, and crossover. To catch up with recent developments in XCS called the classifier *subsumption* mechanism, the "is-more-general" operator has been additionally defined which checks whether the classifier can subsume the other target classifier.

Representation of Classifier Conditions: The representation of the classifier in the CS-Model differs from the original XCSR in the condition part, which

---

[1] The implementation of the XCS part of the CS and LU-Models is based on Butz and Wilson [1].

replaces the bit string with a set of attributes named *interval predicates* by Wilson. The interval predicate is composed of two real values (c_i, s_i) where suffix i denotes the position in the condition part. Each interval predicate represents an interval $[c_i - s_i, c_i + s_i]$ on the real number line, and if the corresponding element of the input (which is a real-valued vector) is included in the interval, matching succeeds. If, and only if, all elements match the corresponding interval predicates in the classifier condition, can matching be considered a success. The domain of attributes c_i and s_i are both set between 0 and 1, which inherit the setting of XCSR in the CS-Model, but is not a necessary requirement for this representation.

Covering Operator: The covering operator creates a new classifier that matches a specified input. When a real-valued vector is denoted as $(x_1, ..., x_i, ..., x_n)$, where n is the dimension of input, each interval predicate of covered classifier condition $(c_1, s_1)...(c_i, s_i)...(c_n, s_n)$ is set as follows.

$$\begin{cases} c_i = x_i \\ s_i = rand(r_0). \end{cases} \quad (1)$$

Here, r_0 is a parameter used to decide the distribution range of the spread of the covering interval, where $rand(x)$ is a function that returns a random value distributed in the interval $0 \leq rand(x) \leq x$. The value of r_0 is set below 1 to maintain the s_i within its domain of $[0, 1]$ inherited from XCSR, but is not a necessary requirement for this operation.

Mutation Operator: The mutation operator mutates the classifier condition by adding delta values Δc_i and Δs_i to interval predicate variables c_i and s_i at the constant possibility of mutation parameter μ at each interval predicate. Each delta value for attributes c_i and s_i are calculated as follows.

$$\begin{cases} \Delta c_i = \pm rand(m_0) \\ \Delta s_i = \pm rand(m_0). \end{cases} \quad (2)$$

Here, m_0 is a parameter used to decide the distribution range of both Δc_i and Δs_i, where $\pm rand(x)$ is a function that returns a random value distributed in interval $0 \leq rand(x) \leq x$ with the sign chosen uniform randomly. If the mutated value exceeds the domain of $[0, 1]$, the value is adjusted to 0 or 1. The setting for this domain is inherited from XCSR, but is not a necessary requirement for this operation.

Crossover Operator: The crossover operator works the same as the crossover in XCS, except that the crossover point is not set between the condition bits but between the interval predicates.

Is-More-General Operator: The is-more-general operator judges whether a classifier condition is more general than another classifier condition. The basic idea of generality is the inclusion of the set of classifier condition's possible matching inputs. If the possible matching inputs of classifier condition X completely include and are larger than the possible matching inputs of classifier condition Y, X is more general than Y. This idea can be realized for real-valued classifier representation by comparing the inclusion of the interval on the real number line for each corresponding interval predicate. For two classifier conditions $X : (c_1, s_1)...(c_i, s_i)...(c_n, s_n)$ and $Y : (c'_1, s'_1)...(c'_i, s'_i)...(c'_n, s'_n)$, if $(c_i - s_i) \leq (c'_i - s'_i)$ and $(c'_i + s'_i) \leq (c_i + s_i)$ for all i except where all attributes are equal, X is more general than Y.

2.2   XCSI-Inspired Classifier Representation (LU-Model)

This subsection proposes the LU-Model with another real-valued classifier representation inspired by XCSI, which is an XCS extended model to deal with integer-valued inputs. XCSI adopts a different design concept over classifier representation, as it specifies the interval by using the value for the lower and upper bounds. This concept can easily be applied to designing real-valued classifier representation that differs from the CS-Model. The details are described below.

Representation of Classifier Condition: The representation of classifier condition in the LU-Model seems to be like that in the CS-Model as its interval predicate is composed of two real values (l_i, u_i), where suffix i denotes the position in the condition part. However, the denoting interval on the real number line differs from the CS-Model. The ith interval predicate simply denotes an interval $[l_i, u_i]$. If the corresponding element of input is included in the interval, matching between the element and the interval predicate succeeds. If, and only if, all elements match the corresponding interval predicates in the classifier condition, can matching be considered a success. The domain of attributes is restricted to $0 \leq l_i \leq u_i \leq 1$. This setting for domain inherits the concept of XCSI, but is not a necessary requirement for this representation.

Covering Operator: The covering operator creates a new classifier that matches a specified input. When a real-valued vector is denoted as $(x_1, ..., x_i, ..., x_n)$, where n is the dimension of the input, each interval predicate of the covered classifier condition $(l_1, u_1)...(l_i, u_i)...(l_n, u_n)$ is set as follows.

$$\begin{cases} l_i = x_i - rand(r_0) \\ u_i = x_i + rand(r_0). \end{cases} \tag{3}$$

Here, r_0 is a parameter used to decide the distribution range of the distance from input value x_i to l_i and u_i, where $rand(x)$ is a function that returns a random value distributed in the interval $0 \leq rand(x) \leq x$. If the covering value exceeds the domain of $0 \leq l_i \leq u_i \leq 1$, l_i and u_i are set to be kept within their domains as the follows: if l_i is smaller than 0, l_i is set to 0; and if u_i exceeds 1, u_i is set to 1.

Mutation Operator: The mutation operator mutates the classifier condition by adding delta values Δl_i and Δu_i to l_i and u_i at the constant possibility of mutation parameter μ at each interval predicate. Each delta value for attributes l_i and u_i is calculated as follows.

$$\begin{cases} \Delta l_i = \pm rand(m_0) \\ \Delta u_i = \pm rand(m_0). \end{cases} \tag{4}$$

Here, m_0 is a parameter used to decide the distribution range of both Δl_i and Δu_i, where $\pm rand(x)$ is a function that returns a random value distributed in the interval $0 \leq rand(x) \leq x$ with the sign chosen uniform randomly. If the mutated value exceeds the domain of $0 \leq l_i \leq u_i \leq 1$, l_i is set to 0 or u_i, and u_i is set to l_i or 1 depending on the following: (1) if the mutated l_i is smaller than 0, l_i is set to 0; (2) if the mutated l_i exceeds u_i, l_i is set to u_i; (3) if the mutated u_i exceeds 1, u_i is set to 1; and (4) if the mutated u_i is smaller than l_i, u_i is set to l_i.

Crossover Operator: The crossover operator works the same as the crossover in the CS-Model, except for the difference between the format of interval predicates, which is of no concern in this operation.

Is-More-General Operator: The is-more-general operator judges whether a classifier condition is more general than another classifier condition. This is achieved for LU-Model classifier representation as follows. For two classifier conditions $X : (l_1, u_1)...(l_i, u_i)...(l_n, u_n)$ and $Y : (l'_1, u'_1)...(l'_i, u'_i)...(l'_n, u'_n)$, if $l_i \leq l'_i$ and $u'_i \leq u_i$ for all i except for where all attributes are equal, X is more general than Y.

2.3 Real-Valued XCS Specific Parameters

While extending XCS to deal with real-valued inputs, new parameters m_0 and r_0 are introduced to XCS for both CS and LU-Models. Although the processes for how m_0 and r_0 are used in each model are different, roughly m_0 controls the upper limit for the random distribution of delta values used in the mutation operator, while r_0 is concerned with the distribution of the spread of covering condition intervals. Originally, the corresponding parameters in XCSR were labeled m and s_0. However, to avoid confusion caused by differences in discussing both models, we unified the names of these corresponding parameters to m_0 and r_0. The correspondence in the names of these parameters are in Table 1, where XCSR's parameters have been renamed m_0 and r_0 in the CS-Model to match the others.

Table 1. Correspondence of names of real-valued XCS specific parameters

XCSR	XCSI	CS-Model	LU-Model
m	m_0	m_0	m_0
s_0	r_0	r_0	r_0

3   Real-Valued 6-Multiplexor Problem

The real-valued 6-multiplexor problem is a sample problem presented by Wilson to validate XCSR, which is a real-valued version of the Boolean 6-multiplexor problem. We also employed this problem for two reasons. The first was that the simplicity of the problem made analysis of the experimental results easier while low computational costs allowed comprehensive experiments to analyze parameter dependence in the model that required a huge number of simulations. The second reason was that it would enable us to refer to Wilson's preceding experiment on XCSR, and further discuss the validity of XCSR, and the CS and LU-Models under the same conditions.

The Boolean 6-multiplexor function took a six-bit string as input and output a truth value of 1 or 0. The function was designed as output that would have a value of $(n + 2)$th bits where n was calculated by interpreting the two leftmost bits as a binary formatted number. For example, the first two bits of the input string "011010" were "01", which denotes the decimal 1 when interpreted as a binary formatted number, so the output value is the third bit of the string, in this case, 1. Alternatively, in disjunctive normal form, the Boolean 6-multiplexor function F_6 is given as follows where b_i stands for the i-th bit of the strings, the over-line negates the bit, and "+" takes a logical sum.

$$F_6 = \overline{b_0}\overline{b_1}b_2 + \overline{b_0}b_1b_3 + b_0\overline{b_1}b_4 + b_0b_1b_5. \tag{5}$$

To modify the Boolean 6-multiplexor problem to the real-valued 6-multiplexor problem, a parameter vector $\theta = (\theta_0, ..., \theta_5)$ is introduced. For each element in the real-valued input vector $x = (x_0, ..., x_5)$, x_i is converted to 0 if $x_i < \theta_i$; otherwise it is 1. In each learning step of simulation, a randomly chosen value from the domain $[0,1]$ is set to each element of vector x and given for input. If the returned output for x is correct, which has the same value as $F_6(x)$, reward r_{imm} is given, where this is a parameter denoting "immediate reward."

4   Experiment

Simulation experiments were done to validate real-valued XCS by applying the CS and LU-Models to the real-valued 6-multiplexor problem. Common conditions for all experiments can be described as follows[2]: N(max population size)= 800, β(learning rate)= 0.2, ϵ_0(error threshold to calculate classifier fitness)= 0.2, ν(power parameter to calculate classifier fitness)= 5, θ_{GA}(threshold to invoke GA)= 12, χ(possibility invoking crossover)= 0.8, μ(possibility invoking mutation) = 0.04. The threshold parameter vector $\theta_i(i = 1, ..., 6)$ for the real-valued 6-multiplexor problem was set as $(0.5, 0.5, 0.5, 0.5, 0.5, 0.5)$. In all simulations, the initial classifier population was set to empty. These settings to evaluate XCSR were the same as those in Wilson experiments.

The simulations for all experiments were evaluated by the average reward and the size of the classifier population. The classifier system was expected to

---

[2] Refer to Wilson [8] for meaning of these parameters.

acquire a population as small as possible to attain high average reward. For both values, the moving average of 50 previous iterations were calculated to check the temporal change. Here, iteration denotes the number of explored problems in the real-valued 6-multiplexor. Ten simulations were done on each case to obtain average variations.

4.1 Preliminary Experiment 1: Parameter Dependency on CS-Model

To examine the dependence of real-valued classifier representation specific parameters m_0 and r_0, we did a preliminary experiment by applying the CS-Model to the real-valued 6-multiplexor problem using four sample combinations of (m_0, r_0) pairs: (a) $(0.1, 1.0)$, (b) $(0.1, 0.5)$, (c) $(0.1, 0.25)$ and (d) $(0.5, 0.5)$. Here, the setting for (a) represents the same conditions as Wilson's experiment. The settings for (b) and (c) were selected to check the effect of change on r_0 parameter compared with (a). The setting for (d) was selected to check the effect of the m_0 parameter.

The results we obtained from the experiments are Fig. 1, which shows the relation between average reward on the vertical axis and iterations on the horizontal axis. Figure 2 shows the relation between population size on the vertical axis and iterations on the horizontal axis. Focusing on the r_0 parameter by comparing (a), (b) and (c), there is a significant difference between (c) and the others. For (c), there is no improvement in average reward as Fig. 1 shows, and the population size remains at a maximum limit of 800 throughout the simulation period in Fig. 2. This implies the existence of a threshold on parameter r_0, which is roughly between the 0.5 used in (b) and the 0.25 used in (c) causing a serious decrease in performance . Focusing on the m_0 parameter by comparing (a) and (d), there are no noticeable differences in average reward as shown in Fig. 1. In terms of population size, (d) converges smaller than (a) as shown in Fig. 2, where the difference is far smaller than that of the previous effect of r_0 between (a) and (c).

4.2 Preliminary Experiment 2: CS-Model vs. LU-Model

To evaluate the differences between the CS and LU-Models, we did another preliminary experiment by applying the LU-Model to the real-valued 6-multiplexor problem. We used 2 settings of (a) $(0.1, 1.0)$ and (c) $(0.1, 0.25)$ for parameters (m_0, r_0). (a) had the same setting as Wilson's original experiment, and (c) yielded a distinctive result in the previous experiment on the CS-Model.

The experimental results for the LU-Model compared with the results of the previous experiment on the CS-Model are in Fig. 3, which reveals the relation between average reward on the vertical axis and iterations on the horizontal axis. Figure 4 has the relation between population size on the vertical axis and iterations on the horizontal axis. By comparing the CS and LU-Models, for (a), there is a difference where the LU-Model records a higher average reward than the CS-Model throughout the period of simulation as Fig. 3 shows, while the population size converges to less than that in the CS-Model in Fig. 4. In the

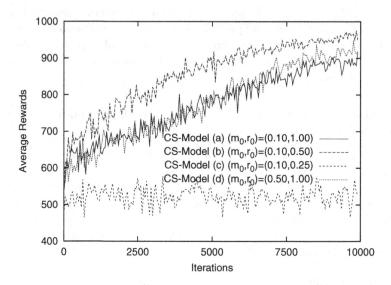

Fig. 1. Experimental results for (a) to (d) on CS-Model: relation between average reward and iterations

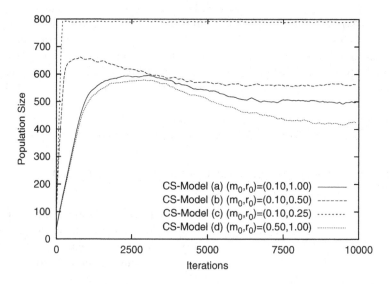

Fig. 2. Experimental results for (a) to (d) on CS-Model: relation between average performance and iterations

(c) of the LU-Model, a similar phenomenon with that of the CS-Model can be observed where there are no improvements in average reward in Fig. 3. and the population size remains at a maximum limit of 800 throughout the simulation

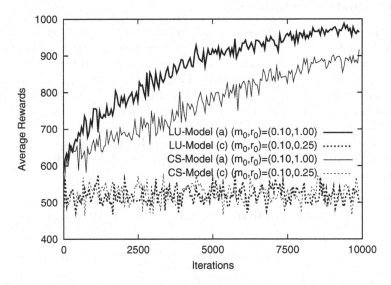

Fig. 3. Comparison of the experiments between CS and LU-Models for (a) and (c): relation between average performance and iterations

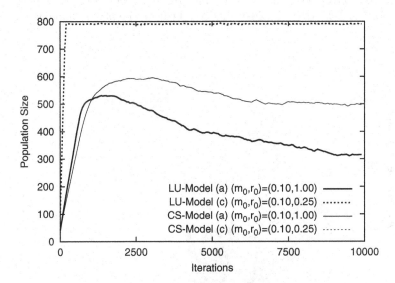

Fig. 4. Comparison of the experiments between CS and LU-Models for (a) and (c): relation between average performance and iterations

period in Fig. 4. From a comparison of both cases, there seems to be a similar performance dependence on the r_0 setting between the CS and LU-Models, which indicates that learning proceeds during a large r_0 value but fails with a small r_0 value.

4.3 Comprehensive Experiment

The results of the two preliminary experiments described in Section 4.1 and Section 4.2 indicate that there seems to be a significant dependence of both average reward and population size on the combination of parameters m_0 and r_0. To reveal the big picture on the landscape for the m_0-r_0 plane, we did a comprehensive experiment on both the CS and LU-Models. In the experiment, 400 combinations of parameters m_0 and r_0 were examined, which covered the m_0-r_0 plane with a series of grid points defined in the following matrix. The grid size was set to 0.05, so that the grid points included four sample combinations (a) $(0.1, 1.0)$, (b) $(0.1, 0.5)$, (c) $(0.1, 0.25)$ and (d) $(0.5, 0.5)$, which were used in the preliminary experiments.

$$
(m_0, r_0) = \begin{bmatrix}
(0.05, 0.05) & (0.05, 0.10) & \cdots & (0.05, 1.00) \\
(0.10, 0.05) & (0.10, 0.10) & \cdots & (0.10, 1.00) \\
\vdots & \vdots & \ddots & \vdots \\
(1.00, 0.05) & (0.10, 0.10) & \cdots & (1.00, 1.00)
\end{bmatrix}
\tag{6}
$$

The results are as follows. Figures 5, 6, 7 and 8 show the average reward for the CS-Model, its population size, and the average reward for the LU-Model, and its population size. In all figures, the x-axis denotes m_0, and the y-axis denotes r_0. Converged values at iterations of 10000 were used to describe the surface for the z-axis. Each grid points' height from the m_0-r_0 plane denotes the

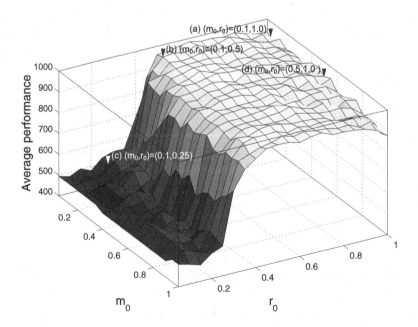

Fig. 5. Relation between average performance and time steps in CS-Model

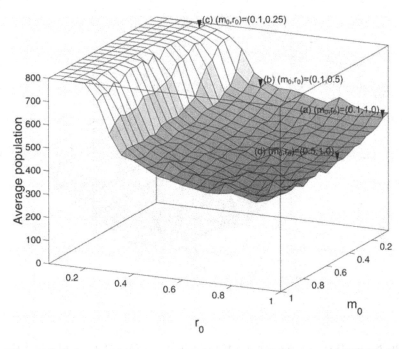

Fig. 6. Relation between average performance and time steps in CS-Model

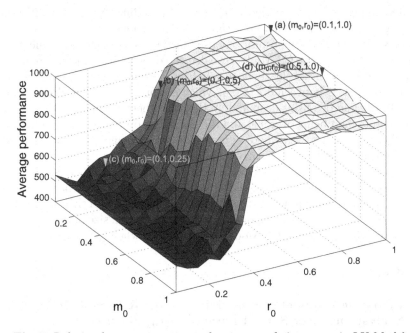

Fig. 7. Relation between average performance and time steps in LU-Model

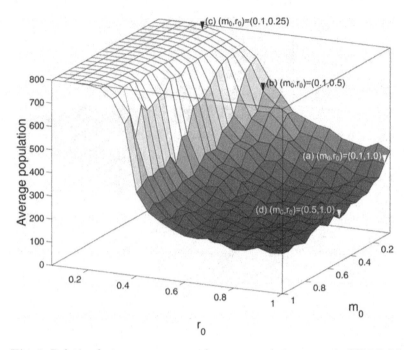

Fig. 8. Relation between average performance and time steps in LU-Model

corresponding values. The viewpoint of Figs. 6 and 8 differs from that of Figs. 5 and 7 presenting the whole surface without hidden regions. The labels (a) to (d) each denote the four sample conditions used in the preliminary experiments.

The simulations on the CS-Model revealed the dependence of performance on the m_0-r_0 plane, which can clearly be seen in Figs. 7 and 8. There is an r_0 threshold where the r_0 value is between 0.2 to 0.5, which is plotted as a sharp drop on each surface. In the middle of this drop and, in the area where r_0 is larger, the average reward is quite high being over 900 and the classifier population size converge towards 300. In the other area where r_0 is smaller, the average reward remains low at 500 and the population size nearly remains at the maximum limit of 800. From the simulations on the LU-Model described in Figs. 5 and 6, we can see that the dependence of performance on the m_0-r_0 plane is quite similar to that of the CS-Model, which can be explained by the similarity in the shapes of their landscapes. However, comparing their corresponding values, population size of the LU-Model at less than 200 is below the CS-Model's of above 300 when r_0 is large.

5   Discussion

5.1   Validating Parameter Sensitivity of XCSR (CS-Model): Analysis of Cover Spread Parameter(r_0) Sensitivity

To validate our evaluation of XCSR, we will discuss the sensitivity of model specific parameters for the CS-Model, especially the serious performance decrease caused by the small r_0 value. We first found this in the first preliminary experiment and verified it through the comprehensive experiment described in Section 4.3. It was not a special case on a specific (m_0, r_0) setting, but a general phenomenon that occurred when r_0 was smaller than a specific threshold that was roughly between 0.2 to 0.5.

As r_0 is a parameter used to decide the distribution range of intervals in covered classifier condition, an assumption concerning the covering process can be made, where the scenario can be described as follows. Consider the entire process of simulation where r_0 was set to be small. In the early stages of simulation, the covering would frequently occur as classifier population was initially set as empty. The population size would soon reach the maximum limit, because the classifiers created by covering could only cover small areas of the input state space as the size of their condition intervals were limited within r_0. Soon, input that was not covered by the present classifier population would arrive, then a new classifiers would be created to cover it while one of the existing classifiers would be deleted. This cycle would be repeated until the state space was covered through simulation. During this period, classifiers would be replaced one after another before they had become experienced, and learning could not be attained.

This assumption explains results such as (c), where no improvements in the average reward can be seen (Fig. 1) and the population size remains at a maximum limit of 800 throughout the simulation period (Fig. 2.) We can check this assumption by calculating the rate the area is covered by the N classifiers over the state space, where N is the maximum limit for population size. In the real-valued 6-multiplexor problem, the area of the state space is 1.0^6. Here, if we assume that all the intervals in N classifier conditions takes a maximum value of r_0, the total area of the covered space would be $N \times r_0^6$ and the coverage rate would be $(N \times r_0^6)/1.0^6$ where r_0 must at least be larger to make $(N \times r_0^6)/1.0^6$ larger than 1.0 to cover the entire state space. In practice, there are overlapping areas between each classifier condition, which require an extra r_0 value to cover the state space. This can be calculated by simple simulation that examines the rate of coverage of the state space by N covered classifiers. The process of simulation is described as follows.

1. Generate a 6-dimensional input vector by setting a random number within a range of [0,1] for each element and check if the randomly generated input vector is covered by any existing classifiers. Repeat until the uncovered input vector is found.
2. Create a classifier by covering operation for the input vector generated in 1 and insert it into the classifier population.
3. Repeat 1 and 2 until the size of the classifier population reaches N.
4. Calculate the rate of coverage of the state space by generating 1000 sample random inputs. The coverage rate is estimated by dividing the number of covered inputs within the number of total sample inputs.

The results are in Fig. 9, indicating the relation between r_0 and the simulated value for state space coverage rate compared with the value of $(N \times r_0^6)/1.0^6$.

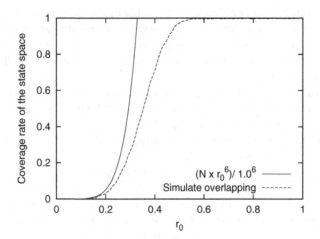

Fig. 9. Relation between coverage rate of the state space and r_0 (N=800)

The horizontal axis denotes r_0 and the vertical axis denotes the coverage rate of state space. When r_0 gets smaller than the threshold roughly between 0.2 and 0.5, the simulated coverage rate quickly decreases to converge to 0. As this curve resembles the curve of the relation between r_0 and average reward, the assumption seems to be valid.

Therefore, the evaluation done by Wilson where XCSR could learn appropriately on the real-valued 6-multiplexor problem should be limited within conditions where the parameter r_0 is set sufficiently large for the covered classifiers conditions to cover the entire input state space. These conditions should be maintained to avoid serious decreases in performance.

5.2 Validating Classifier Representation for Real-Valued XCS: Superiority of LU-Model to CS-Model on Classifier Population Size

Although the CS and LU-Models have a similar tendency towards parameter dependence, by focusing on an absolute value of performance, the LU-model performs well as was found during the second preliminary experiment in Section 4.2, and verified by the comprehensive experiment in Section 4.3 and this is what we will discuss here. The superiority of the LU-Model over the CS-Model is that it requires a smaller classifier population size, while achieving the same average reward where r_0 is large enough to learn, where the threshold is that discussed in the first discussion. This can be explained by analyzing how the difference in the classifier condition expression affects the difficulty of classifier subsumption. The classifier subsumption is a process that suppresses the classifier population size by letting a more general classifier to subsume the other classifier, where the definition of generality is described in Section 2.1 for the CS-Model and in Section 2.2 for the LU-Model.

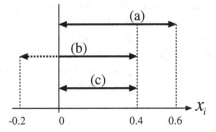

Fig. 10. Interval Graph to Describe Subsumption Difficulty in CS-Model

Intervals expressed by interval predicates (c_i, s_i) are allowed to take ranges which exceed [0,1] in the CS-Model, because the bounds of both c_i and s_i are between 0 and 1. For example, the interval predicate $(c_i, s_i) = (0.1, 0.3)$ denotes [-0.2,0.4] as indicated by (b) in Fig. 10, where c_i and s_i values are both within [0,1] but the denoting interval exceeds [0,1]. Here, the Fig. 10 has three intervals (a) to (c) on the real number line, where the horizontal axis denotes the real number line and the vertical axis is used to distinguish the three intervals. This excess of matching, covering and mutation operations causes no problems, as only the sub part within [0,1] is used for the classifier operations, which is [0,0.4] in this case, described as interval (c) in Fig. 10. However, there is a problem with subsumption. For example, although interval (a) in Fig. 10 denoting [0,0.6] is more general than interval (b) in the effective range, which is equal to (c), interval (a) cannot subsume interval (b). This could occur, in general, to an interval that exceeds the range of [0,1].

However, the LU-Model does not suffer from this problem, as the interval expressed by its interval predicate (l_i, u_i) is limited within the effective range of $0 \leq l_i \leq u_i \leq 1$. For this reason, we found that the LU-Model could successfully subsume not general classifiers which could be alternated by more general classifiers, and this resulted in a smaller classifier population size than in the CS-Model. This presents the possibility of an alternative classifier representation for real-valued XCS, which was adopted and validated in the LU-Model.

6   Conclusion

In this paper, we discussed a validation of XCSR in Wilson's experiment in two respects: (1) we analyzed the settings of real-valued XCS specific parameters to evaluate the model; and (2) we analyzed classifier representation with classifier operators such as covering and mutation. To achieve the latter, we proposed an opponent – the LU-Model and compared it with the original – the CS-Model. We conducted comprehensive experiments by applying the 6-dimensional real-valued multiplexor problem to both models, which revealed the following: (1) there is a critical threshold on covering operation parameter (r_0), which must be considered in setting parameters to avoid a serious decrease in performance; and (2) the LU-Model has an advantage with smaller classifier population size within the same rate of performance as the CS-Model, which demonstrated alternative classifier representation for real-valued XCS.

In future work, we intend to do an intensive analysis on GA operations to validate the discovery of a general and accurate real-valued classifier set. Other classes of problems should then be applied to make the discussion general, as all the results are based on the real-valued 6-multiplexor problem in this paper.

Acknowledgement

The research reported here was supported in part by a contract with the Telecommunications Advancement Organization of Japan entitled, 'Research on Human Communication'.

References

1. M. V. Butz and S. W. Wilson, "An algorithmic description of XCS," in Soft Computing, Vol. 6, pp. 144-153, 2002.
2. J. H. Holland, "Escaping brittleness: the possibilities of general-purpose learning algorithms applied to parallel rule-based systems," in Machine learning, an artificial intelligence approach, Volume II, 1986.
3. T. Kovacs, "XCS classifier system reliably evolves accurate, complete, and minimal representations for boolean functions," in Soft Computing in Engineering Design and Manufacturing, pp. 59-68, 1997.
4. C. Stone and L. Bull, "For Real! XCS with Continuous-Valued Inputs," in Evolutionary Computation, Vol. 11, No. 3, pp. 298-336, 2004. Evolutionary Computation 11(3): 298-336 (2003).
5. R. S. Sutton and A. G. Barto, Reinforcement Learning: an Introduction, MIT Press, Cambridge MA, 1998.
6. S. W. Wilson, "Classifier fitness based on accuracy," in Evolutionary Computation, Vol. 3, No. 2, pp. 149-175, 1995.
7. S. W. Wilson, "Generalization in the XCS classifier system," in Genetic Programming 1998: Proc. of the Third Annual Conference, pp. 665-674, 1998.
8. S. W. Wilson, "Get real! XCS with continuous-valued inputs," in Learning Classifier Systems: from Foundations to Applications, Lecture Notes in Artificial Intelligence (LNAI-1813). Berlin: Springer-Verlag, 2000.
9. S. W. Wilson, "Mining oblique data with XCS," in Lanzi, P. L., Stolzmann, W., and S. W. Wilson (Eds.), Advances in Learning Classifier Systems. Third International Workshop (IWLCS-2000), Lecture Notes in Artificial Intelligence (LNAI-1996). Berlin: Springer-Verlag, 2001.

Use of Learning Classifier System
for Inferring Natural Language Grammar

Olgierd Unold[1] and Grzegorz Dabrowski[2]

[1] Institute of Engineering Cybernetics, Wroclaw University of Technology
Wyb. Wyspianskiego 27, 50-370 Wroclaw, Poland
Phone: (+4871) 320 20 15, Fax: (+4871) 321 26 77
olgierd.unold@pwr.wroc.pl
http://www.ict.pwr.wroc.pl/~unold
[2] IMG Information Management Polska Sp. z o.o.
grzegorz.dabrowski@img.com

Abstract. This paper deals with the use of learning classifier system—LCS—for inferring a nontrivial natural language grammar. In a repeated analysis LCS infers the grammar of a given natural language from an exemplary set of correct and incorrect sentences. A genetic algorithm used periodically strengthens LCS's operation. A context-free grammar is used in the description of language structure.

1 Introduction

One among the key elements in the successful communication of humans with the computer is an exact grammar modelled upon a natural language. However, devising such a model is a difficult and expensive task, and essentially requires an input of expert knowledge into the system. Therefore, grammar search involving learning classifier system, and especially grammatical inference, seems a better solution [4]. Thanks to this approach, it is possible to infer a language structure model by employing a learning set of correct and incorrect sentences. Moreover, having made an adequate algorithm, we can relatively easily adjust the system to infer the grammar of another natural language by introducing changes only in the given learning sentences and certain system parameters. Furthermore, we obtain in this way a dynamic knowledge of a language, thanks to creating an expandable language model that can be adapted to our needs by adding a new learning set.

Still, we should bear in mind that inferring a correct grammar only from a set of positive sentences is impossible [4], and effective learning algorithms exist solely for regular languages [16]. Making algorithms that learn free-context grammars is one of the open and at the same time critical problems of grammatical inference [6]. Free-context grammar inference applies evolution methods among others [2, 3, 10, 12, 13, 14, 15, 17, 19, 21]. This paper is based on the assumption that the searched-for natural language model is described with context - free grammar, and that machine learning is based on genetically modified classifier systems. Section 2 contains a presentation of classifier systems. Section 3 introduces context - free grammars. Section 4 deals

X. Llorà et al. (Eds.): IWLCS 2003-2005, LNAI 4399, pp. 17–24, 2007.

with the architecture of an LCS inferring a natural language grammar. Section 5 shows a selection of experiment results. Section 6 contains conclusions upon the results.

2 Learning Classifier Systems

Classifier systems were created by Holland [7, 8]. According to Goldberg's definition [5], a classifier system learns syntactically simple rules (namely, classifiers) in order to coordinate its own operations in a given environment. Other elements, apart from an environment, are: receptors 'observing' their environment, effectors with which the system influences its environment, and the system itself.

The aforementioned syntactically simple rules are so called product rules (productions), *if <condition> then <action>*. A classic classifying system allows the usage of fixed length rules and simultaneous execution of rules. Thanks to the interaction with its environment, a system can evaluate the strength of particular rules. Rules compete among themselves for the right to match the external messages—posted by the environment—by entering the bid with their offers. Bid entrance fees become the prizes for senders of those messages which have been matched by rules. In this way, a chain connecting receptors and a group of effectors is created. The strength-bid adjustment mechanism depicted above guarantees that the 'good' classifiers, that is the rules which frequently enter the bid and achieve payoff, will survive, and the 'weak' ones will be eliminated. Additionally, a genetic algorithm is employed, which allows mutation and crossing, thanks to which new rules can be created.

It should be noted that apart from the classic classifier systems structure created by Holland, there are proposed their promising mutations, as, for instance, Wilson's XCS [20], Stolzmann's ACS [18], or Holmes's EpiCS [9].

3 Context-Free Grammar

The description of the language structure in the system is based upon context-free grammar, which is conditioned by its following properties:

-   context-free grammar describes well even the highly complex syntactical relationships of any chosen language, including natural languages;
-   machine systems based on context- free grammars usually operate faster than comparable applications employing other solutions;
-   context-free grammar formalism can be relatively easily (by comparison to ATN or systemic grammars) implemented by using any of the majority of popular programming languages.

Formally, context- free grammar is a quadruple:

$$G = (V_N, V_T, P, S) \tag{1}$$

where production rules P are expressed as: $A \rightarrow \alpha$ and $A \in V_N$ and $\alpha \in (V_T \cup V_N)^*$. S is an initial symbol, V_N is a set of nonterminals, and V_T a set of terminals. It is assumed that the searched for grammar is expressed as:

$$G = \{R_1, R_2, \ldots, R_n\} \tag{2}$$

where R_i is the rule expressed as:

$$R_i: N_i: p_{1i} \ldots p_{mi} \tag{3}$$

in which N is a nonterminal from the set V_N, and p is a production in Greibach normal form.

An additional modification which resulted from the analysis of initial experiments was to exclude the S symbol from the grammar, as its function might be taken over by any nonterminal symbol.

On the assumption that the symbol a stands for an adjective, b for an adverb, c for an article, d for an auxiliary verb, e for a full stop, f for a conjunction, g for a noun, h for a numeral, i for a preposition, j for a pronoun, k for a question mark, l for a verb, m for an exclamation mark, n for a comma, and # for any word
the sentence

What do you like doing in your spare time?

can be represented by the string

a d j l g i a a g k.

The example grammar accepted the question mentioned above can be expressed as follows:

$G = \{$
A: aB, lC
B: djl, djlC, djd
C: iD
D: aD, agk
$\}$

where A, B, C, D belong to the nonterminal set.

4 LCS Inferring a Natural Language Grammar

The structure of LCS is based on the so called Michigan approach, where classifying rules form a population [5]. The system generates new rules and tests the effectiveness of the existing ones. Apart from the genetic algorithm module generating new rules, the system contains the following elements: external environment messages input and system-produced messages output module, incoming, outcoming and internal messages list servicing module, rules (population) module upon which the genetic algorithm operates, and—at last—credit assignment module based on 'bucket brigade' algorithm. The messages posted by the environment are the data coming from the outside into the classifier system. These messages activate the classifying rules. The rules that get activated place their messages on the list. New messages may also activate other rules, or send certain messages to the list of output messages. These messages are coded into output messages and are transferred to the environment as the system's response. The system environment estimates the system's answers, and the estimating algorithm updates the estimate of the classifying rules' strength by either increasing or decreasing it if the answer is estimated as—respectively—correct or incorrect.

In the system mentioned, the population of grammars plays the role of system classifiers. Every classifier is of a form: *grammar_production: message*, where *grammar_production* is one of productions of the given grammar, and *message* is a message sent to the list if the respective classifier reacts. Execution of a rule takes place when a production can parse a part of a sentence or a whole sentence that is currently being analysed.

Every classifier has an assigned strength, whose task is to determine the respective classifier's / production's usability for the given grammar. In the beginning all classifiers have the same strength, but in the analysis process they receive positive or negative reward, which is the result of the actions they undertake.

Every sentence of the given natural language is analysed in turns by the whole population. The initial form of the grammars used by the system is randomly generated. The LCS has two aims. The first is checking the adjustment level of the grammars, that is the correctness of analyses of the given sentences in a natural language. The second is paying to the classifiers, or reducing the strength assigned to the classifiers, free-context grammar productions taking part in the analysis. The list of messages is the system memory storing the results of every classifier's action. The first item in the list is always a sentence sent by the system environment. The messages in the list are acted on by the classifiers in turns. These classifiers which are able to match the production to the particular sentence get the right to place their own message in the list. The list is compiled until none of the classifiers can execute its action, that is none of the production matches the sentence part being currently analysed. Every classifier placing its message in the list has to pay a conventional fee, being a part of its strength. The fee is then transferred to a classifier (or distributed among a greater number of classifiers) that placed in the message reinforcing the execution of the current rule. The classifier finishing a sentence analysis receives reward from the system environment. Moreover, all classifiers taking part in the full or partial analysis of a given sentence can be additionally paid a conventional number of strength points. The algorithm described concerns correct sentences. The only modification of the algorithm concerning incorrect sentences is the negative value of the reward.

Genetic algorithm takes the particular grammars as chromosomes, and the grammar productions as genes. Crossing and selection are carried out upon all the grammars (treated as production vectors), whereas mutation modifies (adds, deletes, replaces) singular symbols of the particular productions.

For illustration, we show the process of 2-point crossing of CFG grammars:
before crossing (the symbol $//$ denotes randomly chosen crossing points)

$G_1 = \{$
$A_1: a_{11}, a_{12}, // a_{13}, a_{14}$
$A_2: a_{21}, a_{22}, a_{23},$
$A_3: a_{31} // a_{32}, a_{33}, a_{34}, a_{35}, a_{36}$
$\}$

$G_2 = \{$
$B_1: b_{11}, b_{12}, // b_{13}$
$B_2: b_{21}, b_{22}, b_{23}, b_{24}$
$B_3: b_{31}, // b_{32}$
$\}$

and after exchanging the chromosomes part between the crossing points

$G_1' = \{$
$A_1': a_{11}, a_{12}, // b_{13}$
$A_2': b_{21}, b_{22}, b_{23}, b_{24}$
$A_3': b_{31} // a_{32}, a_{33}, a_{34}, a_{35}, a_{36}$
$\}$

$G_2' = \{$
$B_1': b_{11}, b_{12}, // a_{13}, a_{14}$
$B_2': a_{21}, a_{22}, a_{23},$
$B_3': a_{31} // b_{32}$
$\}$

The symbols a_{ij} and b_{ij} stand for rewriting rules, the symbols A_i and B_i represent nonterminals.

Genetic operators may be used after every full analysis cycle (that is after the analysis of the whole set of correct and incorrect sentences by all the grammars in the population). It is also possible to define the time interval between the genetic algorithm's operations (counted as full analysis cycles).

5 Experiments

Experiments were conducted in two stages. In the first, the system was adjusted with palindromes. In the second, the system operated on the target language grammar, that is English. The correct sentences were purposefully 'tampered with' in order to create incorrect sentences, in a way excluding the possibility of an accidental creation of a correct sentence. Moreover, in the incorrect exemplary set the syntactically wrong structures were placed at the end of most sentences, which aimed at sensitivising the system to the numerous cases of incorrect sentences only slightly differing from the correct ones [1].

About a hundred of correct and thirty incorrect sentences were used in the experiments. The average adjustment strength for all grammar classifiers / productions, and the difference of the number of the analysed correct and incorrect sentences were taken as the fitness function.

Figure 1 illustrates the results of one of the numerous experiments conducted upon English sentences. Figure 1 shows values of the analysed features averaged for all grammars in the given population. Graph A denotes the number of full parse paths performed in a single analysis cycle. Graph B denotes the number of correct sentences fully analysed by the grammars in a single analysis cycle. Graph C denotes the number of full paths of analysed incorrect sentences in a single analysis cycle. Graph D denotes the number of incorrect sentences fully analysed by the grammars in a single analysis cycle. The number of sentences analysed correctly by the evolved grammars reaches 90%. The values of certain parameters were as follows: 5000 generations, 14 terminals, 8 nonterminals, maximally 16 rules for one nonterminal, 8 symbols in a rule, fitness function of the type 'number of correct sentences analysed by the grammar minus number of incorrect sentences analysed', size of the population 30, 3-point crossing, crossing probability 90%, mutation probability 1%, genetic algorithm operation every 10 cycles, reward for a full analysis of a correct sentence 40 points, reward for a partial analysis 25 points, negative reward for full analysis of an incorrect sentence 20 points, negative reward for partial analysis 10 points, 102 correct and 30 incorrect sentences in the learning set.

An example of grammar found is given beneath. The strength of classifiers/productions is shown in curl brackets.

G = {
A: dF(315), lD(30), jA(492), iD(245), cF (872), bA(381), jF(92), lm (200), ebiA(200), gelgC(200), khfkehjB(200), hagB(200), fgiD(200), dk(290)
B: cD(1108), aD(891), b(200), fnjab(200), gamhlb(200), ijB(251), fc(200), lmhefC(200), chhC(200), j#(495), dA(858), idide(200), cdnaF(200)
C: lbB(1038), cA(354), jlnF(200), andg#f(200), keclfaA(200), cjkD(200), dbE(148), ad(200)
D: gD(123), e(339), k(221), eA(540), gnA(200), hhE(200), ae(495), c(200), ageD(200), jgci(200), fehaB(200)
E: h#m(200), kf(200), aF(307), biC(200), d(200), jcE(200), g(200), ad#D(200)
F: bB(459), aE(37), cce(200), dhimE(200), eB(646), fnbnaji(200), aB(736), inC(200), gd(200), fcdf(200), ejdaB(200), aA(477), fja(200), kF(15), jmD(200), jjbmiim(200)
}

Fig. 1. Average values of parameters for evolved grammars

6 Conclusions

The initial assumptions have been proven by the results of experiments. Firstly, the best individuals of the population were able to analyse even 90% of the correct sentences. Secondly, the generalizing properties of the LCSs have been proven. Moreover, it appeared that the genetic algorithm operates in a better way when time intervals between its every operation are longer than a single analysis cycle. This stems from the fact that during several analysis cycles performed on the same learning sets of sentences the differences in strength of the 'very good' (well adjusted) productions /

classifiers and the 'very weak' (badly adjusted) ones start to deepen. Such a situation strengthens the employed selection and mutation operators. Better results can be obtained by employing a greater population of grammars, than by increasing sizes of individuals, that is by increasing excessively the number of productions and their lengths.

References

1. Dabrowski, G.: Use of Learning Classifier System in Natural Language Processing, M.Sc.-thesis, Wroclaw University of Technology, (2001)
2. Dulewicz, G., Unold, O.: Evolving Natural Language Parser with Genetic Programming, In: Abraham, A., Koppen, M. (eds.) Advances in Soft Computing. Hybrid Information Systems, Physica-Verlag, Springer-Verlag Company, Germany, (2002) 361-377
3. Dupont, P.: Regular Grammatical Inference from Positive and Negative Samples by Genetic Search, Grammatical Inference and Application, Second International Colloquium ICG-94, Berlin, Springer, (1994) 236-245
4. Gold, E.: Language Identification in the Limit, Information Control, 10, (1967) 447-474
5. Goldberg, D.E.: Genetic Algorithms in Search, Optimization, and Machine Learning. Addison-Wesley, Reading, Massachusetts, (1989)
6. de la Higuera, C.: Current Trends in Grammatical Inference, In: Ferri, F. J. *at al.* (eds.) Advances in Pattern Recognition, Joint IAPR International Workshops SSPR+SPR'2000, LNCS 1876, Springer, (2000) 28-31
7. Holland, J.: Adaptation in Natural and Artificial Systems, MIT Press, (1975)
8. Holland, J.: Escaping brittleness: The possibilities of general-purpose learning algorithms applied to parallel rule-based systems, In: Michalski, R. S., Carbonell, J.G., Mitchell, T.M. (eds.) Machine Learning Vol.II, chapter 20, Morgan Kaufmann Publishers, Inc., (1986) 593-623.
9. Holmes, J.H.: Evolution-Assisted Discovery of Sentinel Features in Epidemiologic Surveillance, Ph.D.–thesis, Drexel University, (1996)
10. Huijsen, W.: Genetic Grammatical Inference: Induction of Pushdown Automata and Context-Free Grammars from Examples Using Genetic Algorithms, M.Sc.-thesis, Dept. of Computer Science, University of Twente, Enschede, The Netherlands, (1993)
11. Kammeyer, T.E., Belew, R.K.: Stochastic context-free grammar induction with a genetic algorithm using local search. Technical Report CS96-476, Cognitive Computer Science Research Group, Computer Science and Engineering Department, University of California at San Diego, (1996)
12. Korkmaz, E.E., Ucoluk, G.: Genetic Programming for Grammar Induction, Proc. of the Genetic and Evolutionary Conference GECCO-2001, San Francisco Ca, Morgan Kaufmann Publishers, (2001) 180.
13. Lankhorst, M.M.: A Genetic Algorithm for the Induction of Nondeterministic Pushdown Automata. Computing Science Reports CS-R 9502, Department of Computing Science, University of Groningen, (1995)
14. Losee, R.M.: Learning Syntactic Rules and Tags with Genetic Algorithms for Information Retrieval and Filtering: An Empirical Basis for Grammatical Rules, In: Information Processing & Management, (1995)
15. Lucas, S.: Context-Free Grammar Evolution, In: First International Conference on Evolutionary Computing, (1994) 130-135

16. Miclet, L., de la Higuera, C. (eds.): Grammatical Inference: Learning Syntax from Sentences, 3rd International Colloquium, ICGI-96, Montpellier, France, September 25-27, 1996, LNCS 1147, Springer, (1996)
17. Smith, T.C., Witten, I.H.: Learning Language Using Genetic Algorithms, In: Wermter, S., Rilo, E., Scheler, G. (eds.) Connectionist, Statistical, and Symbolic Approaches to Learning for Natural Language Processing, LNAI 1040, (1996)
18. Stolzmann, W.: An Introduction to Anticipatory Classifier Systems, In: Lanzi *at al.* (eds.) Learning Classifier Systems: From Foundation to Application, LNAI 1813, Springer-Verlag, Berlin, (2000) 175-194
19. Unold, O.: An Evolutionary Approach for the Design of Natural Language Parser, In: Suzuki, Y. *at al.* [eds.] Soft Computing in Industrial Applications, Springer Verlag, London, (2000) 293-297
20. Wilson, S.W.: Classifier Systems and the Animat Problem, Machine Learning 2, (1987) 199-228
21. Wyard, P.: Context Free Grammar Induction Using Genetic Algorithms, In: Belew, R.K. Booker, L.B. (eds.) Proceedings of the Fourth International Conference on Genetic Algorithms, San Diego, CA. Morgan Kaufmann, (1991) 514—518

Backpropagation in Accuracy-Based Neural Learning Classifier Systems

Toby O'Hara and Larry Bull

Faculty of Computing, Engineering & Mathematical Sciences,
University of the West of England, Bristol BS16 1QY, U.K.
{Toby.OHara, Larry.Bull}@uwe.ac.uk

Abstract. Learning Classifier Systems traditionally use a binary string rule representation with wildcards added to allow for generalizations over the problem encoding. We have presented a neural network-based representation to aid their use in complex problem domains. Here each rule's condition and action are represented by a small neural network, evolved through the actions of the genetic algorithm. In this paper we present results from the use of backpropagation in conjunction with the genetic algorithm within XCS. After describing the minor changes required to the standard production system functionality, performance is presented from using backpropagation in a number of ways within the system. Results from both continuous and discrete action tasks indicate that significant decreases in the time taken to reach optimal behaviour can be obtained from the incorporation of the local learning algorithm.

1 Introduction

Since their inception Learning Classifier Systems (LCS) [10] have been compared to neural networks, both conceptually [8] and functionally ([6],[7] and [21]). Previously, we have presented a way to incorporate the neural paradigm into the accuracy-based XCS [24], termed X-NCS [4]. Learning Classifier Systems traditionally incorporate a binary rule representation, augmented with 'wildcard' symbols to allow for generalizations. This representation can be limiting in complex domains (e.g., see [20] for early discussions). As a consequence, more sophisticated rule representations have been presented, including integers [26], real numbers [25], messy GAs [14], logical S-expressions [15], and those where the output is a function of the input, including numerical S-expressions [1] and fuzzy logic [22].

We have presented a neural network-based scheme where each rule's condition and action are represented by a neural network, typically using multi-layer perceptrons. The weights of each neural rule being concatenated together and evolved under the actions of the genetic algorithm (GA)[10]. The approach is closely related to the use of evolutionary computing techniques in general to produce neural networks (see [28] for an overview). In contrast to most of that work, an LCS-based approach is coevolutionary, the aim being to develop a number of (small) cooperative neural

X. Llorà et al. (Eds.): IWLCS 2003-2005, LNAI 4399, pp. 25–39, 2007.

networks to solve the given task, as opposed to the evolution of one (large) network. That is, our approach is potentially decompositional in an automatic way. Moriarty and Miikulainen's SANE [17] is most similar, however SANE coevolves individual neurons to form a large network rather than small networks of neurons as rules.

In this paper we investigate ways in which to include backpropagation (BP)[19] within X-NCS in order to increase performance, termed X-NCS(BP). Belew et al. [2] were the first to highlight the potential of combining the two search techniques in the evolution of artificial neural networks, suggesting that "local search performed by backpropagation and other gradient descent procedures is well complemented by the global sampling performed by the GA" (see also ([16], [18]). Using a version of the LCS for function approximation we examine the effects of varying the amount of BP undertaken and which rules to update. Versions of X-NCS(BP) are then investigated for single-step and multi-step tasks with discrete outputs.

2 X-NCS: A Neural LCS

2.1 Discrete Actions

The neural learning classifier system (NCS) used here is based on Wilson's accuracy-based XCS (see [5]) and the majority of its internal mechanisms are unchanged. Each traditional condition-action rule is replaced by a fully connected multi-layer perceptron (MLP). All rules have the same number of nodes in their hidden layers (simplest case [4]) and one more output node than there are possible actions. All weights are randomly initialized in the range [−1.0, 1.0] concatenated together in an arbitrary order and thereafter determined by the GA (and BP here). The system starts with an initial random population, containing the maximum number of classifiers specified in the particular experiment.

The production system cycles through the same input-match-action-update cycle as XCS. However, since all rules explicitly 'see' all inputs, unlike the traditional scheme whereby defined loci can exclude certain rules from certain matchsets, the extra output node is added. This is used to signify membership of a given matchset. After the presentation of an input, each neural network rule produces a value on each of its output nodes in the appropriate manner, i.e., fed forward through sigmoid transfer function nodes. If the extra 'not matchset member' node has the highest output value, the rule does not form part of the resulting matchset. In all other cases the rule forms part of the matchset, proposing the action corresponding to the output node with the highest activation. This matching procedure is repeated for all rules on each cycle.

Rule discovery operates in the same way as usual for XCS with real numbers [25]. Hence the mutation operator is altered to adjust gene values using a normal distribution; small changes in weights are more likely than large changes upon satisfaction of the mutation probability (μ). The cover operator is altered such that when the matchset is empty, random neural networks are created until one gives its highest activation on an action node for the given input. Subsumption is not included.

2.2   Continuous-Valued Actions: Function Approximation

It is well-known that multi-layered perceptrons with an appropriate single hidden layer and a non-linear activation function are universal classifiers (e.g.,[10]). Until recently LCS had not been used to solve tasks of the form $y = f(x)$ since their traditional representation scheme does not lend itself to such classes of problem. Fuzzy Logic LCS (see [3] for an overview) represent, in principle, a production system-like scheme which can be used for such tasks but this remains unexplored. Ahluwalia and Bull [1] presented a simple form of LCS which used numerical S-expressions for feature extraction in classification tasks. Here each rule's condition was a binary string indicating whether or not a rule matched for a given feature and the actions were S-expressions which performed a function on the input feature value. Wilson [27] has presented a form of XCS, termed XCSF, which uses piecewise-linear approximation for such tasks; using only explore trials all matching rules update their parameters, where such trials are run consecutively as a training period.

We have presented a version of X-NCS for tasks of the form $y = f(x)$, where both x and y are real numbers between 0.0 and 1.0. However, unlike the above mentioned work, the system requires very few changes to the design of the standard XCS system:

2.2.1   Processing of Real Numbers
The real number inputs are scaled between 0.4 and 0.8 to accommodate the lack of discrimination of the upper and lower end of the sigmoid function, as is usual in the use of MLPs.

2.2.2   Changes to Error Threshold Processing and System Error
In standard XCS, the error threshold ε_0 is a fixed fraction of the payment range. However with function approximation across a continuous value range, a fixed value may result in very inaccurate classifiers at the bottom end of the input range. It was therefore decided that ε_0 should be variable to enable the accuracy, and hence fitness, of the classifiers across the range to be equivalent. The variable value was chosen as the percentage of the target value at any particular point. The percentage chosen was 1% so, for example under x-squared, if the input was 0.3 the target output value $f(x)$ would be 0.09. Here the required accuracy ε_0 would be 0.0009 and so classifiers that predicted within the range 0.0891 to 0.0909 would be given an accuracy of 1.0. In the same way, when the performance of the system is measured, the system error was calculated by taking the absolute difference between the target value and the prediction of the selected classifier, and dividing this by the target value, i.e., the system error is the percentage error between the target and the prediction value.

2.2.3   Matchset and Action Set Processing
The rule prediction value is taken from one output node of the individual's neural network. The selection of the match set is such that the 'not matchset member' output node merely has to have a positive value, rather than a value less than the primary output node, to indicate a match. In exploration all members of the match set are updated and rule discovery invoked if appropriate as per standard XCS. In XCS, under exploitation, all classifiers that advocate the same action are put into the same

set [A]. The chosen action set is the one that has the highest fitness weighted prediction. For function approximation we are looking for the rule whose prediction is most accurate, i.e., has the least error, and hence taking the classifier with the highest fitness weighted prediction would be inappropriate. Instead, the counterpart of prediction for such tasks is chosen, i.e., rule error, and so we choose the rule with the lowest value of error divided by fitness. This is equivalent to having the chosen action set containing just one macro-classifier.

2.2.4 Rule Updating

The prediction value for each rule is taken as the value of the output of the neural network, i.e., the prediction value of the classifier can change at each iteration. By contrast, the error value of a rule is determined as per standard XCS. Accuracy is determined in the standard XCS way except, as mentioned above, the accuracy criterion is taken as a percentage of the current target value. Fitness is again calculated in the standard XCS way. Thus the output value for a particular rule will change for each different input value. For example, for problem n with input value 0.3 -> prediction 0.0891, but problem $n+1$ with input 0.4 -> prediction 0.160. However the error value for each accurate classifier, although it varies as the predictions can deviate from their respective targets, is a small value that oscillates according to ε_0.

We now examine the effects of incorporating backpropagation into X-NCS, starting with function approximation tasks.

3 Function Approximation

3.1 Backpropagation

In this paper we use standard BP with a learning rate of 0.2 which determines by how much we change the weights at each step and a momentum rate 0.5. If the learning rate is too small, the weights take a long time to converge, conversely, if it is too large, we may end up reverberating around the error surface. Use of a momentum term helps reduce oscillation between iterations, and allows a higher learning rate to be specified without the risk of non-convergence. The learning rate is comparable to the β parameter in XCS.

We have experimented with two training procedures: in the first, only the macroclassifier in the selected actionset, i.e., that providing the system output, is updated under exploit trials; and in the second, **all** rules in the match set have their weights updated using the target value for $f(x)$ under both explore and exploit trails. In both cases offspring are produced using the parent rules' current (BP adjusted) connection weights rather than their original weights; a Lamarckian learning scheme is utilized (e.g., [23]) rather than the Baldwin effect (e.g., [9]). Further, we have experimented with using one, two and three cycles of BP per system cycle.

3.2 Results for $y=x^2$

In this task training consists of (alternating) 20,000 explore trials and 20,000 exploit trials each presenting a random input in the range [0.0, 1.0] scaled as described above.

Figure 1 shows the performance of the accuracy-based neural classifier system on the x-squared function, averaged over ten runs, with a running average over the previous fifty exploit trials. The parameters used were: $N=3200$, $\beta=0.2$, $\phi=0.5$, $\mu=0.15$, $\alpha=0.1$, $\chi=0.8$, $\theta=10$, $\delta=0.1$, $F_I=1.0$, $\varepsilon_I=1\%$. Rules contained five hidden layer nodes.

X-NCS X Squared with backpropagation

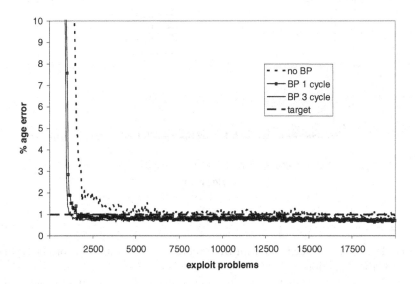

Fig. 1. X-NCS on the x-squared problem

From Figure 1 it can be seen that when not using BP, X-NCS requires around 15,000 problems to solve the task, i.e., for the accuracy of the approximations to fall within 1% of the real f(x). It also can be seen that after 6000 problems the accuracy comes very close to 1% and then gradually reduces to be less than 1%. Analysis of the resulting systems shows that better performance is achieved when one neural network emerges to cover the whole problem space, rather than through cooperative neural networks. The reasons for this appear two-fold: MLPs attempt to form global models by approximating between known data points; and the niche-based scheme of XCS encourages maximally general rules through increased chances to reproduce sets (see [4] for discussions). However, other work (unpublished) on a more difficult real-world problem indicated that, in such circumstances, the problem space is divided into a small number of regions covered by separate neural networks. Additionally, within these discrete regions very accurate classifiers appear, covering very small regions of the problem space. Similar results have been gained from using Radial Basis Function networks on the problems used here [4]. Figure 1 also shows the effects of using one to three cycles of BP, in this case reducing the learning time to less than 2000 problems. It can be seen that BP always reduces the time taken for the system to reach optimal performance but, that three cycles gives no greater benefit

X-NCS X Squared with backpropagation
updating all in match set

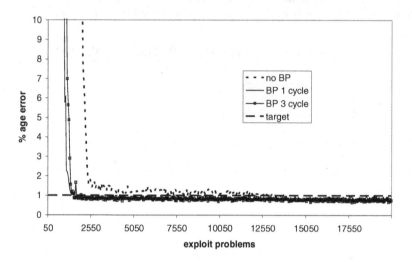

Fig. 2. X-NCS on the x-squared problem using second update scheme

over one. It can also be seen that with BP a more accurate solution is produced than the pre-specified criterion.

Figure 2 shows the results from the same experiments but using the second update scheme, that is, all of the matchset were updated per exploit trial, as happens on explore trials. The parameters are the same as for the previous scheme. It can be seen that worse (slower) performance is obtained than under the previous scheme, and that, again, no benefits are gained from more than one BP cycle. We suggest that the previous update scheme helps XCS separate the more accurate rules from the rest of the match set, by updating them more often.

3.3 Results for Root-Mean-Square

We have also examined the performance of the system on functions which contain more than one variable. Wilson [27] presented a general, multi-dimensional function of the form $y = [(x1^2 + \ldots + xn^2) / n]$ ½. We have used this "root mean squared" function with $n=6$, where training was identical to that of the x-squared task above. The parameters used were: $N=3200$, $\beta=0.2$, $\phi=0.5$, $\mu=0.07$, $\alpha=0.1$, $\chi=0.8$, $\theta=10$, $\delta=0.1$, $F_I=1.0$, $\varepsilon_I=1\%$. Rules contained five hidden layer nodes.

From Figure 3 it can be seen that using the neural representation on its own requires around 14,000 problems to solve the task to an accuracy of 1%. As with the x-squared function, the most accurate solutions came from those in which one classifier covered the whole input range. Again, three cycles of BP gives similar benefit as one. Results from using the second update scheme, updating all the classifiers in the matchset, again gave inferior performance to updating the best macro-classifier (not shown).

Therefore it would appear that the search process of the GA is greatly enhanced by the inclusion of the BP search heuristic within the LCS framework. That is, as discussed by Belew et al. [2], the BP allows the GA to sample the fitness landscape in a way where it need only be possible to reach good solutions via BP rather than consistently produce an exact, high-performance solution. We now investigate applying the approach to a commonly used single-step task with discrete outputs.

X-NCS 6 rms with backpropagation

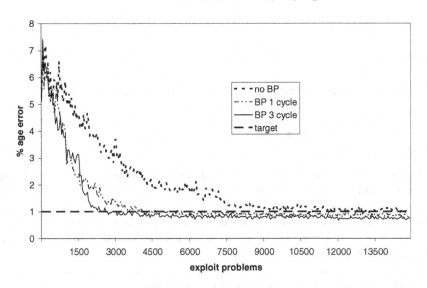

Fig. 3. X-NCS on the 6RMS problem

4 Multiplexer Task

We have tested the scheme on the 6-bit and 11-bit multiplexer problems. These Boolean functions are defined for binary strings of length $l = k + 2^k$ under which the first k bits index into the 2^k remaining bits, returning the indexed bit. A payoff of 1000 is given for a correct answer, otherwise 0.

Two learning schemes have been explored for such tasks. In these schemes we move away from the supervised learning scenario and each rule has three output nodes as described in Section 2.1. In the first (penalty) scheme, when a rule is in an action set proposing a different payoff level to the fittest rule, the value zero is used in a BP cycle(s) on its 'matchset member node' along with its own output values on the other two nodes. In the second (general) scheme, the output values of the fittest rule

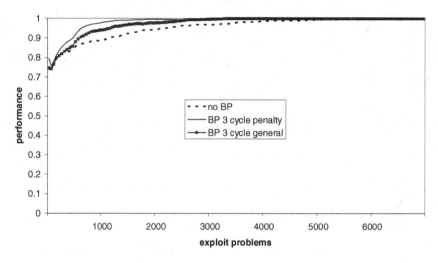

Fig. 4. X-NCS on the 6-bit Multiplexer

(ties broken randomly) are used by the BP as a target for all classifiers to be updated, regardless of a rule's predicted payoff.

Figure 4 shows the results of using X-NCS on the 6-bit version of the single-step problem, averaged over ten runs, with the following parameters: $N=1000$, $\mu=0.01$, $\beta=0.2$, $\phi=0.5$, $\alpha=0.1$, $\chi=0.8$, $\theta=10$, $\delta=0.1$, $pl=0$, $Fl=1.0$, $\varepsilon_1=0.0$. Rules contain five nodes in their hidden layer.

From Figure 4 it can be seen that using the neural representation alone requires around 5000 problems to solve the task. Analysis of the resulting rule-bases shows that, as well as the usual rules which match multiple inputs and propose a single action at a given payoff prediction level, multiple action rules emerge. That is, *for a given prediction level, accurate rules are evolved which suggest different actions depending on the input.* Figure 4 also shows results from the BP penalty training scheme under which rules predicting a different payoff level to the fittest rule are encouraged not to match in that action set in future by receiving a zero input on their match node. It can be seen that this type of BP is beneficial and we have found that three cycles (not shown) gives no greater benefit over two (shown). It can also been seen that using the more general BP update procedure gives some benefit, but not as great as for the penalty form of BP.

Figure 5 shows results for the same versions of the system on the 11-bit multiplexer using the same parameters as before, except: $N=3000$. Again, the penalty form of BP is most beneficial and we have found that no more than one cycle of BP provides any benefit (not shown).

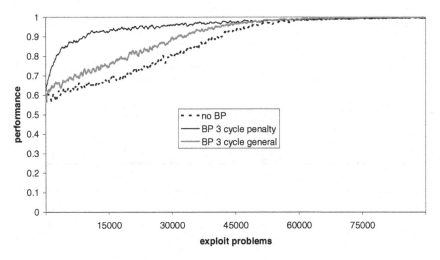

Fig. 5. X-NCS on the 11-bit Multiplexer using backpropagation

5   Multi-step Environments

5.1   Woods 2

Wilson [24] presented the multi-step, and hence delayed reward, maze task Woods 2 to test XCS. Woods 2 is a toroidal grid environment containing two types of food (encoded 110 and 111), two types of rock (encoded 010 and 011) in regularly spaced 3 by 3 cells, and free space (000) (see [ibid]) for full details). The learner is positioned randomly in one of the blank cells and can move into any one of the surrounding eight cells on each discrete time step, unless occupied by a rock. If it moves into a food cell the system receives a reward from the environment (1000) and the task is reset, i.e., food is replaced and the learner randomly relocated. On each time-step the learning system receives a sensory message, which describes the eight surrounding cells, ordered with the cell directly north and proceeding clockwise around it. Here, as in [24], the trial is repeated 10000 times, half explore and half exploit, and a record is kept of the moving average (over the previous 50 exploit trials) of how many steps it takes the system to move into a food cell on each trial. The parameters used for all the following were: N=3200, β=0.2, ϕ=0.5, μ=0.07, α=0.1, χ=0.8, θ=10, δ=0.1, F_I=1.0, ε_I=1%. Rules contained five hidden layer nodes.

Figure 6 shows how X-NCS without BP takes around 3000 trials to solve Woods 2. Analysis of the resulting rule-bases shows that neural rules emerge which have no error and produce different actions depending upon the input. Unlike in the multiplexer problem, we find that for some payoff levels these multi-action rules are more numerous than the equivalent single action rules. We presume that, if left to run

X-NCS Woods2 with with backpropagation

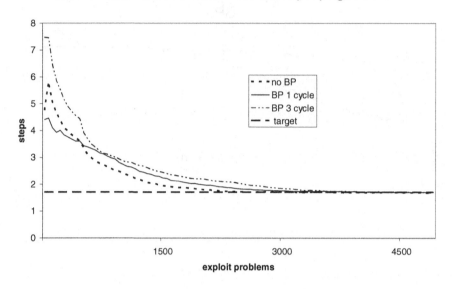

Fig. 6. X-NCS on Woods2

X-NCS Woods2 3 cycles of BP with fitness differentials of 10 15 25

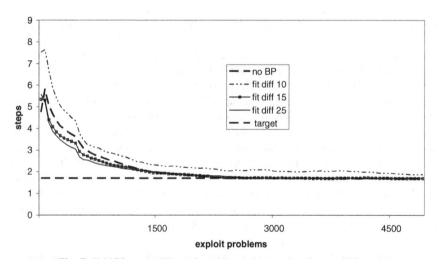

Fig. 7. X-NCS on the Woods2 problem with varying fitness differentials

for longer, the system would converge on a single neural rule for each payoff level; maximal generalizations would be produced in both the condition and action space.

In multi-step tasks the number of payoff levels is potentially large and rules' estimations of their expected payoff may take much longer to stabilize under the temporal difference chains. Hence, the penalty BP scheme described above would not appear to be appropriate in such tasks. Therefore we have only applied the general scheme here. Figure 6 shows how the scheme seems to have a disruptive effect, performing worse than the original system without BP. This was found to be true regardless of the number of BP cycles used. It was therefore decided to try to limit this disruption by only updating those classifiers that were very unfit compared to the fittest classifier. This was achieved by only updating those classifiers whose fitness was smaller than the fittest by a factor Z - the fitness differential. Here, if the fittest classifier had a fitness f, then only those classifiers with a fitness less than f/Z would be updated.

Figure 7 shows the results from using the general BP update procedure with fitness differentials of 10, 15 and 25, and with 3 cycles of BP, as this amount gave slightly better results in Figure 6. As can be seen, a differential of 10 produces a result worse than that of Woods2 without BP, and that with a differential of 15 and 25, the overall performance with BP is only marginally better than without BP, and even then only in the early stages of the trial.

5.2 Mazes 5 and 6

As mentioned above, results for Woods2 showed very little advantage to using BP. However, the Woods2 problem is relatively easy and thus any advantage from using BP might not be significant. To verify this, we tested the BP scheme on two well-known larger environments: Maze5 and Maze6 (Figure 8.a and Figure 8.b)[15].

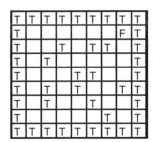 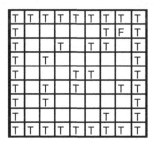

Fig. 8. (a) The Maze5 environment, and (b) the Maze6 environment

Maze6 is derived from Maze5 by moving one obstacle so that the final step to food is reduced from two squares to one. This both increases the optimal solution from 4.75 steps to 5.05 steps and also reduces the number of penultimate and ante-penultimate steps to food. Therefore it is significantly harder than Maze5.

The parameters used were: N=6500, β=0.2, ϕ=0.5, μ=0.07, α=0.1, χ=0.8, θ=10, δ=0.1, F_I=1.0, ε_I=1%. The BP parameter fitness differential Z value was set at 15.

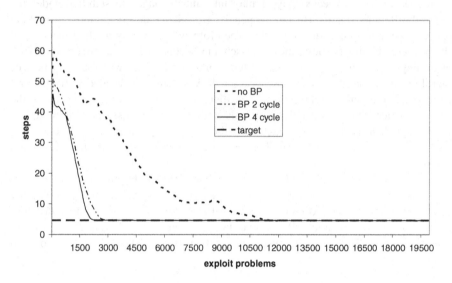

Fig. 9. X-NCS on the Maze5 problem with varying cycles of BP

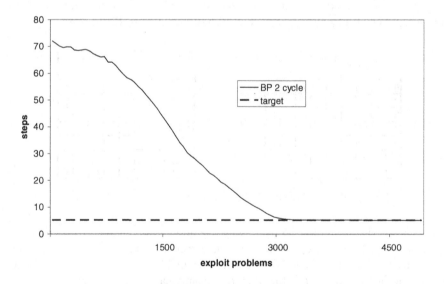

Fig. 10. X-NCS on the Maze6 problem with two cycles of BP

From Figure 9 it can be seen that using the neural representation alone requires around 125,000 problems to solve the task. Figure 9 also shows results from using the general BP update procedure with the fitness differential. As can be seen, a significant change in performance is obtained, although four cycles do not have much more advantage than two.

Figure 10 shows the performance of X-NCS(BP) on the harder Maze 6 task using two cycles of BP and the same parameters as for Maze 5. The original X-NCS has not been tried on this maze simply for time reasons; runs longer than 125,000 problems are expected. As can be seen, it takes approximately 3000 problems to solve the task. Runs with more cycles of BP (e.g., 4) give roughly the same performance (not shown). Therefore, again, the use of backpropagation proves very beneficial to X-NCS on a complex task.

6 Conclusions

In this paper we have presented results from using backpropagation in conjunction with a neural rule representation scheme within an accuracy-based learning classifier system. The effective combination of evolutionary computing and neural computing has long been an aim of machine learning (e.g., see [28] for discussions). It is our aim to exploit the coevolutionary and accuracy processes of XCS to realize such systems for reinforcement learning in complex domains. We have presented mechanisms by which to effectively exploit the local search of BP on top of the traditional GA-based search of LCS in a number of single-step and multi-step tasks.

This version of XCS is a more complex implementation, than standard XCS, and for function approximation at least, does not provide rules that are comprehensible, which is one of the key strengths of LCS. Set against this, is the ease which the embedded neural network can accept all types of input, i.e., binary, integer, real or ordinal in any combination, and importantly that neural networks can be universal approximators, which, in principle, means that they can represent any problem. Furthermore, that unlike the traditional use of neural networks, where the neural network has to cover the whole of the problem space, the XCS classifier itself determines what part of the problem space, or niche, it will cover without reference to external domain expertise. This we feel offsets the disadvantages of lack of comprehension in areas where the problem space is unknown, sparse and noisy.

We are currently examining the use of the hybrid system for more complex single-step tasks, and with discrete or continuous action spaces. In addition are planning to introduce recurrent connections, (a well established neural network architecture) to the evolving neural networks to add memory so that our method can be used in non-Markovian problems.

References

1. Ahluwalia, M. & Bull, L. (1999) A Genetic Programming Classifier System. In W. Banzhaf, J. Daida, A.E. Eiben, M.H. Garzon, V. Honavar, M. Jakiela & R.E. Smith (eds) *Proceedings of the Genetic and Evolutionary Computation Conference – GECCO-99.* Morgan Kaufmann, pp11-18.
2. Belew, R.K., McInerney, J. & Schraudolph, N.N. (1991) Evolving Networks: Using the Genetic Algorithm with Connectionist Learning. In C.G. Langton, C. Taylor, J.D. Farmer & S. Rasmussen (eds) Artificial Life II, Addison-Wesley, pp511-548.

3. Bonarini, A. (2000) An Introduction to Learning Fuzzy Classifier Systems. In P-L. Lanzi, W. Stolzmann & S.W. Wilson (eds) Learning Classifier Systems: From Foundations to Applications. Springer, pp83-106.

4. Bull, L. & O'Hara, T (2002) Accuracy-based Neuro and Neuro-Fuzzy Classifier Systems. In W. Banzhaf, J. Daida, A.E. Eiben, M.H. Garzon, V. Honavar, M. Jakiela & R.E. Smith (eds) *Proceedings of the Genetic and Evolutionary Computation Conference – GECCO-2003*. Morgan Kaufmann, pp11-18.

5. Butz, M. & Wilson, S.W. (2001) An Algorithmic Description of XCS. Advances in Learning Classifier Systems: *Proceedings of the Third International Conference – IWLCS2000*. Springer, pp253-272.

6. Davis, L. (1989) Mapping Neural Networks into Classifier Systems. In J.D. Schaffer (ed.) *Proceedings of the Third International Conference on Genetic Algorithms*, Morgan Kaufmann, pp375-378.

7. Dorigo, M. and Bersini, H. (1994). A comparison of Q-learning and classifier systems. In D. Cliff, P. Husbands, J.-A. Meyer, and S. W. Wilson (eds.), From Animals to Animats 3: *Proceedings of the Third International Conference on Simulation of Adaptive Behaviour* pp. 248-255.

8. Farmer, D. (1989) A Rosetta Stone for Connectionism. Physica D 42:153-187.

9. Hinton, G.E. and Nowlan, S.J. (1987, June). How learning can guide evolution. Complex Systems, pp. 495-502.

10. Holland, J.H.(1975) Adaptation in Natural and Artificial Systems. University of Michigan Press.

11. Holland, J.H. (1986) Escaping Brittleness. In R.S. Michalski, J.G. Carnoell & T.M. Mitchell (eds) Machine Learning: An Artificial Intelligence Approach 2. Morgan Kaufmann, pp48-78.

12. Hornick, K., Stinchombe, M. & White, H. (1989) Multiayer Feedforward Networks are Universal Approximators. Neural Networks 2(5): 359-366.

13. Lanzi, P-L (1997) A Model of the Environment to Avoid Local Learning. In Technical Report N. 97. Politecnico di Milano

14. Lanzi, P-L (1999a) Extending the Representation of Classifier Conditions Part I: From Binary to Messy Coding. In W. Banzhaf, J. Daida, A.E. Eiben, M.H. Garzon, V. Honavar, M. Jakiela & R.E. Smith (eds) *Proceedings of the Genetic and Evolutionary Computation Conference – GECCO-99*. Morgan Kaufmann, pp11-18.

15. Lanzi, P-L (1999b) Extending the Representation of Classifier Conditions Part II: From Messy Coding to S-Expressions. In W. Banzhaf, J. Daida, A.E. Eiben, M.H. Garzon, V. Honavar, M. Jakiela & R.E. Smith (eds) *Proceedings of the Genetic and Evolutionary Computation Conference – GECCO-99*. Morgan Kaufmann, pp11-18.

16. Lee, S.W. (1996) Off-line Recognition of Totally Unconstrained Handwritten Numerals using Multilayer Cluster Neural Networks. IEEE Transactions on Pattern Analysis and Machine Intelligence 18(6):648-652.

17. Moriarty, D.E. & Miikulainen, R. (1997) Forming Neural Networks Through Efficient and Adaptive Coevolution. Evolutionary Computation 5(2): 373-399.

18. Omatu, S. & Yoshioka, M. (1997) Self-tuning Neuro-IPD Control and Applications. In *Proceedings of the 1997 IEEE Conference on Man, Systems and Cybernetics*. IEEE Press, pp1985-1989.

19. Rumelhart, D.E. & McClelland, J.L. (1986) Explorations in Parallel Distributed Processing. MIT Press.

20. Schuurmans, D. & Schaeffer, J. (1989) Representational Difficulties with Classifier Systems. In J.D. Schaffer (ed.) *Proceedings of the Third International Conference on Genetic Algorithms*, Morgan Kaufmann, pp328-333.
21. Smith, R.E. & Cribbs, H.B. (1994) Is a Learning Classifier System a Type of Neural Network? Evolutionary Computation 2(1): 19-36.
22. Valenzuela-Rendon, M. (1991) The Fuzzy Classifier System: a Classifier System for Continuously Varying Variables. In L. Booker & R. Belew (eds) *Proceedings of the Fourth International Conference on Genetic Algorithms*. Morgan Kaufmann, pp346-353.
23. Whitley, D.W., Scott-Gordon, V. & Mathias, K. (1994) Lamarckian Evolution, the Baldwin Effect and Functional Optimisation. In H-P Schwefel & R. Manner (eds) Parallel Problem Solving from Nature – PPSNIII. Springer, pp6-15.
24. Wilson, S.W. (1995) Classifier fitness based on accuracy. Evolutionary Computation 3(2): 149-175.
25. Wilson, S.W. (2000) Get Real! XCS with Continuous-Valued Inputs. In P-L. Lanzi, W. Stolzmann & S.W. Wilson (eds) *Learning Classifier Systems: From Foundations to Applications*. Springer, pp209-222.
26. Wilson, S.W. (2001a) Mining oblique data with XCS In Lanzi, P. L., Stolzmann, W., and S. W. Wilson (Eds.), Advances in Learning Classifier Systems. *Third International Workshop (IWLCS-2000)*, Lecture Notes in Artificial Intelligence (LNAI-1996). Berlin: Springer-Verlag.
27. Wilson, S.W. (2001b) Function Approximation with a Classifier System. In L. Spector. E.D Goodman, A. Wu, W.B. Langdon, H-M. Voigt, M. Gen, S. Sen, M. Dorigo, S. Pezeshk, M. Garzon & E. Burke (eds*) Proceedings of the Genetic and Evolutionary Computation Conference – GECCO-2001*. Morgan Kaufmann, pp974-984.
28. Yao, X. (1999) Evolving Artificial Neural Networks. *Proceedings of the IEEE* 87(9): 1423-1447.

Binary Rule Encoding Schemes: A Study Using the Compact Classifier System

Xavier Llorà, Kumara Sastry, and David E. Goldberg

Illinois Genetic Algorithms Laboratory (IlliGAL),
National Center for Supercomputing Applications,
University of Illinois at Urbana-Champaign, Urbana, IL 61801
{xllora,kumara,deg}@illigal.ge.uiuc.edu

Abstract. Several binary rule encoding schemes have been proposed for Pittsburgh-style classifier systems. This paper focus on the analysis of how maximally general and accurate rules, regardless of the encoding, can be evolved in a such classifier systems. The theoretical analysis of maximally general and accurate rules using two different binary rule encoding schemes showed some theoretical results with clear implications to the scalability of any genetic-based machine learning system that uses the studied encoding schemes. Such results are clearly relevant since one of the binary representations studied is widely used on Pittsburgh-style classifier systems, and shows an exponential shrink of the useful rules available as the problem size increases . In order to be able to perform such analysis we use a simple barebones Pittsburgh classifier system— the compact classifier system (CCS)—based on estimation of distribution algorithms.

1 Introduction

The work of Wilson in 1995 [1] was the starter of a major shift on the way that fitness was computed on classifier systems of the so call Michigan approach. Accuracy became a central element in the process of computing the fitness of rules (or classifiers). With the inception of XCS, the evolved rules targeted became the ones that were maximally maximally general (cover a large number of examples) and accurate (good classification accuracy) . After a decade since the paper published by Wilson, such road has been shown to be a successfull one. However, few attempts have been made to do the same revision exercise on the Pittsburgh-style classifier systems.

This paper revisits Wilson's work and applies some of the his original ideas to Pittsburgh-style classifier systems. We start analyzing how maximally general and accurate rules can be obtained in a barebones Pittsburgh style classifier system. Usually, classifier systems are built around a given knowledge representation and encoding of the rules is not even considered as a variable in the design of most classifier systems. Little research has been done about the relevance of the rule encoding schemes used in Pittsburgh-style systems. Our work propose an alternative of how such rules may be evolved using a simple barebones Pitsburgh style classifier systems—the compact classifier system (CCS)—based on estimation of distribution algorithms.

X. Llorà et al. (Eds.): IWLCS 2003-2005, LNAI 4399, pp. 40–58, 2007.

Such work also analyzed two different binary rule encoding schemes. Surprisingly, one of the most commonly used representation [2,3] inherently posses a bias that challenge the scalability of any system that uses such encoding. In this representation, theory show how the area of meaningful rules shrinks exponentially, leading the learning mechanism into a nail-in-a-haystack situation. Such situation can be corrected using alternating binary encoding schemes, as we show with a simple alternative binary encoding mechanism.

The rest of this paper is structured as follows. Section 2 reviews the binary rule encoding proposed by De Jong & Spears [2], widely used on Pittsburgh-style systems. Section 3 presents how maximally general and accurate rules may be characterized regardless of the representation used. Given the elements presented in the previous sections, theory of exponential shrinking of the number of meaningful rules is presented in section 4 when the De Jong & Spears representation is used. Section 5 presents an alternative encoding and how such exponential shrinking behavior may be avoided. Section 6 presents a first empirical study of the how maximally general and accurate rules may be evolved using a Pittsburgh-style systems. Finally, a summary of the conclusions of the work presented in this paper is presented in section 7.

2   Binary Rule Encoding

Regardless of the Pittsburgh or Michigan approach taken, a wide variety of knowledge representations have been used to describe rules in the genetics-based machine learning community [4,5,6,2,3,7,8,9,10,11]. This paper focuses on the rule representation proposed by De Jong & Spears [2] and later adopted by Janikow [3] in their early works on Pittsburgh-style classifier systems. The main property of such representation is it simple mapping on binary string, when compared to the χ-ary mapping required by the initial Michigan one proposed by Holland [4] and later mainly followed by Goldberg [5] and Wilson [1].

The rule representation proposed by De Jong [2] is based on a finite set of attributes with a finite number of possible values, and a close world assumption. We illustrate the rule encoding representation with the help of a simple example. Let's assume that for a given learning problem objects are described by three different attributes: *color*, *shape*, and *size*. The available colors are **red**, **green**, **blue**, and *white*. Shape comes in two forms, **round** and **square**. Finally, **huge**, **large**, **medium**, and **small** are the possible sizes. Rules describing patterns of objects are encoded as follows

color				*shape*		*size*			
red	green	blue	white	round	square	huge	large	medium	small
1	1	1	1	0	1	0	1	1	0

The previous rule (1111|01|0110) represents all **square** and **large** or **medium** objects. A 0 represents excluding a value, whereas a 1 indicates that such value may be present on the described object. Such simple example shows one of the main differences between this representation and the usual one used in the Michigan approach. This representation holds internal disjunctions among attribute values. The previous rule using the χ-ary alphabet used by traditional Michigan approaches would require two rules {#11, #12}[1] to express the same concept.

The previous rule states that the target object may have any possible *color* by means of allowing all its possible values (red, green, blue, and white). Such approach represents the equivalent of the *don't care* symbol (#) used by traditional Michigan systems. Another important difference is that a rule may never be matched. An example of such situation is the following rule.

color				shape		size			
red	green	blue	white	round	square	huge	large	medium	small
1	1	1	1	0	0	0	1	1	0

The rule 1111|00|0110 describes objects that are not round nor square. Since the objects or the problem may only take two possible values (round nor square), such a rule would never be satisfied.

Finally, a concept can be expressed by the disjunction of several of the aforementioned rules. For instance, the rule set {1111|01|0100, 0010|10|0010} represents all the square and large objects, as well as all blue, round, and medium ones.

The initial proposal by De Jong [2] assumed that rules match positive examples of the concept to be learnt. Any example not matched by a given rule set is, therefore, a negative example of such concept—or close world assumption [12].

3   Maximally General and Accurate Rules

Wilson [1] introduced the concept of classifiers based on accuracy. Besides setting one of the current Michigan approach standards (XCS), Wilson's work emphasized the importance of taking into account the accuracy when evaluating the fitness of classifiers—or rules. Another important concept of such work was the definition of maximally general and accurate rules. Such rules are at the boundary of accuracy and generality–a more general rule will have lower accuracy than the maximally general and maximally accurate one. This section applies some of the Wilson's ideas to Pittsburgh style classifiers. Such an endeavor requires an initial revision of the knowledge representation used and how to introduce the concept of maximally general and accurate rules for a Pittsburgh-style classifier systems.

In order to promote maximally general and maximally accurate rules a la XCS [1], we need to compute the *accuracy* of a rule (α) and its *error* (ε). In a Pittsburgh-style classifier, the *accuracy* may be computed as the proportion of overall examples correctly classified, whereas the *error* is the proportion of incorrect classifications issued by the activation of the rule. For computation simplicity we assume $\varepsilon(r) = 1$ when all the predictions were accurate, and $\varepsilon(r) = 0$ when all were incorrectly issued. Let n_{t+} be the number of positive examples

---

[1] Please refer to Holland (1975) [4], Goldberg (1995) [6], or Wilson (1995) [1] for further details about the equivalent Michigan χ-ary encoding.

correctly classified, n_{t-} the number of negative examples correctly classified, n_m the number of times that a rule has been matched, and n_t the number of examples available. Using this values the *accuracy* and *error* of a rule r can be computed as:

$$\alpha(r) = \frac{n_{t+}(r) + n_{t-}(r)}{n_t} \tag{1}$$

$$\varepsilon(r) = \frac{n_{t+}}{n_m} \tag{2}$$

It is worth to note that the error (equation 2) only take into account the number of correct positive examples classified[2]. This is a byproduct of the close world assumption of this knowledge representation. Once the *accuracy* and *error* of a rule are known, the fitness can be computed as follows.

$$f(r) = \alpha(r) \cdot \varepsilon(r)^\gamma \tag{3}$$

Such fitness favors rules with a good classification accuracy and a low error, or maximally general and maximally accurate rules. Throughout the rest of this paper we assume $\gamma = 1$. Traditional Pittsburgh-style classifier systems mainly relied on some sort of fitness based only on the accuracy (equation 1) [2,3,10]. Such fitness guidance makes no differences between two rules with the same *accuracy* but different errors. Hence, no bias toward maximally general and accurate rules usually exist in the initial Pittsburgh-style systems.

4   Unmatchable Rules: A Representation Side Product

Figures 1, 2, and 3 show the fitness presented in equation 3 for each of the possible rules in three different multiplexer problem (MUX3, MUX6, and MUX11). Figure s 1, 2, and 3 also show another property of the encoding proposed by De Jong & Spears [2], the exponential growth in the number of unmatchable rules. If a rule, because of the close world assumption in a binary classification problem, does not match a positive example, then the example is classified a as negative one. Hence, in a binary problem with a 50% positive and 50% negative examples—such as any multiplexer problem,—an unmatchable rule presents an accuracy $\alpha(r) = 0.5$— since an unmatchable rule classifies all instances as negative due to the close world assumption—and an error $\varepsilon(r) = 1$—since it was never activated and no prediction was issued. Hence the fitness of an unmatched rule is $f(r) = 0.5$.

Comparing figures 1(b), 2(b), and 3(b), a central plateau of unmatched rules grows non linearly. Moreover, the size of such plateau—let $\Phi(\ell)$ be the number of unmatched rules—is theoretically computable. The following calculations assume the binary coding for a binary attribute problem such as the multiplexer problem discussed above. However, such measure is easily extend to any arbitrary χ-ary attribute problem.

---

[2] We also assume that if a rule is never matched, no error is made and, hence, $\varepsilon(r) = 1$.

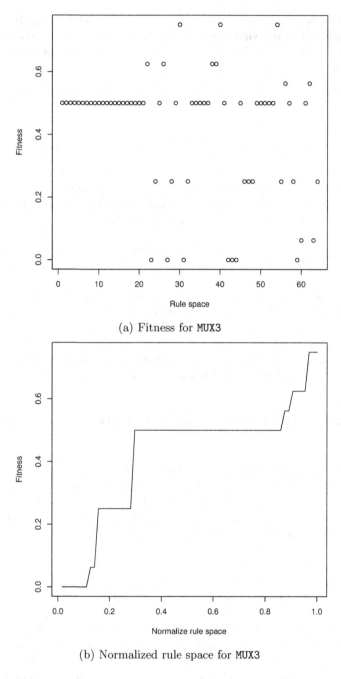

(a) Fitness for MUX3

(b) Normalized rule space for MUX3

Fig. 1. Given the fitness presented in equation 3, figures display the fitness of all possible rules for the 3-input multiplexer (MUX3) using De Jong & Spears representation. The normalized rules space is obtained by sorting all the possible rules according to their fitness.

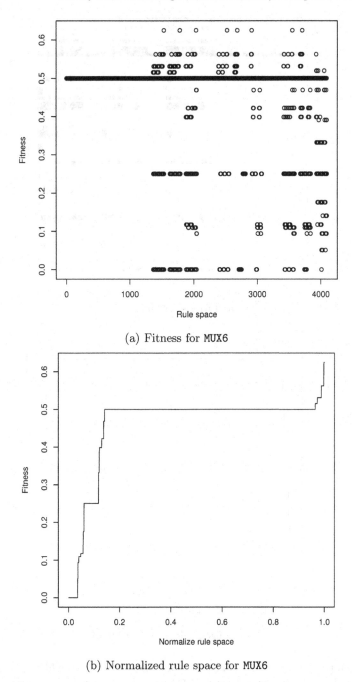

(a) Fitness for MUX6

(b) Normalized rule space for MUX6

Fig. 2. Given the fitness presented in equation 3, figures display the fitness of all possible rules for the 6-input multiplexer (MUX6) using De Jong & Spears representation. The normalized rules space is obtained by sorting all the possible rules according to their fitness.

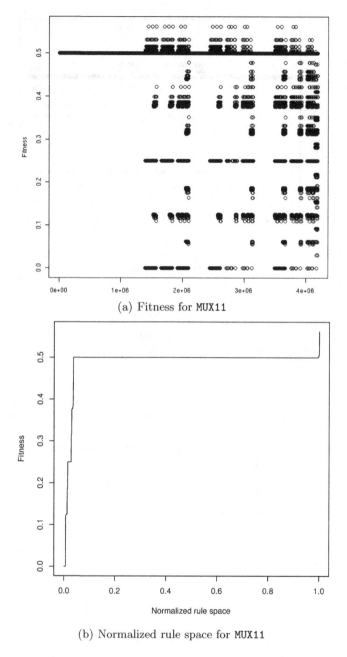

(a) Fitness for MUX11

(b) Normalized rule space for MUX11

Fig. 3. Given the fitness presented in equation 3, figures display the fitness of all possible rules for the 11-input multiplexer (MUX11) using De Jong & Spears representation. The normalized rules space is obtained by sorting all the possible rules according to their fitness.

The total number of possible binary-encoded rules Σ given a given length ℓ is,

$$\Sigma(\ell) = 2^\ell \tag{4}$$

A rule is matchable if it guarantees that for each binary attribute, the two coding bits are not both 0 simultaneously. Thus, for any given binary attribute, four possible combinations are possible (00, 01, 10, and 11), and one (00) needs to be avoided to guarantee that the attribute is matchable. Since the total number of attributes of the rule is $\ell/2$, the number of matchable rules $\Psi(\ell)$—the ones that none of the attributes contain the 00 combination—is:

$$\Psi(\ell) = 3^{\frac{\ell}{2}} \tag{5}$$

Hence, the size of the plateau of unmatchable rules $\Phi(\ell)$, is computed as

$$\Phi(\ell) = \Sigma(\ell) - \Psi(\ell) = 2^\ell - 3^{\frac{\ell}{2}} \tag{6}$$

Figures 1(a), 2(a), and 3(a) display the fitness of all possible rules for three different multiplexer problems, 3-input (MUX3), 6-input (MUX6), 11-input (MUX11). The rule space presented by each figure can be normalized, as figures 1(b), 2(b), and 3(b) show. Such normalization is achieved by sorting the rules according to their fitnes $f(r)$. After the normalization, the grow of the plateau is obvious. Such growth needs to be compared to the growth of matchable rules. The ratio between unmatchable and matchable rules $\rho(\ell)$ shows how scallable such rule encoding is. The $\rho(\ell)$ ratio may be computed using equations 5 and 6 as

$$\rho(\ell) = \frac{\Phi(\ell)}{\Psi(\ell)} = \frac{2^\ell}{3^{\frac{\ell}{2}}} - 1 \tag{7}$$

Equation 7 computes the exact ratio among unmatchable and matchable rules. Since we are interest on how $\rho(\ell)$ grows, such ratio may be approximated as follows ,

$$\rho(\ell) \approx e^{c\ell} \tag{8}$$

where c is given by

$$c = \ln\left(\frac{2}{\sqrt{3}}\right) = 0.143 \tag{9}$$

Figure 4 shows $\rho(\ell)$ and its approximation by equation 8. Unfortunately, equation 8 also shows that $\rho(\ell)$ grows exponentially. That is, such a rule coding procedure produce an exponentially growing number of unmatchable rules, introducing a serious handicap to the scalability of any genetic-based machine learning approach using it.

5 Specificity-Based Rule Encoding: An Alternative

Butz, Pelikan, Llorà, and Goldberg [13,14] introduced an alternative binary encoding as the result of the work for *building-block* identification in XCS. Such

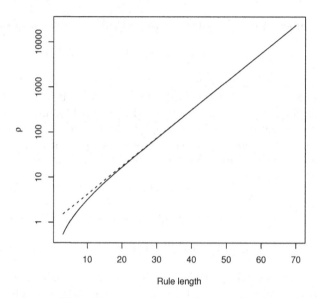

Fig. 4. Figure shows $\rho(\ell)$ and its approximation by equation 8 growing exponentially

endeavor requires transcoding each 3-ary $(0,1,\#)$ rule into a binary represented rule. The reason behind such transcoding is the need of the simplest ECGA and BOA [15] models used to deal with binary strings. Such encoding was proposed for binary-valued attributes $(0,1)$, however the same approach can be applied to χ-ary alphabets without lost of generality. A detailed discussion of this work can be found elsewhere [13,14].

For each attribute, two genes are available in Butz, Pelikan, Llorà, and Goldberg representation. The first gene selects if the condition is a general (equivalent of having a $\#$ in such position) or specific one. If the condition is marked as specific, then the value used is the one represented on the second binary-encoded gene value.

As introduced in the example presented in section 2, let's assume that for a given learning problem objects are described by three different attributes: *color*, *shape*, and *size*. The available colors are **red**, **green**, **blue**, and *white*. Shape comes in two forms, **round** and **square**. Finally, **huge**, **large**, **medium**, and **small** are the possible sizes. A rule describing all *round* and *large* objects will be encoded as (#01) in a Michigan χ-ary alphabet. The equivalent transcoded rule obtained using the procedure mentioned above is

color			shape		size		
specific	$value_0$	$value_1$	specific	$value_0$	specific	$value_0$	$value_1$
0	0	1	1	0	1	0	1

Hence, the equivalent binary transcoded rule for (#01) is $(001|10|101)$.

This encoding presents two interesting properties. The first one is that the relation general:specific is 1:1, removing any bias toward general or specific conditions. The second property is related to the expression of the second gene of

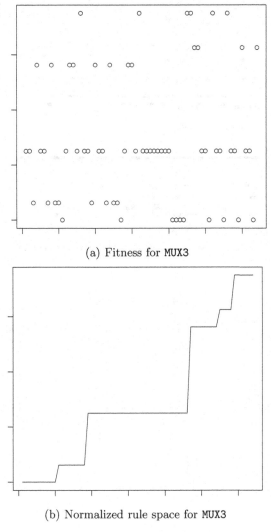

(a) Fitness for MUX3

(b) Normalized rule space for MUX3

Fig. 5. Given the fitness presented in equation 3, figures display the fitness of all possible rules for the 3-input multiplexer (MUX3) using Butz, Pelikan, Llorà & Goldberg representation. The normalized rules space is obtained by sorting all the possible rules according to their fitness.

an attribute and rule redundancy. If an attribute is marked as general then, at least χ binary rules represent the same rule before the transcoding process, and the second gene is not expressed.

Figures 5(a), 6(a), and 7(a) display the fitness of all possible rules using the encoding mentioned above for three different multiplexer problems, 3-input (MUX3), 6-input (MUX6), 11-input (MUX11). The rule space presented by each figure is also normalized, as figures 5(b), 6(b), and 7(b) show. Figures also show a plateau of equally evaluated rules.

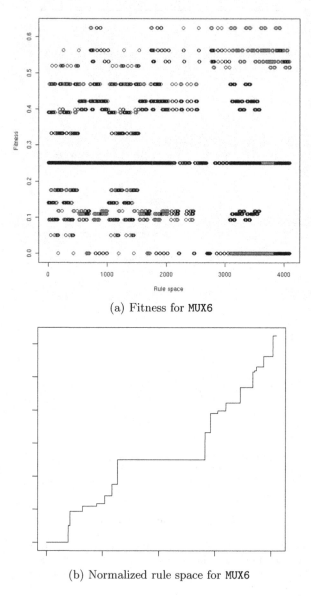

(a) Fitness for MUX6

(b) Normalized rule space for MUX6

Fig. 6. Given the fitness presented in equation 3, figures display the fitness of all possible rules for the 6-input multiplexer (MUX6) using Butz, Pelikan, Llorà & Goldberg representation. The normalized rules space is obtained by sorting all the possible rules according to their fitness.

A close inspection to these rules show a total different scenario than the one proposed by DeJong & Spears encoding. The plateau of equally evaluated rules represent those rules whose prediction is randomly left to the input values. For instance, in the MX3 28 redundant binary transcoded rules form such plateau,

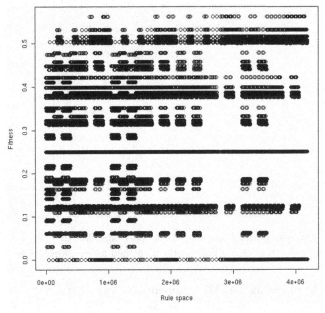

(a) Fitness for MUX11

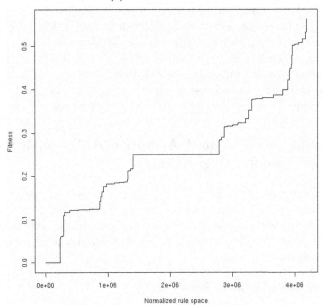

(b) Normalized rule space for MUX11

Fig. 7. Given the fitness presented in equation 3, figures display the fitness of all possible rules for the 11-input multiplexer (MUX11) using Butz, Pelikan, Llorà & Goldberg representation. The normalized rules space is obtained by sorting all the possible rules according to their fitness.

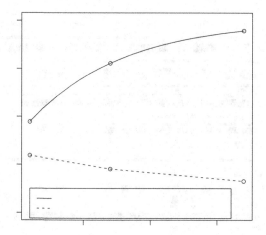

Fig. 8. Ratio between plateau size and rule space size for the two representations studied. Two different tendencies are clearly identified as the problem size grows.

representing the following nine Michigan-encoded rules {###, #01, #10, 0##, 0#0, 0#1, 1##, 10#, 11# }. Each of these rules show how the classification has a fifty-fifty change of being properly issued. Thus, these rules presents the same fitness $f(r) = 0.25$ ($\alpha(r) = 0.5$ and $\varepsilon(r) = 0.5$).

However, this plateau presents a total different behavior. Figure 8 shows how the exponential growth of unmatchable rules in De Jong & Spears representation reduce the space left to useful rules. On the other hand, the plateau of equally evaluated random guessing rules in Butz, Pelikan, Llorà & Goldberg shrinks as the problem size increase. Such shrinking behavior leave more room for interesting rules that may be explore along an evolutionary learning process.

6   Maximally General and Accurate Rules with the Compact Genetic Algorithm

In order to evolve maximally general and accurate rules, we use the compact gentic algorithm (cGA) [16]. It is important to mention here, that such algorithm does not provide any nitching capability and, hence, will only produce one maximally general and accurate rule. How to overcome such limitation is explained elsewhere [17].

6.1   The Compact Genetic Algorithm

The compact genetic algorithm [16], is one of the simplest estimation distribution algorithms (EDAs) [15,18]. Similar to other EDAs, cGA replaces traditional variation operators of genetic algorithms by building a probabilistic model of promising solutions and sampling the model to generate new candidate solutions. The probabilistic model used to represent the population is a vector of probabilities, and therefore implicitly assumes each gene (or variable) to be independent of the other. Specifically, each element in the vector represents the proportion of ones (and consequently zeros) in each gene position. The probability vectors are used to guide further search by generating new candidate solutions variable by variable according to the frequency values.

The compact genetic algorithm consists of the following steps:

1. *Initialization:* As in simple GAs, where the population is usually initialized with random individuals, in cGA we start with a probability vector where the probabilities are initially set to 0.5. However, other initialization procedures can also be used in a straightforward manner.

2. *Model sampling:* We generate two candidate solutions by sampling the probability vector. The model sampling procedure is equivalent to uniform crossover in simple GAs.

3. *Evaluation:* The fitness or the quality-measure of the individuals are computed.

4. *Selection:* Like traditional genetic algorithms, cGA is a selectionist scheme, because only the better individual is permitted to influence the subsequent generation of candidate solutions. The key idea is that a "survival-of-the-fittest" mechanism is used to *bias* the generation of new individuals. We usually use tournament selection [19] in cGA.

5. *Probabilistic model updation:* After selection, the proportion of winning alleles is increased by $1/n$. Note that only the probabilities of those genes that are different between the two competitors are updated. That is,

$$p_{x_i}^{t+1} = \begin{cases} p_{x_i}^t + 1/n & \text{If } x_{w,i} \neq x_{c,i} \text{ and } x_{w,i} = 1, \\ p_{x_i}^t - 1/n & \text{If } x_{w,i} \neq x_{c,i} \text{ and } x_{w,i} = 0, \\ p_{x_i}^t & \text{Otherwise.} \end{cases} \quad (10)$$

Where, $\mathbf{x}_{w,i}$ is the ith gene of the winning chromosome, $\mathbf{x}_{c,i}$ is the ith gene of the competing chromosome, and $p_{x_i}^t$ is the ith element of the probability vector—representing the proportion of ith gene being one—at generation t. This updating procedure of cGA is equivalent to the behavior of a GA with a population size of n and steady-state binary tournament selection.

6. Repeat steps 2–5 until one or more termination criteria are met.

The probabilistic model of cGA is similar to those used in population-based incremental learning (PBIL) [20,21] and the univariate marginal distribution algorithm (UMDA) [22,23]. However, unlike PBIL and UMDA, cGA can simulate a genetic algorithm with a given population size. That is, unlike the PBIL and UMDA, cGA modifies the probability vector so that there is direct correspondence between the population that is represented by the probability vector and the probability vector itself. Instead of shifting the vector components proportionally to the distance from either 0 or 1, each component of the vector is updated by shifting its value by the contribution of a single individual to the total frequency assuming a particular population size.

Additionally, cGA significantly reduces the memory requirements when compared to simple genetic algorithms and PBIL. While the simple GA needs to store n bits, cGA only needs to keep the proportion of ones, a finite set of n numbers that can be stored in $\log_2 n$ for each of the ℓ gene positions. With PBIL's update rule, an element of the probability vector can have any arbitrary precision, and the number of values that can be stored in an element of the vector is not finite.

Elsewhere, it has been shown that cGA is operationally equivalent to the order-one behavior of simple genetic algorithm with steady state selection and uniform crossover [16]. Therefore, the theory of simple genetic algorithms can be directly used in order to estimate the parameters and behavior of the cGA. For determining the parameter n that is used in the update rule, we can use an approximate form of the gambler's ruin population-sizing[3] model proposed by Harik, Cantú-Paz, Goldberg, & Miller [24]:

$$n = -log\alpha \cdot \frac{\sigma_{BB}}{d} \cdot 2^{k-1}\sqrt{\pi \cdot m}, \tag{11}$$

where k is the BB size, m is the number of BBs (note that the problem size $\ell = k \cdot m$), d is the size signal between the competing BBs, and σ_{BB} is the fitness variance of a building block, and α is the failure probability.

6.2   Model Perturbation

However, as mentioned earlier, cGA evolves only one rule at a time. Therefore we propose a modifies cGA to evolve different maximally accurate and maximally general rules not based on niching techniques. We note that the proposed approach is a viable alternative to niching mechanisms. Our approach is based on perturbating the initial probability vector with an uniform noise. Several runs of cGA using different initial perturbated initial probability vectors may lead to different accurate and maximally general rules. Instead of the initial cGA probability vector we used

$$p^0_{x_i} = 0.5 + \mathcal{U}(-0.4, 0.4) \tag{12}$$

Such a perturbation arises another question. Is there any relation among the initial perturbed model and the final rule obtained? Since our approach requires the evolution of rules based on cGA reruns, such relation among the initial perturbated model and the evolved rule is a key element. In order to analyze such a relation, we conducted a simple experiment based on the 3-input multiplexer (MUX3) problem. We randomly generate 1,000 perturbated models using the equation 12 and run cGA. The rule representation used was the on proposed by De Jong & Spears.

About 97% of the runs lead to one of the three accurate and maximally general rules, 10|01|11 (01#), 01|11|01 (1#1), and 11|01|01) (#11). Figure 9, presents pair plots of the initial perturbated probability vector of CGA that leaded to evolve a 01|11|01 (1#1) rule. Figure 9 shows all the initial probability vectors that lead to the rule 01|11|01 (1#1). The pair-wise of the probability vectors

---

[3] The experiments conducted in this paper used $n = 3\ell$.

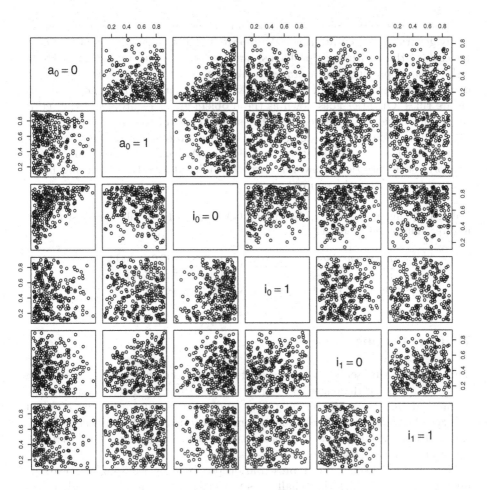

Fig. 9. Pair plots of the initial perturbation of CGA model. The pair-wise plots display the initial probability vectors that evolved the rule 01|11|01 (1#1). The pair-wise plots between ($a_0 = 1, i_0 = 0$) show a clear infeasible region. Such region arises from the rule learned in the multiplexer problem. If a 0 is set in a_0, then i_1 should not contain a 0 in order to evolve the rule 01|11|01 (1#1). This rule was evolved in 31.1% of the successful runs.

that lead to this rule show a clear infeasible region between bits ($a_0 = 1, i_0 = 0$). This region is directly connected to the 3-input multiplexer structure problem. If a 1 is set in a_0, then i_0 should not contain a 0 in order to evolve the rule 01|11|01 (1#1). Such behavior repeated for the other maximally general and accurate rules in MX3. The same behavior also repeated in the other two multiplexer problems MX6 and MX11. The same kind of unfeasible regions appear on the Butz, Pelikan, Llorà, & Goldberg representation. However, in that representation the unfeasibility regions tend to show up more often between the specificity bits of of the different multiplexer attributes.

Another interesting results of these runs is the success rate achieving maximally general and accurate rules. We repeated 10,000 independent runs of cGA for each of the multiplexer problems, and the two representations available as well. Table 1 presents the averaged quality achieved in those runs. Quality is defined as the percentage of the runs that evolved a maximally general and accurate rules. Qualities goes down as the problem grows regardless of the representation. However, preliminary results show a tendency to stabilize much faster in the Butz, Pelikan, Llorà, & Goldberg representation than in the De Jong & Spears one. The exponential growth of ρ for the De Jong & Spears model clearly supports such intuition. However, further empirical investigation using larger multiplexer problems is required.

Table 1. Quality of the rules evolved using cGA using the two available representations in three multiplexer problems

Representation	MX3	MX6	MX11
De Jong & Spears	97%	73.93%	43.03%
Butz, Pelikan, Llorà & Goldberg	95.55%	68.71%	47.97%

7   Conclusions

This paper has presented an analysis of how maximally general and accurate rules, regardless of the encoding, can be evolved Pittsburgh-style systems. The theoretical analysis of maximally general and accurate rules using two different binary rule encoding schemes showed some theoretical results with clear implications to the scalability of any genetic-based machine learning system that uses the studied encoding schemes.

The binary rule-encoding representation proposed by De Jong & Spears [2] inherently posses a bias that challenges the scalability of any system that uses such encoding. In this representation, theory shows how the area of meaningful rules—the ρ ratio—shrinks exponentially, leading the learning mechanism into a nail-in-a-haystack situation. A way for fixing such a behavior would be to turn the unmatching conditions into *dont care* ones. In binary problems, such a change will behave in the same manner as the ternary representation. However, increasing the cardinality beyond binary will drastically chage the balance among generality and specificity distribution of those conditions. Nevertheless, the alternative representation—proposed by Butz, Pelikan, Llorà, & Golberg [13,14]—show that such an exponentially trend is only linked to the encoding scheme used.

Another interesting results after the CCS runs to evolve maximally general and accurate rules is the empirical result that the quality of the rules evolve goes down as the problem grows regardless of the representation used. Such result suggest that a revision of the population size used by the cGA of the CCS is needed. Nevertheless, the preliminary results show a tendency to stabilize the quality drop much faster in the Butz, Pelikan, Llorà, & Goldberg representation than in the De Jong & Spears one. The exponential behavior of ρ for the De Jong & Spears model clearly supports such intuition. However, further empirical investigation using larger multiplexer problems is required. Another open research area is to formally define the ρ ratio for the Butz, Pelikan, Llorà & Goldberg representation, and theoretically computed the expression of ρ.

Acknowledgments

This work was sponsored by the Air Force Office of Scientific Research, Air Force Materiel Command, USAF, under grant F49620-03-1-0129, and by the Technology Research, Education, and Commercialization Center (TRECC), at University of Illinois at Urbana-Champaign, administered by the National Center for Supercomputing Applications (NCSA) and funded by the Office of Naval Research under grant N00014-01-1-0175. The US Government is authorized to reproduce and distribute reprints for Government purposes notwithstanding any copyright notation thereon.

The views and conclusions contained herein are those of the authors and should not be interpreted as necessarily representing the official policies or endorsements, either expressed or implied, of the Air Force Office of Scientific Research, the Technology Research, Education, and Commercialization Center, the Office of Naval Research, or the U.S. Government.

References

1. Wilson, S.W.: Classifier fitness based on accuracy. Evolutionary Computation **3** (1995) 149–175
2. De Jong, K.A., Spears, W.M.: Learning Concept Classification Rules using Genetic Algorithms. In: Proceedings of the Twelfth International Conference on Artificial Intelligence IJCAI-91. Volume 2., Morgan Kaufmann (1991) 651–656
3. Janikow, C.: A Knowledge Intensive Genetic Algorithm for Supervised Learning. Machine Learning **13** (1993) 198–228
4. Holland, J.H.: Adaptation in Natural and Artificial Systems. University of Michigan Press, Ann Arbor, MI (1975)
5. Goldberg, D.E.: Computer-aided gas pipeline operation using genetic algorithms and rule learning. Dissertation Abstracts International **44** (1983) 3174B Doctoral dissertation, University of Michigan.
6. Goldberg, D.E.: Genetic algorithms in search, optimization, and machine learning. Addison-Wesley, Reading, MA (1989)
7. Wilson, S.W.: Get real! XCS with continuous valued inputs. Festschrift in Honor of John H. Holland (1999) 111–121
8. Lanzi, P.L.: Extending the Representation of Classifier Conditions Part I: From Binary to Messy Coding. In: Proceedings of the Genetic and Evolutinary Computation Conference (GECCO'99), Morgan Kauffmann (1999) 337–344
9. Lanzi, P.L., Perrucci, A.: Extending the Representation of Classifier Conditions Part II: From Messy Coding to S-Expressions. In: Proceedings of the Genetic and Evolutinary Computation Conference (GECCO'99), Morgan Kauffmann (1999) 345–352

10. Llorà, X., Garrell, J.M.: Knowledge-Independent Data Mining with Fine-Grained Parallel Evolutionary Algorithms. In: Proceedings of the Genetic and Evolutionary Computation Conference (GECCO'2001), Morgan Kaufmann Publishers (2001) 461–468

11. Llorà, X.: Genetic Based Machine Learning using Fine-grained Parallelism for Data Mining. PhD thesis, Enginyeria i Arquitectura La Salle. Ramon Llull University, Barcelona (February, 2002)

12. Mitchell, T.M.: Machine Learning. McGraw Hill (1997)

13. Butz, M., Pelikan, M., Llorà, X., Goldberg, D.E.: Automated Global Structure Extraction For Effective Local Building Block Processing in XCS. IlliGAL Report No. 2005011, University of Illinois at Urbana-Champaign, Illinois Genetic Algorithms Laboratory, Urbana, IL (2005)

14. Butz, M., Pelikan, M., Llorà, X., Goldberg, D.E.: Extracted Global Structure Makes Local Building Block Processing Effective in XCS. IlliGAL Report No. 2005010, University of Illinois at Urbana-Champaign, Illinois Genetic Algorithms Laboratory, Urbana, IL (2005)

15. Pelikan, M., Lobo, F., Goldberg, D.E.: A survey of optimization by building and using probabilistic models. Computational Optimization and Applications 21 (2002) 5–20 (Also IlliGAL Report No. 99018).

16. Harik, G., Lobo, F., Goldberg, D.E.: The compact genetic algorithm. Proceedings of the IEEE International Conference on Evolutionary Computation (1998) 523–528 (Also IlliGAL Report No. 97006).

17. Llorà, X., Sastry, K., Goldberg, D.E.: The Compact Classifier System: Motivation, Analysis, and First Results. In: Proceedings of the Genetic and Evolutinary Computation Conference (GECCO 2005), ACM press (2005) in press

18. Larrañaga, P., Lozano, J.A., eds.: Estimation of Distribution Algorithms. Kluwer Academic Publishers, Boston, MA (2002)

19. Goldberg, D.E., Korb, B., Deb, K.: Messy genetic algorithms: Motivation, analysis, and first results. Complex Systems 3 (1989) 493–530 (Also IlliGAL Report No. 89003).

20. Baluja, S.: Population-based incremental learning: A method of integrating genetic search based function optimization and competitive learning. Technical Report CMU-CS-94-163, Carnegie Mellon University (1994)

21. Baluja, S., Caruana, R.: Removing the genetics from the standard genetic algorithm. Technical Report CMU-CS-95-141, Carnegie Mellon University (1995)

22. Mühlenbein, H., Paaß, G.: From recombination of genes to the estimation of distributions I. Binary parameters. Parallel Problem Solving from Nature, PPSN IV (1996) 178–187

23. Mühlenbein, H.: The equation for response to selection and its use for prediction. Evolutionary Computation 5 (1997) 303–346

24. Harik, G., Cantú-Paz, E., Goldberg, D.E., Miller, B.L.: The gambler's ruin problem, genetic algorithms, and the sizing of populations. Evolutionary Computation 7 (1999) 231–253 (Also IlliGAL Report No. 96004).

Bloat Control and Generalization Pressure Using the Minimum Description Length Principle for a Pittsburgh Approach Learning Classifier System

Jaume Bacardit[1] and Josep Maria Garrell[2]

[1] Automated Scheduling, Optimisation and Planning research group,
School of Computer Science and IT, University of Nottingham, Jubilee Campus,
Wollaton Road, Nottingham, NG8 1BB, UK
jqb@cs.nott.ac.uk
[2] Intelligent Systems Research Group, Enginyeria i Arquitectura La Salle,
Universitat Ramon Llull, Psg. Bonanova 8, 08022-Barcelona, Catalonia,
Spain, Europe
josepmg@salleURL.edu

Abstract. Bloat control and generalization pressure are very important issues in the design of Pittsburgh Approach Learning Classifier Systems (LCS), in order to achieve simple and accurate solutions in a reasonable time. In this paper we propose a method to achieve these objectives based on the *Minimum Description Length* (*MDL*) principle. This principle is a metric which combines in a smart way the accuracy and the complexity of a theory (rule set , instance set, etc.). An extensive comparison with our previous generalization pressure method across several domains and using two knowledge representations has been done. The test show that the *MDL* based size control method is a good and robust choice.

1 Introduction

The application of Genetic Algorithms (GA) [1] to classification domains is usually known as Genetic Based Machine Learning (GBML), and it has traditionally been addressed from two different points of view: the Pittsburgh approach (or Pittsburgh LCS) and the Michigan approach (or Michigan LCS), early exemplified by LS-1 [2] and CS-1 [3], respectively. Some representative systems of each approach are *GABIL* [4] and XCS [5].

The Pittsburgh approach systems usually evolve variable-length individuals that are complete solutions to the classification problem. This paper deals with the control of the individuals length. This control is a very important issue for two main reasons. The first one is that the evolution of variable-length individuals can lead to solutions growing without control. This phenomenon is usually known as *Bloat* [6] and it has been widely studied in the Genetic Programming field.

The second reason is derived from the fact that usually the fitness of the individuals is only based on their predictive accuracy over the training examples, and doesn't take into account their complexity. Given this fitness function, the

X. Llorà et al. (Eds.): IWLCS 2003-2005, LNAI 4399, pp. 59–79, 2007.

easiest way to increase it is to maximize the probability of correctly classifying the train examples, which is achieved by increasing the size of the individuals. This fact produces solutions that are bigger than necessary, contradicting the *Occam's razor* principle [7] which says that "the simplest explanation of the observed phenomena is most likely to be the correct one". A probable consequence of the "over-complexity" is an over-fitting of the solutions created which can lead to a decrease of the generalization capacity. We observed this problem in our previous work [8].

In this paper we propose a bloat control and generalization pressure method (*GPM*) based on the *Minimum Description Length (MDL)* principle [9]. It is an interpretation of the Occam's Razor principle based on the idea of data compression, that takes into account both the simplicity and predictive accuracy of a theory. Pfahringer [10] did a very good and brief introduction of the principle:

> Concept membership of each training example is to be communicated from a sender to a receiver. Both know all examples and all attributes used to describe the examples. Now what is being transmitted is a theory (set of rules) describing the concept and, if necessary, explicitly all positive examples not covered by the theory (the false-negative examples) and all negative examples erroneously covered by the theory (the false-positive examples). Now the cost of a transmission is equivalent to the number of bits needed to encode a theory plus its exceptions in a sensible scheme. The MDL principle states that the best theory derivable from the training data will be the one requiring the minimum number of bits.

The *MDL* principle is integrated into our Pittsburgh LCS adapting it to two knowledge representations. The classic *GABIL* one [4] for discrete attributes and our own *Adaptive Discretization Intervals (ADI) rule representation* [11] for the real-valued ones. We have also added an adaptive heuristic in order to simplify the task of domain specific parameter tuning. The *GPM* based on the *MDL* principle is compared across several domains with our previous work in this area: The hierarchical selection operator [8], which is explained in section 3.

The paper is structured as follows. Section 2 presents a short description of how the bloat effect affects Pittsburgh *LCS* and also some guidelines about how should be defined the measures used to alleviate the bloat effect. Next, section 3 presents some related work. After the related work we describe the framework of our classifier system in section 4. Our implementation of the *MDL* is explained in section 5. Next, section 6 describes the test suite used in the comparison. The results obtained are summarized in section 7. Finally, section 8 discusses the conclusions and some further work.

2   Bloat Effect in Pittsburgh Approach LCS

In this section we will do a brief and illustrative introduction about how and why the bloat effect affects Pittsburgh approach LCS. We will also show that fixing this problem is not a simple task, showing how bad ways to fix this problem can collapse the learning process.

2.1   What Form Does It Take the Bloat Effect?

Usually the bloat effect is defined as the growth without control of the individuals length, and it is a phenomenon that can affect in general all variable-length representations. In Pittsburgh LCS this effect takes the form of an exponential-rate growing of the number of rules of the individuals. This effect can be illustrated by the first 15 iterations in figure 1, which represents the evolution of the average individual size for the *MX11* problem. If we did not apply any measure to control this, the program would crash from out of memory shortly after.

2.2   Why Do We Have Bloat Effect?

The reason of the bloat effect is well explained in [6]. Its cause is the use of a fitness function which only takes into account the goodness of the solution (accuracy in our case). Having a variable-length representation means that it possible to have several individuals with the same fitness value, and there will be more long representations of a given solution (fitness value) that short ones. So, when the exploration finds new solutions, it is more probable that these solutions will be long than short.

The interpretation of this idea in *LCS* is that, it is more probable to classify correctly more train examples with an individual with a lot of rules that with a short individual. Is this long individual a good solution? Probably no, as this individual is *memorizing* the train examples instead of *learning* them. This shows a side effect of the bloat effect in *LCS*: the generated solutions will probably lack generalization, and its test accuracy will probably be poor.

2.3   How Can We Solve the Bloat Effect?

It is obvious that we need to add to the system some bias towards good but simple solutions, but will any intervention in this sense work? The answer is no. If we introduce too much pressure towards finding simple solutions, we are in danger of collapsing the population into individuals of only one rule, which can not generate longer individuals anymore. With this kind of individuals we can only classify the majority class. Again in figure 1 we can see an example of a too much strong pressure in the MX11, which is activated just after 15 iterations. With only a few iterations, a population of an average of more than 120 rules per individual is reduced to one rule individuals. The bloat control method that created this situation is the same presented in this paper, but bad parametrized (InitialRateOfComplexity=0.5).

So, what is the good way to control the bloat effect? There is not a single answer and, beside the method presented in this paper, in the related work section several methods that achieve this control are described. Intuitively we can say that the best method will be the one finding the best balance between accuracy and complexity.

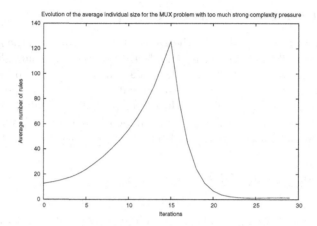

Fig. 1. Illustration of the bloat effect and how a badly designed bloat control method can destroy the population

3   Related Work

The *MDL* principle has been applied as a part of modeling tasks in many different fields. For example, handwriting recognition and robotic arms [12]. The principle has also been widely applied in the Machine Learning field. Some examples are Genetic Programming [13] or c4.5rules [14], where the *MDL* principle is used to select the best subset of rules derived from a c4.5 induced decision tree.

There is extensive prior work in the Evolutionary Computation (EC) field to control the *bloat* effect, specially in Genetic Programming [13,15] where this effect has been more widely studied. However, it has also been studied in other EC paradigms like Genetic Algorithms [8,16] or Evolution Strategies [16]. There is also some work on generalization pressure operators in systems that not suffer the *bloat* effect [17].

4   Framework

In this section we describe the main features of our classifier system. GAssist (*Genetic clASSIfier sySTem*) [18] is a Pittsburgh style classifier system based on GABIL [4]. Directly from GABIL we have borrowed the semantically correct crossover operator.

4.1   General Operators and Policies

Matching strategy. The matching process follows a "if ... then ... else if ... then..." structure, usually called *Decision Lists* [19].

Mutation operators. The system manipulates variable-length individuals, making more difficult the tuning of the classic gene-based mutation probability. In order to simplify this tuning, we define p_{mut} as the probability of mutating an individual. When an individual is selected for mutation (based on p_{mut}), a random gene is chosen inside its chromosome to be mutated.

Policy for missing values. Some of the problems used in the experimentation reproduced in this paper have missing values. A substitution policy has been used. Before starting the learning process all missing values are changed with either the average value of the attribute (for real-valued attributes) or the most frequent value (for symbolic attributes). These averages are not computed using all the train instances, but only the ones belonging to the same class as the instance with a missing values being substituted.

4.2 Bloat Control an Generality Pressure:

We describe briefly our previous work in this area because the *MDL* method presented in this paper will be compared to it in the results section. The bloat control and generalization pressure was achieved by combining the following two techniques:

- *Rule deletion:* This operator deletes the rules of each individual that do not match any training example. This rule deletion is done after the fitness computation and has two constraints: (a) the process is only activated after a predefined number of iterations, to prevent a massive diversity loss and (b) the operator stops when the number of rules of the individual reaches a certain lower threshold.
- *Selection bias using the individual size:* Selection is guided as usual by the accuracy of the individual. However, it also gives certain degree of relevance to the size of the individuals, having a policy similar to multi-objective systems. We use tournament selection because its local behavior lets us implement this policy. The criterion of the tournament is given by our own operator called "hierarchical selection" [8], defined as follows:
  - If $|accuracy_a - accuracy_b| < threshold$ then:
    * If $length_a < length_b$ then a is better than b
    * If $length_a > length_b$ then b is better than a
    * If $length_a = length_b$ then we will use the general case
  - Otherwise, we use the general case: we select the individual with higher fitness.

4.3 Knowledge Representations

The following paragraphs describe the knowledge representations that we use to solve problems with symbolic or real-valued attributes. Some of these representations are well known or have been described in detail elsewhere, but we believe that it is important to describe them again because the *MDL* principle has to be carefully adapted for each of them.

Rule Representations for symbolic or discrete attributes. We will use the **GABIL** [4] representation for this kind of attributes. Each rule consists of a condition part and a classification part: *condition → classification*. Each condition is a Conjunctive Normal Form (CNF) predicate defined as:

$$((A_1 = V_1^1 \vee \ldots \vee A_1 = V_m^1) \bigwedge \ldots \bigwedge (A_n = V_2^n \vee \ldots A_n = V_m^b))$$

Where A_i is the ith attribute of the problem and V_i^j is the jth value that can take the ith attribute.

This kind of predicate can be encoded into a binary string where there is a bit for each value of all attributes of the domain. Attribute values that appear in the CNF predicate have their associated bit set to one. If they not appear in the predicate they have their bit set to 0. An example follows: if we have a problem with two attributes, where each attribute can take three different values {1,2,3}, a rule of the form "If the first attribute has value 1 or 2 and the second one has value 3 then we assign class 1" will be represented by the string 110|001|1.

Rule Representations for real-valued attributes. The representation for real-valued attributes is our own representation called **Adaptive Discretization Intervals** rule representation [18]. Specifically, we will use the second version of the representation (ADI2) [11].

This representation is an evolution of the *GABIL* discrete rule representation. In *GABIL* for each attribute we would use a set of static discretization intervals instead of nominal values. The intervals of the ADI2 representation are not static, but they evolve through the iterations splitting and merging among them (having a minimum size called *micro-interval*). Thus, the binary coding of the *GABIL* representation is extended as represented in figure 2, also showing the split and merge operations.

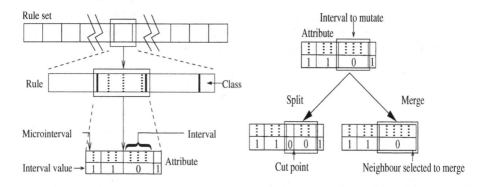

Fig. 2. Adaptive intervals representation and the split and merge operators

The ADI2 representation is defined in depth as follows:

1. Each individuals initial rule and attribute term is assigned a number of "low level" uniform-width and static discretization intervals (called *micro-intervals*).

2. The intervals of the rule are built joining together adjacent *micro-intervals*.
3. Attributes with different numbers of *micro-intervals* can coexist in the population. The evolution will choose the correct number of *micro-intervals* for each attribute.
4. For computational cost reasons, we will have an upper limit in the number of intervals allowed for an attribute, which in most cases will be less than the number of *micro-intervals* assigned to each attribute.
5. When we split an interval, we select a random point in its *micro-intervals* to break it.
6. When we merge two intervals, the state (1 or 0) of the resulting interval is taken from the one which has more *micro-intervals*. If both have the same number of *micro-intervals*, the value is chosen randomly.
7. The number of *micro-intervals* assigned to each attribute term is chosen from a predefined set.
8. The number and size of the initial intervals is selected randomly.
9. The cut points of the crossover operator can only take place in attribute terms boundaries, not between intervals. This restriction takes place in order to maintain the semantical correctness of the rules.
10. The *hierarchical selection* operator uses the length of the individuals (defined as the sum of all the intervals of the individual) instead of the number of rules as the secondary criteria. This change promotes simple individuals with more reduced interval fragmentation.

In order to make the interval splitting and merging part of the evolutionary process, we have to include it in the *GA* genetic operators. We have chosen to add to the GA cycle two special stages applied to the offspring population after the mutation stage. The split and merge operators are controlled by a probability (p_{split} and p_{merge}) defined for each attribute term of each rule. The code for the merge operator probability is represented in figure 3.

```
ForEach Individual i of Population
       ForEach Rule j of Population individual i
              ForEach Attribute k of Rule j of Population individual i
                     If random [0..1] number < p_merge
                            Select one random interval of attribute term k
                                        of rule j of individual i
                            Apply a merge operation to this interval
                     EndIf
              EndForEach
       EndForEach
EndForEach
```

Fig. 3. Code of the application of the merge operator

5   The *MDL* Principle Applied to Generalization Pressure

In this section we describe our proposal of bloat control and generalization pressure based on the *MDL* principle. First, we introduce our implementation of the basic formula of the principle and its adaptation to each of the knowledge representations uses. Finally , we propose a method to adjust automatically the W parameter of the main MDL formula that appears in the introduction section, simplifying the domain-specific adjusting of the principle.

5.1 Basic MDL Formula

As said in the introduction section, the MDL principle is a metric used to evaluate the complexity and accuracy of a theory which is inspired by data compression. The class membership of each training example is to be communicated from a sender to a receiver. This is done by transmitting a theory (set of rules in our case) and, if necessary, transmitting the exceptions to this theory. That is, the misclassified and non-classified examples. The cost of the transmission is equivalent to the number of bits needed to encode the theory plus its exceptions in a sensible scheme. The principle states that the best theory is the one requiring the minimum number of bits. Therefore, the fitness function becomes the minimization of the MDL formula [14]:

$$MDL = W \cdot \text{theory bits} + \text{exception bits} \tag{1}$$

W is a weight that adjust the relation between theory and exception bits. The length of the theory bits (TL) is defined as follows:

$$TL = \sum_{i=1}^{nr} TL_i \tag{2}$$

Where nr is the number of rules of the theory. The definition of the rules for all the knowledge representations used share a common structure: *condition* → *class*. The condition is defined as a conjunction of predicates, where each predicate is associated to an attribute of the problem. Therefore, TL_i is defined as follows:

$$TL_i = \sum_{j=1}^{na} TL_i^j. \tag{3}$$

Where na is the number of attributes of the problem. TL_i^j is the length of the predicate associated to the attribute j of the rule i, and has a specific formula for each knowledge representation used. The reader can see that we have omitted a term in the formula related to the class associated to the rule. As it is a value common for all the possible rules it becomes irrelevant and it has been removed for simplicity reasons.

The exceptions part of the MDL principle (EL) represents the act of sending the class for the misclassified or unclassified examples to the receiver. We implement this idea by sending the number of exceptions plus, for each exception, its index in the examples set (supposing that sender and receiver have the examples organized in the same order) and its class:

$$EL = log_2(ne) + (nm + nu) \cdot (log_2(ne) + log_2(nc)) \tag{4}$$

Where ne is the total number of examples, nm is the number of wrongly classified examples, nu is the number of unclassified examples and nc is the number of classes of the problem. This definition is independant from the knowledge representation.

5.2 Adaptation of the *MDL* Principle for Each Knowledge Representation

The length of the predicate associated to each attribute (TL_i^j) has to be adapted to the type of the attribute and the knowledge representation. While designing the formula to calculate this length we have to remember that the philosophy of the *MDL* principle is to promote simple but accurate solutions. Therefore, we will prefer formula definitions that promote bias towards simpler solutions although there may exist shorter definitions.

MDL Formula for Real-Valued Attributes and ADI2 Rule Representation. The predicate associated to an attribute by this representation is defined as a disjunction of intervals, where each interval is a non-overlapping number of *micro-intervals* and can take a value of either true of false. Therefore, the information to transmit is the number of intervals of the predicate plus, for each interval, its size and value (1 or 0).

$$TL_i^j = log_2(\text{MaxI}) + ni_i^j \cdot (log_2(\text{MaxMI}) + 1) \tag{5}$$

$MaxI$ is the maximum number of intervals allowed in a predicate, ni is the actual number of intervals of the predicate and $MaxMI$ is the maximum allowed number of *micro-intervals* in the predicate.

Given the example of attribute predicate in figure 4, where we have 4 intervals , and supposing that the maximum numbers of intervals and *micro-intervals* are respectively 10 and 25, its MDL size is defined as follows:

$$TL_i^j = log_2(10) + 4 \cdot (log_2(25) + 1)$$

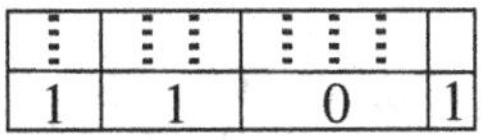

Fig. 4. Example of an ADI2 attribute predicate

MDL Formula for Discrete Attributes and GABIL Representation. The predicate associated to an attribute by this representation is defined as a disjunction of all the possible values that can take the attribute. The simpler way of transmitting this predicate is sending the binary string that the representation uses to encode it. This is the approach used by Quinlan in C4.5rules [14]. However, this definition does not take into account the complexity of the term and does not provide a bias towards generalized solutions.

Therefore, we define a different formula which is very similar to the one proposed for the *ADI2* knowledge representation. In this formula we simulate that we have merged the neighbor values of the predicate which have the same value (true or false):

$$TL_i^j = log_2(nv_j) + 1 + ni_i^j \cdot log_2(nv_j) \qquad (6)$$

nv is the number of possible values of the attribute j and ni is the number of "simulated intervals" that exist in the predicate. The only difference between this formula and the *ADI2* one is that we do not have to transmit the value of all the "simulated intervals", but only the first one (one bit).

If we had an attribute predicate such as "1111100001" we can see that we have 10 values and 3 "simulated intervals" and that the MDL size of the predicate would be:

$$TL_i^j = log_2(10) + 1 + 3 \cdot log_2(10)$$

This approach completely makes sense for ordinal attributes, where there exist an order between values, but not for nominal ones. However, we think that this definition is also useful for nominal attributes because we want to promote generalized predicates, where most of the values are true, and this means having few "simulated intervals".

5.3   Looking for a Parameter-Less *MDL* Principle

If we examine all the formulas of the *MDL* principle we only find one parameter: W which adjusts the relation between the length of the theory and the length of the exceptions. Quinlan used a value of 0.5 for this parameter in *C4.5rules* and reported the following in page 53 of [14]:

> Fortunately, the algorithm does not seem to be particularly sensitive to the value of W.

Unfortunately, our environment of application of the *MDL* principle (a *GBML* system) is quite different and the value of the W parameter is quite sensitive. If the value of W is too high, the population will collapse into one rule individuals, as it can be seen in section *bloat*. If W is too low, the individuals probably will be too much specific.

This problem with the adjusting of W leads to a question: Is it possible to find a good method to adjust automatically this parameter? The completely rigorous answer, being aware of the *No Free Lunch Theorem* [20] and the *Selective Superiority Problem* [21] is no.

Nevertheless, at least we can try to find a way to automatically make the system perform "quite well" in a broad range of problems. In order to achieve this objective we have developed a simple approximation which starts the learning process with a very strict weight (but loose enough to avoid a collapse of the population) and relaxes it through the iterations when the *GA* has not found a better solution for a certain number of iterations. This method can be represented by the code in figure 5.

```
Initialize GA
Ind = Individual with best accuracy from the initial GA population
TL = Theory Length of Ind
EL = Exceptions Length of Ind
NR = Number of rules of Ind
NC = Number of Classes of the domain
TL' = TL · NR/NC
W = InitialRateOfComplexity·EL / (1−InitialRateOfComplexity)·TL'
Iteration = 0
IterationsSinceBest = 0
While Iteration < NumIterations
        Run one iteration of the GA using W in fitness computation
        If a newbest individual has been found then
                IterationsSinceBest = 0
        Else
                IterationsSinceBest = IterationsSinceBest + 1
        EndIf
        If IterationsSinceBest > MaximumBestDelay then
                W = W · WeightRelaxationFactor
                IterationsSinceBest = 0
        EndIf
        Iteration = Iteration + 1
EndWhile
```

Fig. 5. Code of the parameter-less learning process with automatically adjusting of W

$InitialRateOfComplexity$ defines which percentage of the MDL formula should the term $W \cdot TL$ have. Using this supposition and given one individual from the initial population, we can calculate the value of W. We have used a simple policy to select this individual: the one with more train accuracy ($W = \frac{InitialRateOfComplexity \cdot EL}{(1-InitialRateOfComplexity) \cdot TL'}$).

This issue raises a question: is this individual good enough? If we recall section 2, it is more probable that this individual will be long than short. Then, maybe we would be initializing W with a too small value. Therefore, before calculating the initial value of W we do a last step: scaling the theory length of this individual ($TL' = TL \cdot \frac{NR}{NC}$), using as a reference the minimum possible number of rules of an optimal solution: the number of classes of the domain.

We can see that in order to automatically adjust one parameter we have introduced three extra parameters ($InitialRateOfComplexity$, $MaximumBestDelay$ and $WeightRelaxationFactor$). The second parameter is easy to setup if we consider the takeover time for the tournament selection [22]. Given a tournament size of 3 and a population size of 300, the takeover time is 6.77 iterations. Considering that we have both crossover and mutation in our GA, setting $MaximumBestDelay$ to 10 seems quite safe.

Setting $InitialRateOfComplexity$ is also relatively easy: it the value is too high (giving too much importance to the complexity factor of the MDL formula)

Table 1. Tests with the MX-11 domain done to find the values of InitialRateOfComplexity (*IROC*) and WeightRelaxationFactor (*WRF*)

WRF	IROC	Test acc.	Num. of Rules	Iterations until perfect accuracy
	0.05	100.0±0.0	9.3±0.6	301.4±56.8
0.7	0.075	100.0±0.0	9.2±0.5	309.0±62.6
	0.1	100.0±0.0	9.2±0.5	333.3±62.2
	0.05	100.0±0.0	9.3±0.5	331.0±71.5
0.8	0.075	100.0±0.0	9.2±0.3	364.4±75.3
	0.1	100.0±0.0	9.2±0.5	374.3±66.9
	0.05	100.0±0.0	9.2±0.5	428.6±99.7
0.9	0.075	100.0±0.0	9.2±0.4	475.5±95.6
	0.1	100.0±0.0	9.1±0.4	518.4±110.2

the population will collapse. Therefore, we have to find the maximum value of *InitialRateOfComplexity* that lets the system perform a correct learning process. Doing some short tests with various domains we have seen that values over 0.1 are too much dangerous. In order to adjust more finely this parameter and also set *WeightRelaxationFactor* we have done tests using again the *MX-11* domain testing three values of each parameter: 0.1, 0.075 and 0.05 for *InitialRateOfComplexity* and 0.9, 0.8 and 0.7 for *WeightRelaxationFactor*.

The results can be seen in table 1, showing three things: test accuracy and the number of rules of the best individual in the final population and also the average iteration where 100% train accuracy was reached. We can see that all the tested configuration manage to reach a perfect accuracy, and also that the number of rules of the solutions are very close to the optimum 9 ordered rules. The only significant differences between the combinations of parameters tested comes when we observe the iterations needed to reach 100% train accuracy. We can see that as more mild are the parameters used, fewer iterations are needed. This arises the question of how extrapolative to other domains is this behaviour. We have to be aware that *MX-11* is a synthetic problem without noise.

Table 2. Tests with the Wisconsin Breast Cancer domain done to find the values of InitialRateOfComplexity (*IROC*) and WeightRelaxationFactor (*WRF*)

WRF	IROC	Train acc.	Test acc.	Num. of Rules
	0.05	98.2±0.3	95.6±1.5	4.3±1.5
0.7	0.075	98.2±0.3	95.8±1.5	4.1±1.3
	0.1	98.1±0.3	95.9±1.7	3.9±1.2
	0.05	98.1±0.3	95.8±1.5	3.9±1.3
0.8	0.075	98.0±0.3	96.0±1.7	3.7±0.8
	0.1	97.9±0.3	96.0±1.7	3.5±0.9
	0.05	97.8±0.3	95.9±1.7	2.9±0.9
0.9	0.075	97.6±0.3	96.0±1.8	2.3±0.6
	0.1	97.5±0.3	95.9±1.8	2.2±0.5

In order to check how is the system behaving in real problems, we repeated this test with another well-known problem: *Wisconsin Breast Cancer*. The results can be seen in table 2. Iterations are not included in this table because we do not know the ideal solution for this problem. Instead, we have included train accuracy. It will help illustrate the completely different landscape that we have here: Although the differences are not significant, we can see that as more mild are the parameters used, we have more train accuracy, more rules and less test accuracy. It seems quite clear that the system suffers from over-learning if its working parameters are not enough strict. Therefore, we select 0.075 and 0.9 as the values of $InitialRateOfComplexity$ and $WeightRelaxationFactor$ respectively for the rest of this paper. These values seem to be the most stable ones.

Before showing the results for all the datasets tested it would be interesting to see the stability of the W tuning heuristic presented in this section. In figure 6 we can see the evolution of W through the learning process for the *bre* and *tao* problems [1]. The values in the figure have been scaled in relation to the initial W value. These two problems are selected because they show two alternative behaviours due to having very different number of rules in their optimal solutions. We can see that the differences in the evolution of W for different executions shrink through the iterations, showing the stability of the heuristic.

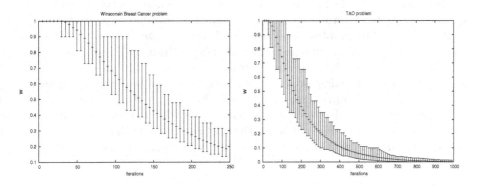

Fig. 6. Evolution of W through the learning process

6   Test Suite

This section summarizes the tests done across several domains in order to evaluate the accuracy and efficiency of the method presented in this paper. We also compare it with our previous proposal.

6.1   Test Problems

The selected test problems for this paper present different characteristics in order to give us a broad overview of the performance of the methods being compared.

---

[1] Datasets are detailed in section 6.

Table 3. Characteristics of the test problems

Dataset	Number of examples	real-valued attributes	discrete attributes	classes
aud	226	-	69	24
bps	1027	24	-	2
bre	699	-	9	2
gls	214	9	-	6
ion	351	34	-	2
irs	150	4	-	3
led	2000	-	7	10
lrn	648	4	2	5
mmg	216	21	-	2
mux	2048	-	11	2
pim	768	8	-	2
prt	339	-	17	22
tao	1888	2	-	2

First we have some synthetic problems: Tao (*tao*) [23], a problem that has non-orthogonal class boundaries, the 11 input multiplexer (*mux*) and LED (*led*), a problem which represents a seven segments display having the represented digit as the class. This problem has a 10% artificially added noise. Second, we also use several real problems provided by the University of California at Irvine (UCI) repository [24]. The problems selected are: Audiology (*aud*), Glass *gls*, Iris *irs*, Ionosphere (*ion*), Pima-indians-diabetes (*pim*), Primary-Tumor (*prt*) and Wisconsin-breast-cancer (*bre*). Finally, we will use three problems from our own private repository. The first two deal with the diagnosis of breast cancer based of biopsies (*bps*) [25] and mammograms (*mmg*) [26] whereas the last one is related to the prediction of student qualifications (*lrn*) [27]. The characteristics of all the datasets are listed in table 3. The partition of the examples into the train and test sets was done using the *stratified ten-fold cross-validation* method [28].

6.2   Experimentation Design

The goal of the tests done in this paper is to evaluate the performance of the implementation of the *MDL* principle described in the prior section. This evaluation includes a comparison of this method our previous generalization pressure methods (GPM): the *Hierarchical Selection* operator [8].

In our previous work, the *Hierarchical Selection* operator was used in combination with the rule deletion operator because it could not control the bloat effect by itself, but only improved the generalization pressure. This fact makes us question if it is necessary to use the rule deletion operator for the *MDL* methods. We performed a short test to answer this question. The test used again *Wisconsin Breast Cancer*. We use the same GA configuration being used in the global tests which is detailed at the end of this section.

The results of this short test are in table 4. We show, for each configuration (GPM with/without rule elimination), the averages and mean deviations of the

Table 4. Test of the effects of the Rule Deletion operator for the *Breast* problem

GPM	Rule Deletion	Test accuracy	Number of Rules			Run Time (s)
			Min.	Max.	Avg.	
Hierar.	No	95.6±1.3	22	2	4.9±3.6	92.8±25.6
Hierar.	Yes	95.8±1.6	6	2	2.4±0.7	57.4±2.5
MDL	No	95.9±1.7	6	2	2.4±0.7	58.9±3.4
MDL	Yes	96.1±1.8	4	2	2.4±0.7	55.8±1.9

test accuracy, the number of rules of the final solution and the run time in seconds (using a Pentium IV at 1.5GHz). We can see that the use of the rule deletion operator improved the accuracy for all the *GPM*. Also, there is a reduction in the average number of rules (for the Hierarchical *GPM*) and run time. The rule set size reduction does not seem very big in average, but the differences are considerable if we look at the maximum and minimum sizes for the Hierarchical method, reflecting that it sometimes cannot control the bloat effect.

Other domains showed similar results. As it seems there does not exist a bad interaction between the GPM and the rule elimination operator, we have decided to use the operator for the rest of the tests.

In order to allow the replication of our results we show the detailed configuration of our tests in table 5. This table is divided in two parts: common and domain-specific parameters.

The value of the initialization probability (p_1) is greater than the usual 0.5 value for some problems. All these problems share a common trait: a high number of attributes. In this environment, a regular initialization policy can lead to a situation where very few (or none) train examples are matched by the individuals. This situation can lead to a collapse of the population towards one rule individuals, because accuracy becomes an insignificant part of the fitness computation.

7   Results

In this section we present the results obtained. The aim of the tests was to determine the performance of the *GPM* tested in three aspects: accuracy and size of the solutions as well as computational cost. For each method and test problem we show the average and standard deviation values of: (1) the cross-validation accuracy, (2) the size of the best individual in number of rules and (3) the execution time in seconds. The tests were executed in an AMD Athlon 1700+ using the Linux operating system, C++ language and GCC v3.2.2 compiler.

The results can be seen in table 6. The results were also analyzed using paired two-sided statistical t-test [29] in order to determine if the *MDL* method outperform our previous approach with a significance level of 1%. No significant outperformances were detected.

Table 5. Common and problem-specific parameters of the GA

Parameter	Value
General parameters	
Crossover probability	0.6
Selection Algorithm	Tournament
Tournament size	3
Population size	300
Probability of mutating an individual	0.6
Number of seeds for each experiment	15
MDL Weight heuristically adjusting	
InitialRateOfComplexity	0.075
MaximumBestDelay	10
WeightRelaxationFactor	0.9
rule deletion operator	
Iteration of activation	40
Minimum number of rules before disabling the operator	$numClasses + 3$
Hierarchical Selection	
Iteration of activation	40
ADI rule representation	
Maximum number of intervals per attribute	10
Possible size in *micro-intervals* of an attribute	5,6,7,8,10,15,20,25
p_{split}	0.05
p_{merge}	0.05

Code	Parameter
#iter	Number of GA iterations
p_1	Probability of value 1 in initialization
d_{comp}	Threshold parameter in Hierarchical Selection

Problem	Parameter		
	#iter	p_1	d_{comp}
aud	1500	0.9	0.005
bps	300	0.75	0.015
bre	250	0.5	0.010
gls	1100	0.5	0.010
ion	450	0.75	0.010
irs	200	0.5	0.010
led	1000	0.5	0.001
lrn	700	0.5	0.010
mmg	275	0.75	0.010
mux	1000	0.5	0.001
pim	225	0.5	0.010
prt	1000	0.9	0.005
tao	900	0.5	0.001

As an external reference of the results, in table 7 the accuracy of the two above methods is compared to *IB1* [30], *C5.5* [14] [2] and *XCS* [5] [3]. We can see that, as usual, each method is the best in some problems but all of them perform similarly in average.

What can we observe in the results? First of all we can see that for the *mux* problem, the *MDL* method manages to generate solutions more near to the optimum rule set than the Hierarchical Selection method. Also, from a global point of view the results tell us that the *MDL* method has achieved our objective of developing a robust and easier to adjust *GPM*. It has managed to outperform

---

[2] Using the Weka [29] implementations.
[3] Results taken from [31].

Table 6. Results of the comparative tests. Bold entries show the method with best results for each test problem.

Problem	Configuration	Accuracy	Number of Rules	Time (s)
aud	Hierar.	60.0±4.2	11.2±2.2	89.1±12.4
	MDL	**63.5±3.9**	10.6±2.9	121.2±20.2
bps	Hierar.	**80.2±2.9**	3.4±0.8	218.0±17.3
	MDL	**80.2±2.9**	3.3±1.0	218.9±13.4
bre	Hierar.	95.8±1.6	2.4±0.7	44.6±1.9
	MDL	**96.1±1.8**	2.4±0.7	43.4±1.5
gls	Hierar.	64.4±3.6	7.2±1.4	71.1±7.9
	MDL	**64.8±3.0**	8.7±1.1	74.2±8.2
ion	Hierar.	90.7±2.8	4.0±1.2	177.6±20.8
	MDL	**91.3±2.9**	5.0±1.6	173.7±17.5
irs	Hierar.	95.1±2.1	4.8±1.0	5.3±0.3
	MDL	**95.6±3.0**	4.6±0.8	5.4±0.3
led	Hierar.	74.4±1.7	18.0±2.0	344.3±13.4
	MDL	**74.6±1.7**	19.3±2.2	332.7±8.2
lrn	Hierar.	**68.2±4.6**	7.1±1.6	82.9±5.9
	MDL	68.1±4.1	9.6±2.0	85.4±5.5
mmg	Hierar.	**66.3±4.5**	5.1±1.1	39.9±4.7
	MDL	64.4±6.4	5.3±1.1	38.6±4.7
mux	Hierar.	**100.0±0.0**	10.9±1.1	519.0±36.7
	MDL	**100.0±0.0**	9.2±0.4	474.2±14.4
pim	Hierar.	**75.0±3.4**	4.5±1.3	57.8±5.0
	MDL	74.8±3.4	3.9±0.9	57.4±4.7
prt	Hierar.	46.9±5.3	10.2±2.6	39.4±5.3
	MDL	**47.1±5.2**	14.9±3.5	47.1±5.5
tao	Hierar.	**94.9±1.1**	18.1±3.9	461.3±46.5
	MDL	94.7±0.9	15.1±4.6	414.1±33.0
average	Hierar.	77.8±15.9	8.2±5.0	165.4±164.7
	MDL	78.1±15.8	8.6±5.0	160.5±148.5

(in average) our previous work, the Hierarchical Selection method, in two ways: accuracy and reduction of the computational cost.

Nevertheless, the differences in the results do not seem to be much significant, but the way to reach these results, the internal behaviour of each method, is very different for both methods. We can observe this fact looking at the evolution of the accuracy average individual size (in rules) through the iterations. It figure 7

Table 7. Accuracy of Hierar. and MDL methods compared to IB1, C4.5 and XCS. Bold entries show the method with best results for each test problem.

Problem	Hierar.	MDL	IB1	C4.5	XCS
aud	60.0±4.2	63.5±3.9	76.0±6.3	**79.0±6.2**	41.6±8.1
bps	80.2±2.9	80.2±2.9	**83.2±3.0**	80.1±4.5	**83.2±2.9**
bre	95.8±1.6	96.1±1.8	96.0±1.4	95.4±1.5	**96.4±2.4**
gls	64.4±3.6	64.8±3.0	66.3±10.4	65.8±9.9	**70.8±8.1**
ion	90.7±2.8	**91.3±2.9**	86.9±4.6	89.8±4.	—
irs	95.1±2.1	**95.6±3.0**	95.3±3.1	95.3±3.1	94.7±5.0
led	74.4±1.7	74.6±1.7	56.5±1.7	**75.0±2.1**	74.5±1.9
lrn	68.2±4.6	68.1±4.1	61.4±5.8	**68.6±4.4**	—
mmg	**66.3±4.5**	64.4±6.4	63.5±11.5	64.8±6.0	64.3±6.4
mux	**100.0±0.0**	**100.0±0.0**	78.6±3.8	99.9±0.2	**100.0±0.0**
pim	75.0±3.4	74.8±3.4	70.3±3.3	73.1±5.0	**75.4±4.4**
prt	46.9±5.3	**47.1±5.2**	37.8±5.3	44.1±5.8	39.9±6.6
tao	94.9±1.1	94.7±0.9	**96.1±1.1**	95.1±1.9	89.9±1.2

we can see this evolution for the *bps,bre*, *mux* and *tao* problems. The plot for the iterations before the rule deletion activation have been removed from the graph because they introduce a high distortion.

The Hierarchical Selection method uses a specific-to-general policy. In the early iterations of the learning process it frequently finds new solutions that outreach the previous best accuracy by more than d_{comp}. In this situation the number of rules of the individuals is irrelevant. But as the learning curve stabilizes, the differences in accuracy between the bests individuals of the population become smaller than d_{comp}. Then, the smaller individual are mostly selected and, as a consequence, the average individual size slowly decreases.

On the other hand the *MDL* method, because of the behaviour of the *W* control heuristic, starts the learning process giving much importance to the size of the individual, and relaxes this importance through the iterations as dictated by the heuristic. Therefore, the behaviour is general-to-specific.

In figure 7 we can also see the main problem of the *MDL* method, which is the over-relaxation of the *W* weight. The philosophy of the algorithm we have proposed to tune *W* is that we relax this weight when it is too much strict, that is, when the *GA* cannot find a better individual for a certain number of iterations. This condition is sometimes difficult to control, and maybe if the

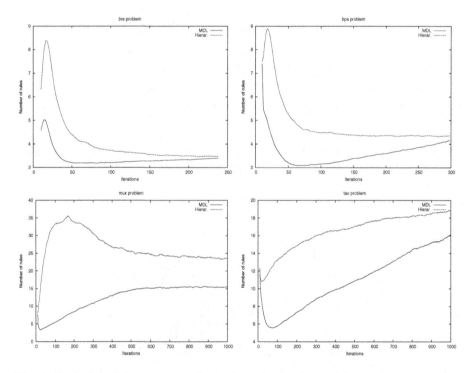

Fig. 7. Evolution of the average individual size for the *bre* and *bps* problems and *ADI* representation

system was given more iterations, the test performance in some domains would decrease. On the other hand we can see in figure 7 and in table 6 the reverse situation for the *tao* problem: The accuracy obtained by the *MDL* method is below the Hierarchical Selection one because the rule set is too much simple. With some more iterations this method probably would increase its accuracy.

Figure 7 can also help explain the notable computational cost difference between the tested methods in some domains (*mux* and *tao*). Smaller individuals are faster to evaluate. Therefore, the notable differences in the average individual size have their consequence in the overall computational cost.

8   Conclusions and Further Work

In this paper we have proposed a generalization pressure method for Pittsburgh Approach Classifier Systems based on the *MDL* principle. This technique proposes a fitness function which combines in a smart way the accuracy and the complexity of the individual being evaluated. The complexity measure is not based only on the size of the individual (number of rules) but also on the content of the rules. This if one of the main differences between this method and others found in the literature. Having a bloat control method that takes into account the semantical content of the rules can help explore better the search space, beside managing the size of the individual.

Extensive tests comparing the *MDL* method with our previous proposal have been done. These tests show that the technique performs slightly better (although not in a significant way based on Student t-tests) and runs also slightly faster. Beside its good results, the *MDL* method has another interesting feature, compared to our previous work in *GPM*, which is that it does not need a specific adjustment for each problem being solved. This is due to an adaptive adjustment of the W parameter. This adjustment is done by an heuristic process that we have developed. The adjusting of the W parameter is critical because applying too much or too little generalization pressure in the population can lead to an incorrect learning process. In the first case the population can collapse into individuals which are too simple. On the other hand, too little pressure can lead to over-fitted solutions. The tests have show that the adjustment of W is good, although it could be better controlled.

Therefore, as further work, other methods of adjusting W (like a specific-to-general policy) or maybe a stop criterion for the current method (leaving the value of W fixed after a certain point of the learning process) should be studied. Also, it could be interesting to extract other measures from the performance of the *GPM* tested, like the degree of diversity existing in the population.

Also, it would be very interesting to compare this bloat control method with recent Pareto-based Multi-Objective techniques like *MOLCS* [16]. This method is completely parameter-less, which was one of our goals. However, this means that the pressure applied to the complexity objective cannot be adjusted and, probably, it will be too strong or too mild for certain problems. If we know how the system will behave, having the possibility of some fine-tuning if quite desirable.

Acknowledgments

The authors acknowledge the support provided under grant numbers 2001FI 00514, TIC2002-04160-C02-02, TIC 2002-04036-C05-03 and 2002SGR 00155. Also, we would like to thank Enginyeria i Arquitectura La Salle for their support to our research group. The Wisconsin breast cancer databases was obtained from the University of Wisconsin Hospitals, Madison from Dr. William H. Wolberg. The primary tumor domain was obtained from the University Medical Centre, Institute of Oncology, Ljubljana, Slovenia. Thanks go to M. Zwitter and M. Soklic for providing the data.

References

1. Holland, J.H.: Adaptation in Natural and Artificial Systems. University of Michigan Press (1975)
2. Smith, S.F.: Flexible learning of problem solving heuristics through adaptive search. In: Proceedings of the Eighth International Joint Conference on Artificial Intelligence, Los Altos, CA, Morgan Kaufmann (1983) 421–425
3. Holland, J.H.: Escaping Brittleness: The possibilities of General-Purpose Learning Algorithms Applied to Parallel Rule-Based Systems. In: Machine learning, an artificial intelligence approach. Volume II. (1986) 593–623
4. DeJong, K.A., Spears, W.M.: Learning concept classification rules using genetic algorithms. Proceedings of the International Joint Conference on Artificial Intelligence (1991) 651–656
5. Wilson, S.W.: Classifier fitness based on accuracy. Evolutionary Computation **3** (1995) 149–175
6. Langdon, W.B.: Fitness causes bloat in variable size representations. Technical Report CSRP-97-14, University of Birmingham, School of Computer Science (1997) Position paper at the Workshop on Evolutionary Computation with Variable Size Representation at ICGA-97.
7. Mitchell, T.M.: Machine Learning. McGraw-Hill (1997)
8. Bacardit, J., Garrell, J.M.: Métodos de generalización para sistemas clasificadores de Pittsburgh. In: Proceedings of the "Primer Congreso Espaol de Algoritmos Evolutivos y Bioinspirados (AEB'02)". (2002) 486–493
9. Rissanen, J.: Modeling by shortest data description. Automatica **vol. 14** (1978) 465–471
10. Pfahringer, B.: Practical uses of the minimum description length principle in inductive learning (1995)
11. Bacardit, J., Garrell, J.M.: Evolving multiple discretizations with adaptive intervals for a pittsburgh rule-based learning classifier system. In: Proceedings of the Genetic and Evolutionary Computation Conference - GECCO2003, (in press), LNCS, Springer (2003)
12. Gao, Q., Li, M., Vitnyi, P.: Applying mdl to learn best model granularity. Artificial Intelligence **121** (2000) 1–29
13. Iba, H., de Garis, H., Sato, T.: Genetic programming using a minimum description length principle. In Kinnear, Jr., K.E., ed.: Advances in Genetic Programming. MIT Press (1994) 265–284
14. Quinlan, J.R.: C4.5: Programs for Machine Learning. Morgan Kaufmann (1993)

15. Luke, S., Panait, L.: Lexicographic parsimony pressure. In: GECCO 2002: Proceedings of the Genetic and Evolutionary Computation Conference. (2002) 829–836
16. Llorà, X., Goldberg, D.E., Traus, I., Bernadó, E.: Accuracy, Parsimony, and Generality in Evolutionary Learning System a Multiobjective Selection. In: Advances in Learning Classifier Systems: proceedings of the 5th International Workshop on Learning Classifier Systems, (in press), LNAI, Springer (2002)
17. Bernadó, E., Garrell, J.M.: Multiobjective learning in a genetic classifier system (MOLeCS). Butlletí de l'Associació Catalana l'Intel.ligència Artificial 22 (2000) 102–111
18. Bacardit, J.: Pittsburgh Genetics-Based Machine Learning in the Data Mining era: Representations, generalization, and run-time. PhD thesis, Ramon Llull University, Barcelona, Catalonia, Spain (2004)
19. Rivest, R.L.: Learning decision lists. Machine Learning 2 (1987) 229–246
20. Wolpert, D.H., Macready, W.G.: No free lunch theorems for search. Technical Report SFI-TR-95-02-010, Santa Fe, NM (1995)
21. Brodley, C.: Addressing the selective superiority problem: Automatic algorithm /model class selection (1993)
22. Goldberg, D.E., Deb, K.: A comparative analysis of selection schemes used in genetic algorithms. In: Foundations of Genetic Algorithms, Morgan Kaufmann (1991) 69–93
23. Llorà, X., Garrell, J.M.: Knowledge-independent data mining with fine-grained parallel evolutionary algorithms. In: Proceedings of the Third Genetic and Evolutionary Computation Conference, Morgan Kaufmann (2001) 461–468
24. Blake, C., Keogh, E., Merz, C.: Uci repository of machine learning databases (1998) (www.ics.uci.edu/mlearn/MLRepository.html).
25. Martínez Marroquín, E., Vos, C., et al.: Morphological analysis of mammary biopsy images. In: Proceedings of the IEEE International Conference on Image Processing. (1996) 943–947
26. Martí, J., Cufí, X., Regincós, J., et al.: Shape-based feature selection for microcalcification evaluation. In: Imaging Conference on Image Processing, 3338:1215-1224. (1998)
27. Golobardes, E., Llorà, X., Garrell, J.M., Vernet, D., Bacardit, J.: Genetic classifier system as a heuristic weighting method for a case-based classifier system. Butlletí de l'Associació Catalana d'Intel.ligència Artificial 22 (2000) 132–141
28. Kohavi, R.: A study of cross-validation and bootstrap for accuracy estimation and model selection. In: IJCAI. (1995) 1137–1145
29. Witten, I.H., Frank, E.: Data Mining: practical machine learning tools and techniques with java implementations. Morgan Kaufmann (2000)
30. Aha, D.W., Kibler, D.F., Albert, M.K.: Instance-based learning algorithms. Machine Learning 6 (1991) 37–66
31. Bernadó, E., Garrell, J.M.: Accuracy-based learning classifier systems: Models, analysis and applications to classification tasks. Special Issue of the Evolutionary Computation Journal on Learning Classifier Systems (in press) (2003)

Post-processing Clustering to Decrease Variability in XCS Induced Rulesets

Flavio Baronti, Alessandro Passaro, and Antonina Starita

Dipartimento di Informatica, Università di Pisa
Largo B. Pontecorvo, 3 — 56127 Pisa, Italy
{baronti, passaro, starita}@di.unipi.it

Abstract. XCS is a stochastic algorithm, so it does not guarantee to produce the same results when run with the same input. When interpretability matters, obtaining a single, stable result is important. We propose an algorithm which applies clustering in order to merge the rules produced from many XCS runs. Such an algorithm needs a measure of distance between rules; we then suggest a general definition for such a measure. We finally evaluate the results obtained on two well-known data sets, with respect to performance and stability. We find that stability is improved, while performance is slightly impaired.

1 Introduction

Randomness of the search process is one of the chief characteristics of evolutionary algorithms (EA), and indeed one of their strong points. It is randomness which allows them to escape local optima, and ensures a broader portion of the search space to be explored.

This non-deterministic behaviour is however a double-edged weapon; in fact, for non-trivial problems, it is likely that many repetitions of the algorithm will produce many different final solutions. For some kinds of tasks, the problem can be solved by simply picking the solution with best fitness: if we are solving the Travelling Salesman Problem with EA, the solution with lower cost is the one to choose, and the others can just be discarded. This is typically the case with strictly single-objective tasks. More often this is not possible; for instance, when employing EAs to train neural networks, many networks with similar accuracy might be found. In such a situation, since there is no obvious way to choose one, a viable alternative is to keep all of them, and set up a voting mechanism to take every one into account.

There are occasions however in which both choosing a single solution and voting are unviable or undesirable. Voting in fact has an important side-effect: it heavily obfuscates the reason why a certain decision was taken, distributing the responsibility among the set of classifiers.

XCS [7] too suffers from the variance problem; even after employing a ruleset reduction algorithm (like CRA [8]), on non-trivial problems many runs will produce different rulesets, each one with its own rules, with similar performance but

X. Llorà et al. (Eds.): IWLCS 2003-2005, LNAI 4399, pp. 80–92, 2007.

no evident way to build (or to choose) a single merged set. This causes one of the most appealing properties of XCS to be lost: interpretability of the classifiers appears much less interesting, if these classifiers continuously change among different executions. The problem is exacerbated by the fact that the same rule can come in many slightly variated forms, from whence the need to define a measure of similarity between rules.

Stability of the results is a main concern in all the settings where finding a set of interpretable classifiers is more the starting point of further research, than the arrival point; this happens for instance in the knowledge discovery in databases framework. One typical case of such a situation is medical research. Medical researchers collect general data, and would like to find associations and rules within it. When a "good" set of rules is found, effort must be spent in order to correctly interpret it, possibly involving the set up of specific medical trials. This makes stability a primary requirement, on par with performance.

We present a post-processing algorithm which tries to solve this problem, or at least mitigate it. The basic assumption is that good rules will be preferred by XCS, and will then appear more often in the output sets, although with slight variations. We repeat an XCS experiment a number of times; then a clustering algorithm is performed on all the resulting rules, putting together similar ones. Bigger clusters contain more frequent rules, so a representative from each of the biggest clusters is chosen; finally, a reduced version of XCS is executed again on this set of rules, in order to train them together and to set their working parameters (accuracy, fitness, etc.).

Section 2 will present all the definition that will be used in the rest of the paper. Subsection 2.1 will show the definitions necessary to understand and run the post-processing algorithm; subsection 2.2 will instead contain the definitions necessary to evaluate the effectiveness of the algorithm. The algorithm itself is outlined in section 3. Section 4 reports the benchmarks performed to evaluate the algorithm. Sections 5 and 6 describe and discuss the obtained results. Summary and conclusions are drawn in section 7.

2 Definitions

Some definitions are necessary before presenting the variability reduction algorithm: they are described in the following subsection. Subsection 2.2 reports instead the definitions necessary to evaluate the actual effectiveness of the algorithm; they are not necessary to simply run it.

2.1 Algorithm Definitions

In the following, we will make use of a measure of distance $\mathcal{D}$, and will assume that a corresponding measure of similarity $\mathcal{S}$ can be defined (which decreases as distance increases). A formal definition of distance and similarity, along with a survey of common measures, can be found in [9]. For our purposes, it will suffice to say that the measure $\mathcal{S}$ must lie in the range from 0 (maximally different items) to 1 (equal items). If the distance measure $\mathcal{D}$ provides values in the $[0, 1]$ range, a corresponding $\mathcal{S}$ can be simply obtained as $\mathcal{S} = 1 - \mathcal{D}$. If instead $\mathcal{D}$ can return any positive value, one possible family of valid definitions of $\mathcal{S}$ given $\mathcal{D}$ is

$$\mathcal{S} = (1 + \mathcal{D})^{-\alpha} \tag{1}$$

where α can be any positive number.

The basic function required by the algorithm is a measure of distance between rules. In general, the shape of rules is completely problem-dependent, so this measure has to be problem-dependent too. We suggest however a way to define it which makes it independent, as long as the problem provides a set of input data to be learned; this is not always the case, as XCS works by reinforcement learning, where the existence of such a set is just a particular situation.

In case this set exists, we define the *S-signature*[1] of a rule r as the set of input patterns the rule applies to. We then define two rules to be similar if they apply mostly to the same inputs (with a maximum when their signatures are equal); they are defined to be diverse instead when they apply to different inputs (with a minimum when the signatures do not have common elements). A suitable measure of distance can then be the Jaccard coefficient [4]:

$$D(r_1, r_2) = 1 - \frac{|S(r_1) \cap S(r_2)|}{|S(r_1) \cup S(r_2)|} \tag{2}$$

$$= 1 - \frac{|S(r_1) \cap S(r_2)|}{|S(r_1)| + |S(r_2)| - |S(r_1) \cap S(r_2)|} \tag{3}$$

This measure ranges from 0 (maximally different rules) to 1 (equal rules), and has been demonstrated to be a metric [5]. The corresponding similarity function is simply $S(r_1, r_2) = 1 - D(r_1, r_2)$.

An equivalent way to view a rule signature is as a Boolean vector. Since the input set size is a fixed value n, the B-signature of a rule r could be represented as a vector $\mathbf{b} \in \{0, 1\}^n$, where $\mathbf{b}_i = 1$ iff rule r applies to input pattern i. Jaccard distance becomes then

$$D(r_1, r_2) = 1 - \frac{\sum_{i=1}^{n} B(r_1)_i \wedge B(r_2)_i}{\sum_{i=1}^{n} B(r_1)_i \vee B(r_2)_i} \tag{4}$$

It is important to note that this measure is not strictly related to the intuitive notion of similarity between two rules. In fact, it bears no notion of how the rules actually *appear*: two rules could look completely different, and still apply to the same set of input patterns. This has experimentally been found to be common when rules are "tailored" to pick a very small subset of inputs — for instance, an outlier. In that situation, the data set probably offers many possible ways to match only that particular pattern with a number of conditions on its values; this will produce many different-looking rules, which for the system actually

---

[1] Note that this S is for "set", and is not to be confused with the $\mathcal{S}$ in Eq. 1, which was for "similarity".

have the same meaning. Consider for instance a setting where a single particular condition is sufficient to make a classifier match one pattern only (say *age>90*, in a data set where only one person is that old). All the other conditions of the classifier can then vary freely, as long as they continue to cover the pattern; it is then easy to build two different-looking classifiers, which maintain the same signature — that is, the same meaning to the system.

Another advantage of this definition of distance is that it does not require to choose a weighting strategy for attributes. If we had to compute similarity on the rule appearance (that is, on its conditions), we should decide how much importance to give to mismatches in the different attributes. This becomes more challenging when attributes do not have the same type. To exemplify, consider again a dataset containing people, with *age* and *gender* attributes among the others. Is the classifier (*gender*=M ∧ *age>25*) closer to (*gender*=F ∧ *age>25*), or to (*gender*=M ∧ *age>75*)?

As last remark, choosing a measure of similarity can create unexpected difficulties also within the same attribute. When modelling a real value for instance, it is entirely possible that its distribution is not uniform in the whole range of validity. Then, a little variation where the values are more frequent should be weighted more than a larger variation in areas where values are few. Recalling the previous example, if the only person above 90 is 95, the two conditions *age>90* and *age>94* appear equal to the system — while *age>50* and *age>54*, although differing of the same amount, probably describe quite different pattern sets.

2.2 Evaluation Definitions

In order to evaluate the goodness of a method to reduce variability, it is first necessary to provide a quantitative definition of variability. The basic block for such a definition is a measure of similarity between results of a classification algorithm. Since the results of our algorithms are rule sets, we need to define a measure of similarity between sets of rules — that is, sets of items which have themselves a similarity measure. This is what Jaccard coefficient does, using the simple *equals* relationship. We then extended Jaccard coefficient to this more general setting; to the best of our knowledge, this extension has not been proposed before.

The only thing which needs to be redefined in (3) for this extension is the size of the intersection. Notice that maximizing this value will minimize set distance (thus maximizing set similarity). We can put the items of the two sets in a fully connected bipartite graph, where each edge is weighted with the similarity between the two nodes. We then define the size of the intersection as the value of the maximal matching in the bipartite graph. This means that we assign each item from one set to at most one item from the other set; the size of intersection is then measured by taking the sum of similarities between the matched items, and by picking the matching which maximizes this sum.

Once we can measure the distance between two results, we must assess the variability contained in a set of such results. In the classical setting where a single result is a real number, the most widespread measure of variability is variance, defined as the average squared difference between each result and the average of the results.

Unfortunately, since in our setting each result is a ruleset, to calculate the classical variance we should be able to define an "average" ruleset. This is exaclty the goal of our algorithm, but we cannot use it to evaluate itself. We provide then two ways to evaluate variability of a set of rulesets:

- The mean and the standard deviation of the pairwise distance of the rulesets in the set. If we have ρ rulesets, this will calculate the mean and the standard deviation of $\frac{\rho(\rho-1)}{2}$ distances.
- A plot of the distribution of the $\frac{\rho(\rho-1)}{2}$ distances. This is not quantitative, but can provide more insight on the dynamics of variability.

3   The Algorithm

We now describe the algorithm we propose to join the results of many runs of XCS. We recall that the underlying assumption is that more important rules will be discovered and reported by XCS more often. This means in turn that, when joining all the rules generated by many runs, they will be more numerous than less important ones. Clustering all these rules, in order to put similar rules together, should thus yield bigger clusters for more important rules, and smaller clusters for less useful rules[2]. The algorithm is sketched in Fig. 1.

After each XCS run, a ruleset reduction algorithm is performed, in order to pick the most important rules. We applied Wilson's CRA [8], with two slight modifications. A sketch of the modified algorithm is in Fig. 2. The symbols follow Wilson's definition.

The first modification is the performance target: in the original CRA, step (a) is executed while maintaining performance at 100%. Our full rulesets did not reach that level of performance, so we simply executed step (a) maintaining the performance value of the original ruleset.

The second modification is in step (b). On the $\mathbf{M}_{n^*}$ set, we remove in turn each classifier c_i, but starting with $i = n^*$ and working back to $i = 0$. If the performance on the reducing set has dropped, we add the classifier again to the set. There are two differences from the original algorithm. First of all, this procedure recalculates performance at each removal, and cancels the removal if there is a negative variation: this ensures the original performance level is maintained. The second difference is the backwards order: we believe that, in this way, we first remove the less important classifiers — which are also more likely to suffer from higher variance.

After ruleset reduction, the Boolean signature $B(r)$ of all the m rules resulting from the runs is computed. This produces m Boolean vectors, which will be

---

[2] Notice that *bigger* and *smaller* refer to the number of items in each cluster. We suppose that the clustering algorithm produces clusters with a similar value of dispersion.

partitioned with a clustering algorithm (like k-means [3], or ROCK [2]) according to the chosen rule distance. Following the basic assumption — the frequency of a rule is a mark of its importance — the cluster sizes sort rules over their importance; the l biggest clusters will then describe the l most important rules. From them, one representative rule must be chosen (in the k-means implementation, we picked the rule closest to the cluster centroid). Finally, the l chosen rules should be trained again to work together; this can be done running XCS again with a fixed population consisting of them alone, and allowing only the performance, error, and fitness values to vary.

1. Run XCS R times; collect the resulting R rulesets.
2. Execute a ruleset reduction algorithm on each of the R rulesets.
3. Collect all the rules from each reduced ruleset; ignore all the parameters (fitness, numerosity, ...). Call m the number of resulting rules.
4. Cluster the m rules, obtaining k clusters.
5. Sort the clusters by descending size (number of rules inside a cluster).
6. Pick a representative rule from the first l clusters.
7. Train these l rules again with fixed-population XCS.

Fig. 1. Sketch of the general variability reduction algorithm

(a) Find the smallest n^* such that $\mathbf{M}_{n^*}$ achieves the same performance ν as the full ruleset.
(b) Remove from $\mathbf{M}_{n^*}$ each classifier c such that the performance of $\mathbf{M}_{n^*} \setminus \{c\}$ is not below ν. Check backwards, from the end of $\mathbf{M}_{n^*}$, up to the start.
(c) Same as Wilson: sort the classifiers in descending order of marginal contribution.

Fig. 2. Details of our implementation of Wilson's CRA

4. Apply k-means to cluster the m rules:
   (a) Calculate the boolean signature of the m rules.
   (b) Calculate distances between rules according to Eq. 4.
   (c) Apply k-means, with $k = \frac{4}{3}\frac{m}{R}$
5. Discard clusters containing less than $R/3$ rules.
6. From the remaining clusters, choose the rule closest to the centroid as representative.

Fig. 3. Details of the particular implementation of steps 4–6 of the general algorithm

4 Test Setting

We tested the whole algorithm on the well-known *WBC* and *mushroom* data sets, from the UCI repository [6]. The first aim was to evaluate the effective reduction in variability provided by the algorithm. 300 10-fold cross-validated XCS runs

Table 1. Summary of XCS parameters. Naming of the variables follows [1].

β	0.15	μ	0.04	ϵ_I	10
α	0.1	θ_{del}	10	F_I	0.01
ϵ_0	10	δ	0.1	θ_{mna}	2
ν	5	θ_{sub}	20	p_{explr}	0.5
γ	N/A	$P_\#$	0.333	p_{GAsub}	1.0
θ_{ga}	40	p_I	500	p_{ASsub}	1.0
χ	0.8				

were executed for each data set; then, the original variability was calculated, by taking the distance between each pair of results obtained from the same fold. This produces $10 \cdot \frac{1}{2} \cdot 300 \cdot 299 = 448500$ distances, whose distribution is plotted on the top-left graph in Figs. 5 and 6. We now applied the clustering algorithm as follows: chosen a number R, we clustered together the results of R runs of the original algorithm, and applied retraining. We finally evaluated distances again, obtaining $10 \cdot \frac{1}{2} \cdot \frac{300}{R} \cdot (\frac{300}{R} - 1)$ distances, which again are plotted on the graph. Average and standard deviation of the distance are calculated as well, and reported on the last column of Tables 2 and 3.

The second aim in testing was to assess if, and how much, the clustering process affected performance. As we already stated, taking apart rules structured to work together and putting them with rules coming from different sets could disrupt their cooperative effect, and impair performance. This effect was evaluated through the accuracy value in the test set obtained through cross validation.

The parameters of XCS common to both problems are reported in Table 1: they are the typical default values. WBC could appear a simpler problem than Mushroom, since its input space is 10^{10}, compared to Mushroom's which is $\approx 10^{15}$. However, the search space complexity is the opposite: while Mushroom has to find good classifiers in a $\approx 10^{17}$ space, WBC must work in a $\approx 10^{65}$ space. We decided then to employ a population size of 200 for Mushroom and 400 for WBC; evolution was allowed to run for $150k$ generations in Mushroom, compared to $250k$ in WBC. These figures, although not very high, were sufficient for the ruleset to reach a stable accuracy.

As regards the clustering procedure (see Fig. 3 for a sketch of the algorithm), we employed k-means with $k = \frac{4}{3}\frac{m}{R}$, where m is the total number of classifiers being clustered, and R is again the number of basic XCS runs these classifiers were produced by. $\frac{m}{R}$ is the average number of classifier produced by a run; if all the runs produced the same classifiers, we should obtain exactly $\frac{m}{R}$ clusters, with R rules each. Since this is generally not the case, we added the $\frac{4}{3}$ coefficient. Then, the clusters with less than $R/3$ rules were discarded; approximately, this means that we keep only the rules which appeared at least once out of three runs. From each cluster, the classifier closest to the cluster centroid was taken as the representative rule.

5   Results

Results of the tests are summarized in Tables 2 and 3, with increasing number R of basic XCS experiments clustered together. Figures 5 and 6 show instead the density function of the distance between two rulesets produced by clustering, again with increasing number of basic experiments joined.

Eye inspection of accuracy, specificity and sensitivity values of the test set for both data sets shows a moderate decrease in performance after clustering, substantially independent of the R value. Statistical analysis confirms these first impressions. On Mushroom, the Kruskal-Wallis test on the accuracy, specificity and sensitivity values did not reveal any significant differences between all the clustered results (p-values: .582, .274 and .320 respectively). On the other side, a Kolmogorov-Smirnov test between the original results and all the clustered results showed a significant difference (all the p values are .000 for accuracy, specificity and sensitivity)[3]. Figures for the WBC problem are similar: no significant difference between clustered results (p-values: .398, .667, .953), significant difference between them and the original results (again the p-values are all .000).

While the two data sets appeared of comparable difficulty with respect to accuracy, results look very different for the distance of produced rules. In fact, in the *mushroom* problem the original rulesets have a quite broad range of distances, centered around the 0.5 value. In the *WBC* problem the distribution curve of distances appears radically different, with a high peak around 0.9, and a long, mostly low tail towards 0. As regards the effects of clustering on distance, small values of R immediately create a radical shift in the distribution, filtering out the tails (on both sides) and lowering the average. t-tests performed on consecutive pairs of results (e.g. $R = 1$ with $R = 5$, $R = 5$ with $R = 10$ and so on) show that for Mushroom there is not much improvement in clustering more than 12 results, while for WBC improvement stops around 25 clustered results (values reported in Tables 2 and 3).

The mushroom data set after clustering produces a distribution with at least three distinct peaks. This peculiar shape could be explained by the resulting rulesets being roughly divided into two groups; the distance within each group is represented by two of the peaks, while the third is the distance across the groups. Further investigation is however necessary in order to confirm this hypothesis.

6   Discussion

The test setting was designed to check whether the clustering algorithm disrupted the original performance, and whether it could reduce variability of the results.

As regards performance, the decrease is not too marked (around 4%), but is significant. The increase in standard deviation is a hint of what's actually happening: a check on the distribution of accuracy values (Fig. 4) reveals that the clustered results have a left tail of sporadic, unsuccesful experiments. Further

---

[3] For the p-values, we use ".000" as a shorthand for "$< .001$".

Table 2. Summary of results on the Mushroom data set. Accuracy, specificity and sensitivity are computed on the test set. The first row reports the results without clustering. The last column shows the p-value of a two-tailed t-test with the distances on the previous row.

R value	Accuracy	Specificity	Sensitivity	Rules no.	Distance	p-value
—	.98 ± .01	.99 ± .01	.97 ± .01	11 ± 3	.48 ± .21	—
5	.95 ± .11	.95 ± .19	.95 ± .11	12 ± 1	.28 ± .10	.000
10	.94 ± .12	.95 ± .19	.94 ± .11	11 ± 1	.25 ± .08	.000
12	.95 ± .10	.95 ± .19	.95 ± .08	11 ± 1	.23 ± .08	.000
15	.95 ± .12	.94 ± .20	.95 ± .10	11 ± 1	.24 ± .08	.042
20	.96 ± .08	.97 ± .13	.95 ± .10	11 ± 1	.24 ± .07	.031
25	.95 ± .10	.94 ± .20	.96 ± .06	11 ± 1	.24 ± .07	.245
30	.94 ± .13	.94 ± .21	.94 ± .13	11 ± 1	.24 ± .07	.316

Table 3. Summary of results on the WBC data set. Accuracy, specificity and sensitivity are computed on the test set. The first row reports the results without clustering. The last column shows the p-value of a two-tailed t-test with the distances on the previous row.

R value	Accuracy	Specificity	Sensitivity	Rules no.	Distance	p-value
—	.93 ± .04	.88 ± .10	.96 ± .03	7 ± 6	.69 ± .23	—
5	.88 ± .13	.84 ± .19	.91 ± .20	6 ± 2	.62 ± .10	.000
10	.89 ± .14	.85 ± .17	.91 ± .21	6 ± 1	.55 ± .09	.000
12	.87 ± .17	.83 ± .21	.89 ± .25	6 ± 1	.54 ± .10	.000
15	.90 ± .12	.85 ± .19	.92 ± .18	6 ± 1	.52 ± .09	.000
20	.89 ± .13	.84 ± .20	.92 ± .19	6 ± 1	.49 ± .08	.000
25	.88 ± .14	.85 ± .19	.90 ± .23	6 ± 1	.47 ± .09	.002
30	.89 ± .15	.86 ± .19	.91 ± .21	6 ± 1	.47 ± .09	.474

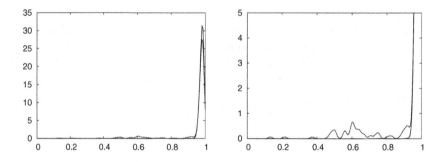

Fig. 4. Distribution of accuracy values for $R = 1$ (solid line) and for $R = 5$ (dashed line) in the WBC experiment. The graph on the right is a detail of the left one.

analysis on the accuracy distribution of WBC reported that 10% of the clustered results were off the unclustered average by more than 3 times the unclustered standard deviation. Most of the time the algorithm does not impair performance; however, the possibility of a failure must then be taken into account in the

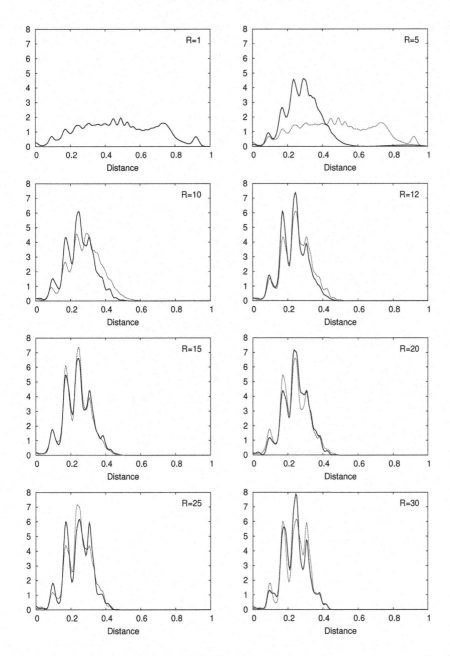

Fig. 5. Distribution of distances for the *mushroom* data set as the number of clustered experiments R grows. The X axis displays distance; the solid curve is the density function of distance for the R value displayed; the dashed curve is the density function for the previous R value.

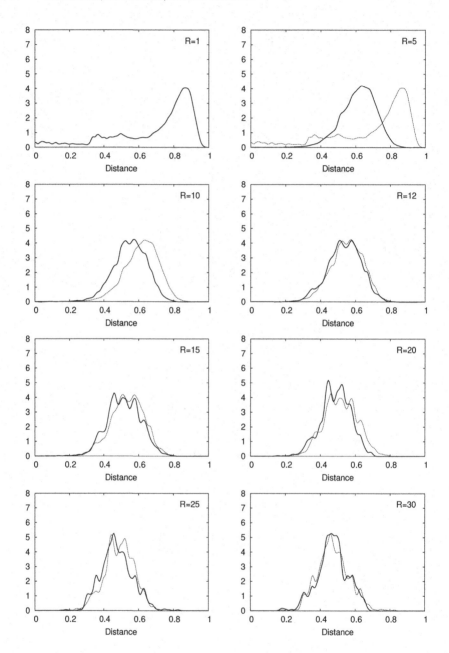

Fig. 6. Distribution of distances for the *WBC* data set as the number of clustered experiments R grows. The X axis displays distance; the solid curve is the density function of distance for the R value displayed; the dashed curve is the density function for the previous R value.

implementation (for instance, repeating the experiment if the accuracy obtained after clustering is much lower than the average of the original accuracies).

Moving to the analysis of distances, the flat distribution for Mushroom values can be interpreted as XCS generally producing a common subset of similar rules, with some slight modifications which account for the higher distance levels. The high peak at 0.9 on WBC is instead a mark of XCS producing very different rulesets, with little in common. The high variance in the results, paired with the low variance in accuracy, points out that the problem is complex enough to offer a fitness surface with many local optima, equivalent from the accuracy point of view, but radically distant in the search space. On this data the algorithm is more effective, moving the distribution of distances towards a 0.5 average.

Finally, while performance was affected in the same way for all R levels, distance gets significantly lower as R increases. This effect however decreases in magnitude as R increases, until it stops: further increases of the number of results clustered together do not promote further decreases on variance.

Our method is not able to produce "the" ruleset for a certain data set. It does instead give a way to merge the sets of rules resulting from many repetitions of experiments, obtaining a single set with comparable performance and higher stability. Very often experiments are already run many times; then, one is choosen either by visual inspection or by higher performance. Running the clustering algorithm in this situation has a low computational cost, and the benefit of lowering variability.

7   Conclusions

This paper tackled the problem of variability in the rules resulting from XCS runs. We presented a measure of distance between rules, and used it in a clustering algorithm to find recurring rules between different runs. Bigger clusters should contain more frequent rules, and thus more important and stable ones. We tested the algorithm on two data sets from the UCI repository, namely *mushroom* and *WBC*, and evaluated the distance between the resulting rulesets through a novel measure of distance. Results show the algorithm is effective, enhancing stability of the results while maintaining performance. Stability however does not appear to be improvable over a certain level, although incrementing the number of basic runs being clustered together (and thus the quantity of data available to clustering).

Future work will involve the application of different, possibly non-parametric clustering algorithms, in order to remove the dependency on a fixed number of clusters k, and to more easily isolate outliers. Another interesting point is the evaluation of the measure of distance between rules against other possible measures. Finally, a different approach could focus on building ruleset reduction algorithms (like CRA) specifically geared towards obtaining low-variance results.

References

1. Martin V. Butz and Stewart W. Wilson. An algorithmic description of XCS. In P. L. Lanzi and et al., editors, *IWLCS 2000*, volume 1996 of *LNAI*, pages 253–272. Springer-Verlag, 2001.
2. Sudipto Guha, Rajeev Rastogi, and Kyuseok Shim. ROCK: A robust clustering algorithm for categorical attribute. *Information Systems*, 25(5):345–366, 2000.
3. Zhexue Huang. Extensions to the k-means algorithm for clustering large data sets with categorical values. *Data Mining and Knowledge Discovery*, 2(3):283–304, Sep 1998.
4. Anil K. Jain and Richard C. Dubes. *Algorithms for Clustering Data*. Prentice Hall, NJ, 1998.
5. Alan H. Lipkus. A proof of the triangle inequality for the Tanimoto distance. *Journal of Mathematical Chemistry*, 26(1–3):263–265, March 1999.
6. D. J. Newman, S. Hettich, C. L. Blake, and C. J. Merz. UCI repository of machine learning databases [http://www.ics.uci.edu/~mlearn/MLRepository.html], 1998.
7. Stewart W. Wilson. Classifier fitness based on accuracy. *Ev. Comp.*, 3(2), 1995.
8. Stewart W. Wilson. Compact rulesets from XCSI. In P. L. Lanzi and et al., editors, *IWLCS 2001*, volume 2321, pages 197–210. Springer-Verlag, 2001.
9. Rui Xu and Donald Wunsch. Survey of clustering algorithms. *IEEE Transactions on Neural Networks*, 16(3):645–678, May 2005.

LCSE: Learning Classifier System Ensemble for Incremental Medical Instances

Yang Gao[1], Joshua Zhexue Huang[2], Hongqiang Rong[3], and Da-qian Gu[1]

[1] State Key Laboratory for Novel Software Technology, Nanjing University,
Nanjing, 210093, China
gaoy@nju.edu.cn
[2] E-business Technology Institute, The University of Hong Kong, China
jhuang@eti.hku.hk
[3] Department of Computer Science, The University of Hong Kong, China
hqrong@cs.hku.hk

Abstract. This paper proposes LCSE, a learning classifier system ensemble, which is an extension to the classical learning classifier system(LCS). The classical LCS includes two major modules, a genetic algorithm module used to facilitate rule discovery, and a reinforcement learning module used to adjust the strength of the corresponding rules after the learning module receives the rewards from the environment. In LCSE we build a two-level ensemble architecture to enhance the generalization of LCS. In the first-level, new instances are first bootstrapped and sent to several LCSs for classification. Then, in the second-level, a simple plurality-vote method is used to combine the classification results of individual LCSs into a final decision. Experiments on some benchmark medical data sets from the UCI repository have shown that LCSE has better performance on incremental medical data learning and better generalization ability than the single LCS and other supervised learning methods.

1 Introduction

Medical data analysis is aimed at discovery and refinement of concepts, or rules, that exist in medical databases. Supervised machine learning systems have been proved useful tools for medical data analysis in several medical domains. Because of special characteristics of medical data, the current learning systems need to be further enhanced. Firstly, sufficient samples for batch training are often difficult to collect in a short time. This requires the learning system to be able to incrementally learn from gradually incoming medical cases. Secondly, the positive and the negative cases are not always identifiable. The conclusions often depend on the historical experience of some rewards (payoff or penalty) received from the environment. In other words, the reinforcement learning paradigm must be adapted to the learning process. Thirdly, in order to understand the learnt knowledge, the process and results of the learning system should be easily interpretable. Many supervised machine learning methods do not have the ability for

X. Llorà et al. (Eds.): IWLCS 2003-2005, LNAI 4399, pp. 93–103, 2007.

incremental reinforcement learning. Although some are able to learn incrementally, such as the artificial neural network systems, the learning process and rules are not comprehensible because of the black box nature of these techniques.

The Learning Classifier System(LCS) is a machine learning technique that combines reinforcement learning, evolutionary computing and heuristics into an adaptive system. Each learning classifier system is a rule-based system in which the rules are in the form of "If conditions THEN action". Evolutionary computing techniques and heuristics are used to search for the possible rules, whilst the reinforcement learning techniques are used to assign utility to existing rules, thereby guiding the search for better rules. Bonelli et al.[1] have demonstrated that the learning classifier system is suitable in three medical domains. Holmes applied LCS to knowledge discovery in the clinical research databases and achieved some fruitful results in estimation of disease risk and epidemiologic surveillance [2][3]. Wilson [4][5] proposed the XCSR technique to adapt real-value attributes in LCS and used XCS in the oblique data set from the Wisconsin Breast Cancer data. Bernadó et al.[6] selected several data sets from the UCI repository, such as Pima-indians etc., to compare the performances of XCS and GALE. Similar works have been done by Bacardit et al. on comparison of XCS and GAssist[7].

The above research has shown that LCS works well on medical data mining. However, LCS tends to over-fit the data for smaller data sets[7]. Other problems are noisy data (incorrect measured values) and missing values which often occur in medical data sets. To solve these problems with LCS, we need to improve it in terms of generalization to avoid over-fitting and to increase accuracy of classification. So far, the ensemble method is one of most interesting and attractive learning systems with strong generalization ability. Ensemble learning refers to a collection of methods that learn a target function by training a number of individual learners and combing their predictions[8]. Breiman[9] proved that generating multiple versions of a predictor and using them to get an aggregated predictor can improve the prediction accuracy of a weak predictor. Usually, multiple learners are generated by training individual classifiers on different datasets obtained by resampling a common training set(such as bagging or boosting). Then, the multiple learners are integrated to form the best learner using a combination strategy(such as voting, averaging the counts, etc.) The ensemble technology based on supervised machine learning has been studied and applied to several domains, such as medical diagnosis, face recognition, scientific image analysis [10], etc. However, combination of LCS with ensemble learning has not yet been well tested.

In this paper, we propose a two-level LCS ensemble system(LCSE) and discuss its applications in knowledge discovery of incremental medical data. In the first-level of the ensemble, new instances are first bootstrapped and sent to several LCSs for classification. Then, in the second-level, a plurality-vote method is used to combine the classification results of individual LCSs into a final decision. Experiments on some benchmark medical data sets from the UCI repository have shown that LCSE has better performance on incremental medical data learning and better generalization ability than the single LCS and other supervised learning methods.

The paper is organized as follows. In Section 2, we introduce the architecture of XCSR which is one of the popular learning classifier systems used to deal with real-value attributes. In Section 3, a two-level learning classifier system ensemble is presented and the learning process is discussed in detail. We present some test experiments on the Pima Indians Diabetes data set and investigate the performance of the respective approaches in Section 4. Finally, we draw some conclusions and outline future works in Section 5.

2   The Architecture and Process Cycle of XCSR

The traditional classifier system(CS) is designed for batch data training. Holland et al. introduced the reinforcement component to the CS to improve the system performance from incremental instances. The new CS was named the Learning Classifier System (LCS)[11]. LCS consists of two important components. One is the genetic algorithm module for LCS to create new classifier rules. The other is the reinforcement learning module that receives payoff from the environment, distributes the message and adjusts the classifier's strength. Currently, there are three types of learning classifier systems – ZCS, XCS and ACS. Considering there is non-causality between the sequence medical instances, we use XCS in the learning classifier system ensemble in this paper. The basic architecture of LCS will be explained using the advanced learning classifier system, XCSR[4].

The condition of rule is often composed of the conjunction of different binary attributes, which can easily match the discrete-valued inputs in LCS. In order to deal with the continuous-valued variables such as temperature, blood pressure etc., XCSR changed the rule representation's structure to use the interval representations instead of a ternary representation $(0, 1, \#)$[4]. The real-value attribute is represented as a half-open interval in the form of (centre-value, width). Based on this representation, XCSR can be broadly applied to the problems in continuous-valued environments. The framework of XCSR is presented in Fig.1.

The XCSR includes two interface modules, the sensor and the effector. The sensor receives outside input and the effector specifies an action. When applied to the data mining domain, the sensor module receives a new instance (without the categorical label) whilst the effector module outputs the classification of the respective input.

There are four important data structures in XCSR. The first one is the population of classification rules, $[P]$. The jth rule C_j in $[P]$ is represented as a 5-tuple of (condition, action, predictive reward p_j, predictive reward's error e_j, fitness value f_j). The condition of a rule uses an interval representation. The action represents a possible category. If C_j is a rule that correctly predicts the reward, then the predictive reward p_j is positive. Otherwise, p_j is negative or zero. The predictive reward's error e_j measures the difference between p_j and the actual receiving reward R as shown in equation(1).

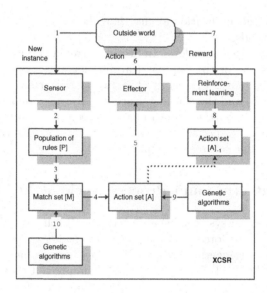

Fig. 1. The Schematic of XCSR

Equation(2) defines the accuracy κ_j and the relative accuracy κ'_j. That the initial fitness f_j is 0 reflects the adaption capacity. f_j is calculated from κ'_j in the reinforcement learning module. In general, the learning system intends to reserve the rules which have higher accuracy and higher relative accuracy. We use the parameter P to measure the size of $[P]$. If we increase the value of P, the system's performance will be improved but the computation cost is increased distinctly.

$$e_j = |R - p_j| \tag{1}$$

$$\begin{cases} \kappa_j = 0.1(e_j/e_0)^{-n} & if \ e_j > e_0 \\[2mm] \kappa_j = 1 & if \ e_j \le e_0 \\[2mm] \kappa'_j = \kappa_j / \sum_i \kappa_i \end{cases} \tag{2}$$

The second data structure is the match set $[M]$. In the 3rd process of Fig. 1, the system checks the input against the conditions of the rules in $[P]$. The rules that meet the conditions form the match set $[M]$. In implementation, we add a flag for each rule in $[P]$ in order to reduce the storage complexity of the system. When the flag of a rule is set true, it indicates that the rule is in $[M]$.

The action set $[A]$ is the third data structure. In $[M]$, each rule has an action(or category). By balancing the exploration and exploitation, we use the roulette wheel selecting algorithm to choose an action in process 4. All rules that have an action the same as the selected action form the action set $[A]$. The action set $[A]_{-1}$ is the action set in the last learning step.

The most important modules in XCSR are the reinforcement learning module and the genetic algorithm module. When the system outputs an action in $[A]$, the environment returns a corresponding payoff in process 7. The system will get the positive reward R if it outputs the correct category. Otherwise, the system will get a negative value. In process 8, according to the payoff R, the system adjusts the performance parameters of all rules in the action set $[A]_{-1}$ using equations(2), (3) and (4) in the reinforcement learning module. Of course, the previous action set $[A]_{-1}$ in Fig. 1 is replaced with the current action set $[A]$ when the problem is single-step. While the system is running, the genetic algorithm module is activated at every regular interval. Using the roulette wheel selection method, two rules that have the great fitness value are chosen from the action set $[A]$ and crossed over two-point with the probability χ and mutated with the probability μ per allele in the 9th process. The newly generated rules are inserted into the population set $[P]$ and two low-fitness rules are removed from $[P]$. If all rules in the population set $[P]$ could't match the incoming instance, the cover module is activated and generates rules in the 10th process.

$$p_j \leftarrow p_j + \beta(R - p_j) \tag{3}$$

$$f_j \leftarrow f_j + \beta(\kappa_j' - f_j) \tag{4}$$

3   Learning Classifier System Ensemble

The learning classifier system ensemble or LCSE we propose in this paper combines the learning classifier system with ensemble learning in order to improve the system generality. Ensemble learning refers to a collection of methods that learn a target function by training a number of individual learners and combining their predictions. Dieterich showed that the uncorrelated errors of individual classifiers can be eliminated through averaging method, so the ensemble machine may approximate the desired target function[8]. Currently, major methods for constructing ensembles include subsampling the training examples, manipulating the input features, manipulating the output targets and modifying the learning parameters of the classifier. In this paper, we choose the first method to construct the ensemble learner.

Fig. 2 shows the system architecture of LCSE. Besides several sub LCSs, a bootstrap module and a voting module are added. The bootstrap module is used to distribute the inputs to different sub-LCSs and the voting module is used to combine all classification results of sub-LCSs to produce the final system output. When subsampling the training set, two different sampling techniques are used, bagging and boosting. Bagging is the sampling-with-replacement procedure where each learner in ensembles is trained on the average probability of the training examples[9]. Boosting takes a different resampling approach than bagging, which maintains a constant probability of $1/N$ for selecting each individual example[12]. In our current research, we investigate the bagging method

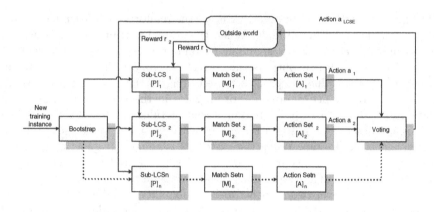

Fig. 2. The Architecture of Two-level Learning Classifier System Ensemble

to construct the ensemble learner. The boosting method will be considered in our future work.

At initialization of the system, the sub-LCSs random initialize their population sets. In each episode, the bootstrap module inputs new samples to every sub-LCSs randomly with respect to the probability λ given in equation(5) where LCS_i denotes the ith sub-LCS. In other words, the bootstrap module executes the bagging procedure. Given a dataset D containing S samples, the bootstrap module generates N bootstrap data sets by drawing some samples from D for N times. Because all samples for training did not be obtained in batch in LCSE, we design the bootstrap module of LCSE as equation(5). The sampling probability λ is set as 63.2% because each sample has a probability of $1 - (1 - 1/S)^S$ of being selected at least once in the S samples of a dataset D in the bagging procedure[13]. When $S \to \infty$, the sampling probability converges to 0.632. According to equation(6), the ith sub-LCS receives current input if the random number generated by the ith sub-LCS is not greater than λ. Or else, the ith sub-LCS rejects current input.

After the ith sub-LCS receives an input, it constructs the match set $[M]_i$. Then the roulette wheel selecting algorithm is used to get the action set $[A]_i$. Finally, the ith sub-LCS outputs its classification result a_i. Hence, multiple different learning classifier systems with the different accuracy on the data are constructed. In order to find the best or near optimal learning classifier system, LCSE combines all sub-LCSs that belong to the same classification result. Usually, the combined result can be obtained by voting or averaging to reduce the overall expected error and improve the generalization performance of LCSE. Since LCSE aims at classifying the unknown medical instances correctly, the majority voting is used to select the final action of each instance. So, the voting module ensembles the sub-LCSs' outputs according to equation(6) to get the final system's classification result a_{LCSE}. The voting module uses the basic plurality voting method.

$$\begin{cases} if\ rand_i() \leq \lambda & bootstrap(LCS_i) \leftarrow TRUE \\ \\ else & bootstrap(LCS_i) \leftarrow FALSE \end{cases} \tag{5}$$

$$a_{LCSE} \leftarrow argmax_a \sum_i vote(a_i) \tag{6}$$

We must emphasize that each sub-LCS that participates in voting will receive a respective payoff r_i from the environment with respect to its input a_i. The sub-LCS will get a reward of positive value r if its output is correct. Otherwise, the sub-LCS will pay the penalty with the negative value $-r$. Then, every sub-LCS processes reinforcement learning of the rule strength and rule discovery based on its payoff. In Fig. 2, the solid line represents that the ith sub-LCS is activated. The dashed line indicates that the respective sub-LCS is not activated in the learning episode.

Essentially, the bootstrap module in LCSE aims at developing multiple sub-LCSs with different classification performances by means of inputting different samples, initiating different population sets, and undertaking different rule learning and discovery processes. In other words, the classification results of every sub-LCS may be different even though the input sample is same. Therefore, the generality of the learning classifier system is improved when combining multiple sub-LCSs with ensemble learning.

Table 1 gives the detail process of LCSE and Table 2 lists the parameters of LCSE.

Table 1. The process of LCSE

---

1. Create populations of each sub-LCS in the system and initialize;
2. While not reach the maximum learning step T {
    2.1 Distribute the input to sub-LCSs randomly using (5);
    2.2 If the ith sub-LCS is activated, form the match set $[M]_i$. Otherwise, do nothing;
    2.3 If the ith sub-LCS is activated, construct the action set $[A]_i$ using the roulette wheel selecting algorithm and generate the classification output a_i. Otherwise, do nothing;
    2.4 Vote the system output a_{LCSE} with (6);
    2.5 If the ith sub-LCS is an activated sub-LCS, receive its own payoff r_i depending on a_i. Otherwise, do nothing;
    2.6 If the ith sub-LCS is an activated sub-LCS, adjust rule parameters using (1), (2), (3) and (4). Otherwise, do nothing;
    2.7 When T reaches the regular steps, do GA on its own action set of every sub-LCS;
    }
3. Return;

---

4 Experimental Results and Analysis

The Pima Indians Diabetes data set from National Institute of Diabetes and Digestive and Kidney diseases was used in a benchmark performance test. The

Table 2. The description and values of parameters in LCSE

Para.	Description	Value
N	Number of sub-LCSs	3, 5, 7 or 9
P	Population set's size of each sub-LCS	800
T	Total training steps	2000
α	Learning rate of reinforcement learning	0.2
χ	Probability of crossover operator	0.8
μ	Probability of mutation operator	0.04
θ_{GA}	The number of steps when activating GA process	12
$\theta_{Covering}$	Threshold for covering process	5
ϵ_0	Threshold of computing error of a classifier	10
s_0	The parameter of create random scope of a condition in classifier	1.0

diabetes dataset consist of eight condition attributes. They are (1) the number of pregnancy times, (2) plasma glucose concentration in 2 hours in an oral glucose tolerance test, (3) diastolic blood pressure (mm Hg), (4) triceps skin fold thickness (mm), (5) 2-hour serum insulin (mu U/ml), (6) body mass index (weight in kg/(height in m)^2), (7) diabetes pedigree function, and (8) age (years). The output value 0 represents the benign state (500 instances, 65.1%) and 1 represents the malignant state (268 instances, 34.9%).

4.1   Experiment 1

In order to highlight the over-fitting problem, the stratified tenfold cross-validation method was applied to the diabetes dataset in the first experiment. We randomly divided the diabetes dataset into 10 subsets (or folds). In each subset all actions had the same likelihood to be taken. In each trial, we used one subset for testing and the remaining nine subsets for learning. Table 2 gives the parameter settings.

Results of the stratified cross-validation test of LCSE with seven sub-LCSs are shown in Table 3. Table 3 shows the comparison results of the average fraction correct between LCSE and other learning methods. Among these methods, we used the decision tree and neural network methods in the Weka toolbox from the University of Waikato in New Zealand [14]. In Table 4, detailed results of stratified cross-validation test on the benchmark dataset is exhibited.

From Table 3 and Table 4, we can see that most LCSE configurations performed better than LCS(XCSR). And, we can also see from Table 3 that the average fraction accuracy increases as the number of sub-LCSs increases in LCSE. For LCSE with 7 sub-LCSs, its average prediction accuracy reached 77.34%. For LCSE with 9 sub-LCSs, this value is 77.47%. Because the difference between them is not obvious, we think that there is a limit in which the method approaches the maximal accuracy and further addition of sub-LCSs will no longer have effect. But, the LCSE is not enough stable since their standard deviations are greater than XCSR, decision tree and neural network. So, we expect we can decrease the standard deviation of LCSE in the future work.

Table 3. Comparing Performance of LCSE with XCSR, Decision Tree and Neural Network

Learning Method	Correct Classification	Misclassification	Average Fraction Accuracy	Standard Deviations
LCSE with 9 sub-LCSs	595	173	0.7747	0.00240
LCSE with 7 sub-LCSs	594	174	0.7734	0.00189
LCSE with 5 sub-LCSs	577	191	0.7513	0.00192
LCSE with 3 sub-LCSs	564	204	0.7344	0.00333
XSCR	548	220	0.7135	0.00107
Decision Tree(J48)	554	214	0.7213	0.00110
Neural Network	585	183	0.7617	0.00049

Table 4. Detailed results of stratified cross-validation test on the benchmark dataset (Correct calssification/Misclassification)

	XCSR	LCSE with 3 sub-LCSs	LCSE with 5 sub-LCSs	LCSE with 7 sub-LCSs	LCSE with 9 sub-LCSs	DT(J48)	NN
Trial 1	52/25	52/25	54/23	58/19	66/11	60/17	58/19
Trial 2	56/21	61/16	62/15	59/16	57/20	57/20	58/19
Trial 3	53/24	57/20	53/24	54/23	61/16	54/23	58/19
Trial 4	50/27	49/28	56/21	57/20	59/18	57/20	58/19
Trial 5	55/22	56/21	57/20	58/19	56/21	57/20	63/14
Trial 6	58/19	56/21	59/18	61/16	59/18	53/24	60/17
Trial 7	54/23	54/23	60/17	63/14	62/15	51/26	57/20
Trial 8	57/20	65/12	61/16	64/13	61/16	53/24	59/18
Trial 9	56/20	54/22	54/22	56/20	62/14	56/20	56/20
Trial 10	57/19	60/16	61/15	64/12	52/24	56/20	58/18

4.2   Experiment 2

We conducted the second experiment on the Pima Indians Diabetes data set to investigate the on-line learning performance. In the second experiment, we also split the whole dataset into 10 subsets. Firstly, we used the first subset for learning and the second subset for testing. Then, we used the first two subsets for learning and the third subset for testing. This process continued till the step that the first nine subsets were used for learning and the last subset for testing. In fact, since both the learning classifier system LCS and the learning classifier system ensemble LCSE are on-line learning methods, it is unnecessary to separate the testing process from the learning process. The purpose of the above learning and testing stages was to compare LCSE with decision tree and neural network methods.

We conducted 10 trials on each learning method. The results are shown in Fig.3 and Fig.4. We can see that the prediction accuracy has been improved when the systems learnt more instances. LCSE with 7 sub-LCSs performed better in the on-line learning than other LCSEs and LCS in almost all learning stages.

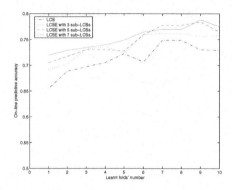 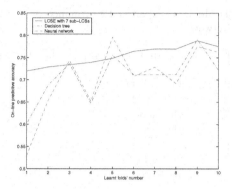

Fig. 3. Comparison between LCSE and LCS

Fig. 4. Comparison between LCSE and Neural Network and Decision Tree

5   Conclusions and Future Work

This paper has presented a Learning Classifier System Ensemble (LCSE) which combines the learning classifier system with ensemble learning to improve the generality of the single learning classifier system. The experiment results have shown that LCSE with sub-LCSs performs better than other learning methods such as decision tree and neural networks on the diabetes data set. Furthermore, the LCSEs with more sub-LCSs outperformed the LCSEs with less sub-LCSs as well as LCS. These initial results have demonstrated the advantages of the LCSE learning system. But, we know this work is a first step to using ensembles of Learning Classifier System to improve classification accuracy. In our future work, we plan to investigate other ensemble methods and benchmark LCSE on other data mining domains.

Acknowledgement

The paper is supported by the Natural Science Foundation of China (No.60475026), the National Outstanding Youth Foundation of China (No.60325207) and the National Grand Fundamental Research 973 Program of China (No.2002CB312002).

References

1. P. Bonelli, A. Parodi, An Efficient Classifier System and Its Experimental Comparison with Two Representative Learning Methods on Three Medical Domains. In R.K. Belew and L.B. Booker, editors, Proceedings of the fourth international conference on Genetic algorithms(ICGA-4), pages 288-295. San Mateo, CA:Morgan Kaufmann, 1991.

2. John H. Holmes, Learning Classifier Systems Applied to Knowledge Discovery in Clinical Research Databases. In: Pier Luca Lanzi, Wolfgang Stolzmann, Stewart W. Wilson, editors, Learning Classifier Systems. From Foundations to Applications. LNAI 1813, pages 243-261, Springer-Verlag, Berlin, 2000.
3. John H. Holmes, Applying a Learning Classifier System to Mining Explanatory and Predictive Models from a Large Database. In: Pier Luca Lanzi, Wolfgang Stolzmann, Stewart W. Wilson, editors, Advances in Learning Classifier Systems. LNAI 1996, pages 103-113, Springer-Verlag, Berlin, 2001.
4. Stewart W. Wilson. Get Real! XCS with continous-valued inputs. In P. L. Lanzi, W. Stolzmann and S. W. Wilson(eds.), Learning Classifier Systems. From Foundations to Applications. LNAI 1813, pages 209-219, Springer-Verlag, Berlin, 2000.
5. Stewart W. Wilson, Mining Obilque Data with XCS. In: Pier Luca Lanzi, Wolfgang Stolzmann, Stewart W. Wilson, editors, Advances in Learning Classifier Systems. LNAI 1996, pages 158-174, Springer-Verlag, Berlin, 2001.
6. Ester Bernadó, Xavier Llorà, Josep M. Garrell, XCS and GALE: A Comparative Study of Two Learning Classifier Systems on Data Mining. In: Pier Luca Lanzi, Wolfgang Stolzmann, Stewart W. Wilson, editors, Advances in Learning Classifier Systems. LNAI 2321, pages 115-132, Springer-Verlag, Berlin, 2002.
7. Jaume Bacardit, Martin V. Butz, Data Mining in Learning Classifier Systems: Comparing XCS with GAssist, in this volume.
8. Dietterich, T. G. Ensemble Learning. In The Handbook of Brain Theory and Neural Networks, Second edition, Cambridge, MA: The MIT Press, 405-408, 2002.
9. Leo Breiman, Bagging Predictors, Machine Learning, Vol.24, No.2, pp.123-140, 1996.
10. Z.-H. Zhou, J. Wu, and W. Tang. Ensembling neural networks: many could be better than all. Artificial Intelligence, 137(1-2): 239-263, 2002.
11. L. Booker, D.E. Goldberg, J.H. Holland, Classifier systems and genetic algorithms. Artificial Intelligence. vol.40(1-3), pp. 235-282, 1989.
12. Y. Freund and R. Schapire. A short introduction to boosting, Journal of Japanese Society for Artificial Intelligence , 14(5):771-780, 1999.
13. E. Bauer, R. Kohavi, An empirical comparison of voting classification algorithms: Bagging, boosting, and variants. Machine Learning, 36(1-2), 105–139, 1999.
14. http://www.cs.waikato.ac.nz/ ml/weka/, Last visit at 22, Dec., 2004.

Effect of Pure Error-Based Fitness in XCS

Martin V. Butz[1], David E. Goldberg[2], and Pier Luca Lanzi[2,3]

[1] Department of Cognitive Psychology,
University of Würzburg, 97070 Würzburg, Germany
butz@psychologie.uni-wuerzburg.de
[2] Illinois Genetic Algorithms Laboratory,
University of Illinois at Urbana-Champaign, Urbana, Illinois, USA
{deg,lanzi}@illigal.ge.uiuc.edu
[3] Dipartimento di Elettronica e Informazione
Politecnico di Milano
Milano, Italy
lanzi@elet.polimi.it

Abstract. The accuracy-based fitness approach in XCS is one of the most significant changes in comparison with original learning classifier systems. Nonetheless, neither the scaled accuracy function, nor the importance of the relative fitness approach has been investigated in detail. The recent introduction of tournament selection to XCS has shown to make the system more independent from parameter settings and scaling issues. The question remains if relative accuracy itself is actually necessary in XCS or if the evolutionary process could be based directly on error. This study investigates advantages and disadvantages of pure error-based fitness vs. relative accuracy-based fitness in XCS.

1 Introduction

Recent advances in XCS understanding have shown that the accuracy-based fitness approach can guide the evolutionary process to the discovery of accurate, maximally general classifiers [7]. Additionally, with the introduction of tournament selection, XCS gained a more reliable and persistent pressure towards accuracy [9]. However, it did not become clear why accuracy needs to be scaled nor why fitness is derived from the *relative* accuracy.

This study investigates the fitness approach in XCS. The relative accuracy-based fitness approach underlies several peculiar parameter choices which need to be investigated and clarified. Moreover, although XCS's fitness approach was successful in many different investigations (e.g. [2,11,4]), it is not clear if the additional accuracy bias is necessary for a successful evolutionary process in XCS. In fact, it seems possible that XCS selection with tournament selection could be solely based on minimizing error instead of maximizing accuracy. In this way, the additional accuracy bias would become irrelevant and parameter estimations should reach less noisy values faster.

The remainder of this study is structured as follows. The next section gives a short overview over the XCS system with the relevant parameter initialization

X. Llorà et al. (Eds.): IWLCS 2003-2005, LNAI 4399, pp. 104–114, 2007.

method and update methods. Next, we study the effect of basing selection directly on error instead of accuracy-based fitness. Summary and conclusions conclude the study.

2 XCS Overview

XCS is a very general learning mechanism that combines gradient-based optimization of predictions with evolutionary-based space partitioning. The partitions evolve to enable maximally accurate predictions. While XCS was also successfully applied in multi-step problems [22,15,16,1], we restrict this study to classification problems to avoid the additional problem of reward propagation. However, the insights of this study should readily carry over to multi-step problems. This section introduces XCS as a pure classification system providing the necessary details to comprehend the remainder of this work. For a more complete introduction to XCS the interested reader is referred to the original paper [22] and the algorithmic description [10].

We define a classification problem as a problem that consists of problem instances $s \in \mathcal{S}$ that need to be classified by XCS with one of the possible classifications $a \in \mathcal{A}$. The problem then provides scalar payoff $R \in \Re$ with respect to the made classification. The goal for XCS is to choose the classification that results in the highest payoff. To do that, XCS is designed to learn a complete mapping from any possible $s \times a$ combination to an accurate payoff value. To keep things simple, we investigate problems with Boolean input and classification, i.e. $\mathcal{S} \subseteq \{0,1\}^L$ where L denotes the fixed length of the input string and $\mathcal{A} = \{0,1\}$.

XCS evolves a population $[P]$ of rules, or *classifiers*. Each classifier in XCS consists of five main components. The condition $C \in \{0,1,\#\}^L$ specifies the subspace of the problem instances in which the classifier is applicable, or *matches*. The "don't care" symbol $\#$ matches in all input cases. The action part $A \in \mathcal{A}$ specifies the advocated action, or classification. The payoff prediction p approaches the average payoff encountered after executing action A in situations in which condition C matches. The prediction error ε estimates the average deviation, or error, of the payoff prediction p. The fitness reflects the average relative accuracy of the classifier with respect to other overlapping classifiers.

XCS iteratively updates its knowledge base with respect to each problem instance. Given current input s, XCS forms a *match set* $[M]$ consisting of all classifiers in $[P]$ whose conditions match s. If an action is not represented in $[M]$, a covering classifier is created that matches s (#-symbols are inserted with a probability of $P_\#$ at each position). For each classification, XCS forms a *payoff prediction* $P(a)$, i.e. the fitness-weighted average of all reward prediction estimates of the classifiers in $[M]$ that advocate classification a. The payoff predictions determine the appropriate classification. After the classification is selected and sent to the problem, payoff R is provided according to which XCS updates all classifiers in the current action set $[A]$ which comprises all classifiers in $[M]$ that advocate the chosen classification a. After update and possible GA invocation, the next iteration starts.

Prediction and prediction error parameters are update in $[A]$ by $p \leftarrow p + \beta(R - p)$ and $\varepsilon \leftarrow \varepsilon + \beta(|R - p| - \varepsilon)$ where β ($\beta \in [0, 1]$) denotes the *learning rate*. The fitness value of each classifier in $[A]$ is updated according to its current scaled relative accuracy κ':

$$\kappa = \begin{cases} 1 & \text{if } \varepsilon < \varepsilon_0 \\ \alpha \left(\frac{\varepsilon_0}{\varepsilon}\right)^\nu & \text{otherwise} \end{cases} \qquad \kappa' = \frac{\kappa}{\sum\limits_{x \in [A]} \kappa_x} \qquad (1)$$

$$F \leftarrow F + \beta(\kappa' - F) \qquad (2)$$

The parameter ε_0 ($\varepsilon_0 > 0$) controls the tolerance for prediction error ε; parameters α ($\alpha \in (0, 1)$) and ν ($\nu > 0$) are constants controlling the rate of decline in accuracy κ when ε_0 is exceeded. The accuracy values κ in the action set $[A]$ are then converted to set-relative accuracies κ'. Finally, classifier fitness F is updated towards the classifier's current set-relative accuracy. All parameters except for fitness F are updated using the *moyenne adaptive modifée* technique [19]. This technique sets parameter values directly to the average of the so far encountered cases as long as the experience of a classifier is still less than $1/\beta$. Each time the parameters of a classifier are updated, the experience counter *exp* of the classifier is increased by one.

A GA is invoked in XCS if the average time since the last GA application on the classifiers in $[A]$ exceeds threshold θ_{ga}. The GA selects two parental classifiers using roulette-wheel selection [22] or the recently introduced tournament selection [9]. Two offspring are generated reproducing the parents and applying crossover and mutation. Parents stay in the population competing with their offspring. We apply free mutation in which each attribute of the offspring condition is mutated to the other two possibilities with equal probability. Parameters of the offspring are inherited from the parents, except for the experience counter *exp* which is set to one, the numerosity *num* which is set to one, and the fitness F which is multiplied by 0.1. In the insertion process, *subsumption deletion* may be applied [23] to stress generalization.

The population of classifiers $[P]$ is of fixed size N. Excess classifiers are deleted from $[P]$ with probability proportional to an estimate of the size of the action sets that the classifiers occur in (stored in the additional parameter *as*). If the classifier is sufficiently experienced and its fitness F is significantly lower than the average fitness of classifiers in $[P]$, its deletion probability is further increased.

3   Error-Based Selection

Although an error-based selection method still pursues the XCS goal of evolving a complete and accurate reward map of a problem several differences can be identified. This section discusses these differences and experimentally investigates error-based fitness in XCS.

3.1 Major Differences

As mentioned in the XCS overview, selection is usually based on the set-relative accuracy derived fitness estimate of a classifier. In offspring classifiers this fitness is usually derived from the parents (sometimes also from the average fitness in the population) and multiplied by 0.1 to be pessimistic about the offspring quality. Dependent on the learning rate β, the moyenne adaptive modifée (MAM) technique, the experience counter, and the accuracy scaling, more accurate offspring reaches a fitness value higher than the parental value after a certain amount of updates. Only then the more-accurate offspring has the chance to outperform its parents and take-over the specific environmental niche it covers. The number of influences suggest that complex interactions of different factors can occur.

Similar to the fitness approach, though, it seems also possible to base selection directly on the prediction error estimate of a classifier. While accuracy-based fitness needs to be maximized, error-based fitness needs to be minimized. Additional effects are expectable, though, since the error estimate is directly derived from the parental value (without a pessimistic increase) and the error estimate is not set relative, effectively disabling fitness sharing. While the former factor should have the effect that offspring sometimes causes additional disruption, the latter factor might result in weaker niche support pressure. These factors are investigated in our experimental study.

Interestingly, though, due to the lack of fitness sharing, additionally, overlapping classifiers are enabled in this framework. The relative-accuracy-based fitness approach in the original XCS causes the evolution of non-overlapping niches that cover the whole reward map of a learning problem (see e.g. [13,14] for further analyses). Error-based fitness will cause the evolution of a similar complete reward map but allows overlapping classifiers. This might be advantageous in unevenly overlapping niches, but has the drawback that more classifiers need to be sustained to continuously cover the whole problem space. The additional classifiers also undergo additional competition due to the unrestricted population-wide deletion technique.

3.2 Implementation

Error-based selection is realized applying tournament selection. Instead of maximizing the fitness estimate of a classifier, the error estimate is minimized. Thus, the classifier wins in the current tournament in an action set that has the lowest reward prediction error estimate. Parameter updates are not changed.

Additionally, to free XCS completely from the fitness evaluation, the prediction array needs to be formed with respect to a classifier's error estimate and not to its fitness estimate. Since the error estimate in young classifiers is very noisy, the reward prediction estimate is less trusted than in elder classifiers. Widrow & Stearns (1985) formalized how the reward prediction error can be expected to vary with respect to the number of encountered reward prediction updates [20].

Assuming the encountering of a perfect signal P and the initial estimate p_i, then the error of the actual estimate $p(t)$ can be determined as follows [20]:

$$p(t) = P + (1 - \beta)^t (p_i - P) \tag{3}$$

Assuming the worst-case initialization error $\epsilon_{wc} = \max\{p_i; P_{max} - p_i\}$ as the initial error and assuming furthermore a perfect reward signal from then on, the maximal difference from the actual average encountered reward can be estimated as follows:

$$\Delta p = (1 - \beta)^{exp} \epsilon_{wc} \tag{4}$$

Since the prediction error of a classifier can reach on average half the maximal reward prediction $P_{max}/2$ (temporarily it might also lie a little above this value), the maximal error in the reward prediction error can be determined as follows denoting $\epsilon_{\epsilon wc}$ as the maximum possible error of the error and assuming a perfect signal.

$$\Delta \epsilon = (1 - \beta)^{exp} \epsilon_{\epsilon wc} \tag{5}$$

The actual error of a classifier can now be estimated somewhat pessimistic as the actual error estimate plus the worst-case differing amount (assuming a perfect signal).

$$\epsilon' = \epsilon + \Delta \epsilon \tag{6}$$

The prediction array may now be weighted according to the estimated error ϵ' in conjunction with the actual reward prediction value p:

$$PA(a) = \frac{\sum_{cl \in [M] \wedge cl.A = a} cl.p \cdot 1/cl.\epsilon' \cdot cl.num}{\sum_{cl \in [M] \wedge cl.A = a} 1/cl.\epsilon' \cdot cl.num} \tag{7}$$

This prediction array determination consequently ignores fitness but weights the reward estimates according to the actual inverse error estimate. Finally, deletion cannot be biased on the fitness estimate of a classifier as originally proposed and investigated in [12]. Consequently, deletion is proportional to the action set size estimate alone as in the original XCS implementation [22]. The next section investigates the impact of these modifications in several typical Boolean function problems.

3.3   Experimental Investigation

Several questions need to be investigated in the new approach. First, the question is if in fact overlapping, accurate classifiers evolve. Next, the speed of evolution will show if the direct error dependence allows a faster or slower detection of the relevant environmental niches and thus, if performance speed increases or decreases. Finally, due to the additional overlapping classifiers, the support of each environmental niche needs to be investigated. Will XCS be able to sustain the representation of the complete problem with the same number of classifiers?

To answer these questions, we apply XCS to the multiplexer function [22,6] and the count ones problem [6,8].

Table 1 shows the typical difference in the classifier lists of relative-accuracy-based fitness and error-based fitness in the six multiplexer. While in the relative case mainly non-overlapping classifiers evolve, that explicitly identify each environmental niche, in the error-based case those classifiers evolve as well as classifiers that overlap two niches. For example, niches 01*0**-1 and 11***0-1 (specifying accurately the incorrect class) would be represented perfectly by the similar classifiers substituting don't care symbols for the star symbols. These evolve in the relative-accuracy-based case. However, in the error-based case, also the overlapping classifiers gain a high numerosity value such as classifier #1#0#0-1. Note that in the exemplar runs, the maximal population size was set to $N = 2000$ so that niche support was not a problem in this case. The overlapping classifiers gain a similar numerosity (on average) as the non-overlapping ones do. Hardly any pressure towards the non-overlapping classifiers can be detected.

Table 1. Typical resulting classifier list for relative-accuracy-based fitness and error-based fitness in the 6-multiplexer problem

Relative-Accuracy Based							Error Based					
C	A	p	ϵ	F	num	exp	C	A	p	ϵ	F	num	exp
11###0	1	0	0	0.836775	85	5134	01#0##	1	0	0	0.900787	106	5994
11###1	1	1000	0	0.792768	73	5478	#1#1#1	1	1000	0	0.637662	91	4973
10##0#	1	0	0	0.702730	67	5847	1###00	1	0	0	0.587410	82	5651
10##1#	1	1000	0	0.653202	59	5270	#01#1#	1	1000	0	0.532509	81	2592
01#0##	1	0	0	0.471205	49	5306	0#11##	1	1000	0	0.429461	65	3552
01#1##	1	1000	0	0.418793	38	5306	000###	1	0	0	0.712403	63	5175
01#0##	1	0	0	0.252941	28	1976	#00#0#	1	0	0	0.435288	62	4410
001###	1	1000	0	0.301881	28	5726	#1#0#0	1	0	0	0.369763	46	5853
#00#0#	1	0	0	0.242931	27	4925	11###1	1	1000	0	0.504067	41	4982
000###	1	0	0	0.328251	27	5529	10##10	1	1000	0	0.412228	38	325
01#01#	1	0	0	0.234058	26	2557	10##0#	1	0	0	0.491715	36	4408
0010##	1	1000	0	0.272719	25	2095	01#1##	1	1000	0	0.409011	32	1174
10##10	1	1000	0	0.256431	24	2269	11###0	1	0	0	0.445064	32	4524
10##01	1	0	0	0.232770	24	2481	10##1#	1	0	0	0.288221	28	1054
01#10#	1	1000	0	0.242531	22	2570	1##11	1	1000	0	0.270195	27	5872
01#0#1	1	0	0	0.210961	22	2636	001##1	1	1000	0	0.239064	25	466
01#11#	1	1000	0	0.222898	20	2651	001###	1	1000	0	0.270045	19	1891
000##1	1	0	0	0.230527	20	2740	0#11#0	1	1000	0	0.139190	13	97
001#0#	1	1000	0	0.204827	20	2786	#00#00	1	0	0	0.049742	8	75
001##0	1	1000	0	0.198300	19	1849	0001##	1	0	0	0.053201	5	67
01#1#0	1	1000	0	0.214924	19	2692	#01#11	1	1000	0	0.043135	5	102
000#1#	1	0	0	0.222182	19	2667	#1#1#0	1	405	501	0.000000	4	1654
001##1	1	1000	0	0.202509	19	2867	1###0#	1	161	302	0.000000	3	191
#1#0#0	1	0	0	0.170386	18	5351	00#0##	1	519	509	0.000000	3	316

Further experiments in the larger 20 multiplexer problem are displayed in Figure 1 showing the normal multiplexer problem, and the problem with additional Gaussian noise (adding a Gaussian Noise with standard deviation $\sigma = 300$ on the provided reward reflecting noise in the fitness evaluation function).

All runs show that XCS with error-based fitness is able to solve the problem as well. The evolutionary speed is slightly decreased, that is, perfect performance is reached after a larger number of steps in comparison to relative-accuracy-based fitness. Part of the explanation for this decrease in learning speed can be attributed to the larger number of classifiers that is evolved. Additionally,

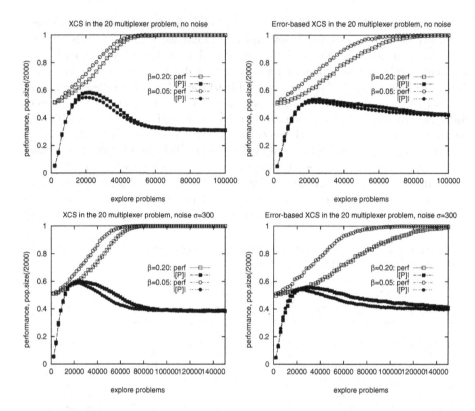

Fig. 1. Slight advantages due to relative-accuracy based selection can be observed in the multiplexer problem. However, the simplicity of error-based selection remains appealing.

parameter initialization issues appear relevant. Since fitness is decreased by 0.1 in offspring classifiers and fitness is updated by the Widrow-Hoff rule from the beginning, disruption by young classifiers appears to be prevented better in the fitness-based selection case than in the error-based selection case.

The relative fitness approach also results in a slightly stronger generalization pressure. The pressure appears to be mainly due the initial decrease in offspring fitness. The decrease in fitness assures that similarly accurate parents win the tournament against their offspring. More specialized classifiers undergo parameter updates less frequently so that the more specialized a classifier the longer it takes for it to exceed its parent's fitness. Thus, more generalized similarly accurate classifiers reach higher fitness values faster.

Besides the multiplexer problem, we experiment with the count ones problem in which overlapping niches need to be sustained to ensure the representation of a complete problem solution. Besides Gaussian noise, we also added alternating noise, in which the incorrect reward is provided with a probability of 0.15, reflecting noise with incorrect classification cases [6]. Similar performance

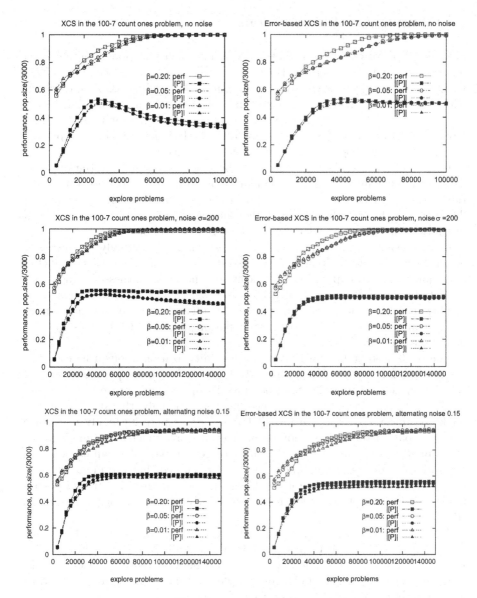

Fig. 2. Differences in the count ones problem with string length 100 and 7 relevant bits are minor. Lower β rates decrease convergence speed but increase accuracy.

observations can be made. The error-based approach again suffers more from more offspring disruption so that a reliable 100% performance is reached slighlty slower. However, particularly in the noisy problem cases, XCS with error-based selection reaches a slightly higher performance level. Thus, fitness sharing may cause disruption in problems in which overlapping niches actually need to be sustained for a complete problem solution.

4   Summary and Conclusions

Summing up, we could show that an XCS in which selection is purely based on error is able to solve similar Boolean function problems as the normal XCS. Hereby, performance was slightly worse than in the original XCS approach but with the gain of having less parameters per classifier and depending less on learning rate β. Moreover, the peculiar accuracy-scaling function is not necessary anymore. Parameter estimates can now be directly inherited from the parents. An additional parameter estimate decrease is not necessary.

Despite the successful application, it became also clear that the approach deserves further investigation. A similar error estimate as done for the prediction array calculation might be useful in the selection mechanism to prevent disruption in the error-based case. Other mechanisms are imaginable to prevent disruption but still ensure detection of better classifiers fast.

In conclusion, the results show that fitness sharing is actually not necessary in the XCS framework. Niching is assured due to the niche-based reproduction in conjunction with population-wide deletion. Thus, while fitness-sharing is very likely to be mandatory in other LCS frameworks, such as the ZCS system [21,3], niching in XCS is accomplished by the niche-based reproduction mechanism. This niching effect and its impact on population sizing in XCS is investigated in detail elsewhere [5,4].

The study also points out that parameter initialization in offspring classifiers is still in its infancy. Proper mathematical approaches to the parameter estimations are necessary to understand possible disruption and ensure fast detection of more accurate (or lower error) classifiers. Additionally, the MAM technique might be questioned because initial, large updates may be highly disruptive. However, parameter initialization becomes even more crucial once pure Widrow-Hoff updates are applied (since then an incorrect initial value can cause strong disruption). More resent modifications of XCS showed that the prediction part in XCS is generally very flexible enabling the estimation of linear and polynomial predictions approximated with recursive least squares or the pseudo inverse [17,18].

The results herein show that in strongly overlapping problems, the fitness sharing approach may be reconsidered or may actually be obsolete. Future analyses on this matter will be relevant not only for the XCS classifier system but also for LCSs in general since all (Michigan-style) classifier systems rely on iterative parameter updates and thus noisy fitness estimates.

Acknowledgments

We are grateful to the whole IlliGAL lab for their help and the useful discussions.

This work was sponsored by the Air Force Office of Scientific Research, Air Force Materiel Command, USAF (F49620-03-1-0129). The US Government is authorized to reproduce and distribute reprints for Government purposes notwithstanding any copyright notation thereon. Butz's contribution received additional funding from the the Computational Science and Engineering graduate option program (CSE) at the University of Illinois at Urbana-Champaign, the German research foundation (DFG) under grant DFG HO1301/4-3 as well as from the European commission contract no. FP6-511931.

The views and conclusions contained herein are those of the authors and should not be interpreted as necessarily representing the official policies or endorsements, either expressed or implied, of any of the organizations mentioned above.

References

1. Alwyn Barry. A hierarchical XCS for long path environments. *Proceedings of the Third Genetic and Evolutionary Computation Conference (GECCO-2001)*, pages 913–920, 2001.
2. Ester Bernadó, Xavier Llorà, and Joseph M. Garrell. XCS and GALE: A comparative study of two learning classifier systems and six other learning algorithms on classification tasks. In P. L. Lanzi, W. Stolzmann, and S. W. Wilson, editors, *Advances in Learning Classifier Systems (LNAI 2321)*, pages 115–132. Springer-Verlag, Berlin Heidelberg, 2002.
3. Larry Bull and Jacob Hurst. ZCS redux. *Evolutionary Computation*, 10(2):185–205, 2002.
4. M. V. Butz. *Rule-Based Evolutionary Online Learning Systems: A Principled Approach to LCS Analysis and Design*. Studies in Fuzziness and Soft Computing. Springer-Verlag, Berlin Heidelberg, 2005.
5. Martin V. Butz, David E. Goldberg, Pier Luca Lanzi, and Kumara Sastry. Bounding the population size to ensure niche support in XCS. IlliGAL report 2004033, Illinois Genetic Algorithms Laboratory, University of Illinois at Urbana-Champaign, 2004.
6. Martin V. Butz, David E. Goldberg, and K. Tharakunnel. Analysis and improvement of fitness exploitation in XCS: Bounding models, tournament selection, and bilateral accuracy. *Evolutionary Computation*, 11:239–277, 2003.
7. Martin V. Butz, Tim Kovacs, Pier Luca Lanzi, and Stewart W. Wilson. How XCS evolves accurate classifiers. *Proceedings of the Third Genetic and Evolutionary Computation Conference (GECCO-2001)*, pages 927–934, 2001.
8. Martin V. Butz, M. Pelikan, X. Llorà, and D.E. Goldberg. Extracted global structure makes local building block processing effective in XCS. *GECCO 2005: Genetic and Evolutionary Computation Conference: Volume 1*, pages 655–662, 2005.
9. Martin V. Butz, Kumara Sastry, and David E. Goldberg. Tournament selection in XCS. *Proceedings of the Fifth Genetic and Evolutionary Computation Conference (GECCO-2003)*, pages 1857–1869, 2003.
10. Martin V. Butz and Stewart W. Wilson. An algorithmic description of XCS. In P. L. Lanzi, W. Stolzmann, and S. W. Wilson, editors, *Advances in learning classifier systems: Third international workshop, IWLCS 2000 (LNAI 1996)*, pages 253–272. Springer-Verlag, Berlin Heidelberg, 2001.
11. Phillip W. Dixon, David W. Corne, and Martin J. Oates. A preliminary investigation of modified XCS as a generic data mining tool. In P. L. Lanzi, W. Stolzmann, and S. W. Wilson, editors, *Advances in learning classifier systems: Fourth international workshop, IWLCS 2001 (LNAI 2321)*, pages 133–150. Springer-Verlag, Berlin Heidelberg, 2002.

12. Tim Kovacs. Deletion schemes for classifier systems. *Proceedings of the Genetic and Evolutionary Computation Conference (GECCO-99)*, pages 329–336, 1999.

13. Tim Kovacs. Strength or Accuracy? Fitness calculation in learning classifier systems. In Pier Luca Lanzi, Wolfgang Stolzmann, and Stewart W. Wilson, editors, *Learning classifier systems: From foundations to applications (LNAI 1813)*, pages 143–160. Springer-Verlag, Berlin Heidelberg, 2000.

14. Tim Kovacs. Towards a theory of strong overgeneral classifiers. *Foundations of Genetic Algorithms 6*, pages 165–184, 2001.

15. Pier Luca Lanzi. An analysis of generalization in the XCS classifier system. *Evolutionary Computation*, 7(2):125–149, 1999.

16. Pier Luca Lanzi. An extension to the XCS classifier system for stochastic environments. *Proceedings of the Genetic and Evolutionary Computation Conference (GECCO-99)*, pages 353–360, 1999.

17. Pier Luca Lanzi, Daniele Loiacono, Stewart W. Wilson, and David E. Goldberg. Extending XCSF beyond linear approximation. *GECCO 2005: Genetic and Evolutionary Computation Conference: Volume 2*, pages 1827–1834, 2005.

18. Pier Luca Lanzi, Daniele Loiacono, Stewart W. Wilson, and David E. Goldberg. Generalization in XCSF for real inputs. IlliGAL report 2005023, Illinois Genetic Algorithms Laboratory, University of Illinois at Urbana-Champaign, 2005.

19. Gilles Venturini. Adaptation in dynamic environments through a minimal probability of exploration. *From Animals to Animats 3: Proceedings of the Third International Conference on Simulation of Adaptive Behavior*, pages 371–381, 1994.

20. Bernard Widrow and Samuel D. Stearns. *Adaptive Signal Processing*. Prentice-Hall, Englewood Cliffs, New Jersey, 1985.

21. Stewart W. Wilson. ZCS: A zeroth level classifier system. *Evolutionary Computation*, 2:1–18, 1994.

22. Stewart W. Wilson. Classifier fitness based on accuracy. *Evolutionary Computation*, 3(2):149–175, 1995.

23. Stewart W. Wilson. Generalization in the XCS classifier system. *Genetic Programming 1998: Proceedings of the Third Annual Conference*, pages 665–674, 1998.

A Fuzzy System to Control Exploration Rate in XCS

Ali Hamzeh and Adel Rahmani

Department of Computer Engineering, Iran University of Science and Technology
Tehran, Iran
{hamzeh, rahmani}@iust.ac.ir

Abstract. Exploration/Exploitation dilemma is one of the most challenging issues in reinforcement learning area as well as learning classifier systems such as XCS. In this paper, an intelligent method is proposed to control the exploration rate in XCS to improve its long-term performance. This method is called Intelligent Exploration Method (IEM) and is applied to some benchmark problems to show advantages of adaptive exploration rate for XCS.

Keywords: Learning Classifier Systems, XCS, Exploration, Exploitation, Adaptive Exploration Rate.

1 Introduction

Assume an agent tries to learn the environment E. At each time step, the agent receives the environmental state as the vector X, then carries out the action a and receives the reward r from the environment. In this model, the agent tries to satisfy two objectives: (i) to learn which actions consequence to achieve more reward in its life time to upgrade its performance and (ii) to reach its predefined goal(s).

When an agent tries to learn to act in an unknown environment, using reinforcement learning paradigm, it always faces a critical problem. It must choose between acting based on its previous experiences and trying some new actions expecting to achieve new knowledge about the environment. The rationale behind the latter approach is to gain some useful information which may help the agent to improve its performance in that environment. The problem of creating a balance between achieving new information and acting based on previous ones is usually called Exploration/Exploitation Dilemma (EED). Holland was one of the first to discuss the dilemma in connection with adaptive systems [1]. He summarizes: "[obtaining] more information means a performance loss, while exploitation of the observed best runs the risk of error perpetuated". The EED is described in the reinforcement learning framework in the following:

All machine-learning approaches which are based on the reinforcement learning paradigm, suffers from this issue. One of these approaches is Learning Classifier Systems (LCS). LCSs are a kind of reinforcement learning algorithms first proposed by Holland and later modified and extended by Wilson [2]. Wilson's Extension is called Accuracy Based Classifier System (XCS) and now is the mostly used extension of LCS.

X. Llorà et al. (Eds.): IWLCS 2003-2005, LNAI 4399, pp. 115–127, 2007.

One of the most challenging issues in XCS as well as other reinforcement learning approaches is to develop a proper EED strategy. In this paper, we propose a new method to develop this balance between the exploration and exploitation rates in XCS's life cycle using Fuzzy Logic.

The rest of the paper is organized as follows: at first we describe some important researches about EED, and then we describe XCS in brief and describe its original approach to create EED balance. Then we introduce our proposed method and describe it in details and finally some benchmark problems are considered, and our experimental results are presented and discussed.

2 Related Works on EED

In this section, we summarize some important researches related to EED issue to show the state of the EED research.

If a system has a well-defined way of making the Explore/Exploit decision at each time-step, we say it has an EED strategy. In [3], the author categorized some famous EED strategies. He introduced two general categories: Global Strategies and Local Strategies. The first category includes the policies where the rate of exploration is based on some features which are independent of the system itself, such as time or the measure of the system's overall performance. These strategies do not change with respect to the time or the problem's experience. As an example of these policies, we can refer to static probability policy, variable probability with respect to prediction error and so on. The local category of policies includes the policies which their exploration rate is determined with respect to the response of the system to the current input of the environment.

In the other words, the global strategies use the property of the whole system to adapt the exploration rate but the local strategies determine the exploration rate based on the current input of the system. These strategies are based on this idea that in some iteration, learning may not be necessary or may be even disruptive. There are many exploration strategies which lay in this category such as roulette wheel selection.

In [4], the authors proposed an EED strategy based on some meta-knowledge about the environment. Their strategy reacts to the performance of the agent. To validate their approach, they applied it to some economic systems and compare it to two adaptive methods: one local and one global.

And finally in [5], the authors develop a model, based on the combination of exploration and exploitation in a common framework. They defined a new concept called 'degree of exploration'. This value is defined for a particular state as the entropy of the probability distribution on the set of acceptable actions. This value is controlled by the user of the system. In [5], the EED problem was seen from another perspective: a global optimization problem. The problem is to find a policy that minimizes the expected cost with a fixed degree of entropy. This formulation leaded the authors to a set of nonlinear updating rules. The authors showed that in specific circumstances, there are some non-linear equations which can easily be solved to obtain the desired policy. Some reported simulations showed that their model works as expected.

3   XCS in Brief

In this section, we briefly describe XCS. XCS contains a population of the classifiers which is called [P]. This population can be empty in the beginning of the experience, or be filled randomly. Each classifier in [P] is made up different parts. These parts are: a condition from the alphabet {0, 1, #}, an action which is an integer, and a set of associated parameters. These parameters are (1) payoff prediction P_j, which estimates the payoff which the system will receive when its action is applied to the corresponding environment; (2) the prediction error ε_j, (3) the fitness F_j, and some other parameters such as *exp*, *num* and etc.

3.1   A Life Cycle of XCS

When XCS receives the environmental state, it forms the related match set [M]. This set includes the classifiers which their condition parts matches the current environmental state. If no classifiers match, the covering operator creates a predefined number of classifiers which match the current input and insert them into the population and into the [M]. If the size of population grows over a predefined size N, due to covering, then some other classifiers are eliminated from the population regarding their fitness and experience.

Then, for each action a_k, which is proposed by the classifiers in [M], the system computes a fitness weighted average P'_k using this equation: $P'_k = \frac{\sum_j F_j P_k}{\sum_j F_j P_j}$, where P_k is the prediction of the classifiers which propose action a_k, this value is used as the bid of corresponding action to win the exploitation phases. Then, XCS chooses an action from those proposed in [M] regarding its EED strategy. Finally, an action set [A] is formed consisting of the subset of [M] having the chosen action.

After that, the selected action is applied to the environment and a reward R was received from the environment which may be zero. Then, the parameters of the classifiers in [A] are re-estimated according to that reward. At first, the predictions are updated as follows: $P_j \leftarrow P_j + \beta(R - P_j)$. Next, the errors are re-estimated using this equation: $\varepsilon_j = \varepsilon_j + \beta(|R - P_j|)$. Then the relative accuracy for the corresponding classifier is calculated as follows: $k_j = 0.1\frac{\varepsilon_0}{\varepsilon_j}^{-v}$ for $\varepsilon_j > \varepsilon_0$, else 1.0. The parameter ε_0 is termed the error threshold and v is a positive integer: both of them are initiated at the beginning of the experiments.

Then, this raw accuracy is used to calculate the relative accuracy for each classifier using the following equation: $k'_j = \frac{k_j}{\sum_j k_j}$. And at last, this relative accuracy is used to update the classifier's fitness: $F_j = F_j + \beta(|k_j - F_j|)$.

These updated values are used in another important component of XCS: The discovery component. On a regular basis depending on the parameter θ_{ga}, the genetic algorithm is applied to the classifiers in [A]. As usual, applying GA consists of three phases: selection, crossover and mutation. In the selection phase, two classifiers are selected with proportionate selection operator regarding their fitness. The crossover operator, usually one-point crossover operator, is applied on the two selected parents at the rate of χ. Then, at the rate of μ, each allele of the generated offspring is mutated. The resulting offspring are inserted into the population and two classifiers are deleted to keep the population size constant.

4   EED in XCS

We encounter the EED in the action selection procedure of XCS. Due to the online performance measuring in XCS, the question is how to create the desired balance between selecting the winner action with respect to agent's previous experiments or let the agent to explore its environment in order to gain more useful information.

It seems that agent must create a balance between these two choices; but the major question is how to create this balance. Yet another issue to be addressed is how to select the winner action during exploration phase. Should this decision be purely random or must incorporate some previously gained knowledge?

In this paper, we try to provide answers by inspecting the current XCS implementation based on [6] by M. Butz and S.W. Wilson. In this implementation the balance between Exploration and Exploitation is created using a P_{exp} constant. P_{exp} is used to determine the probability of exploring environment. This probability remains unchanged during the agent's life cycle and commonly is set equal to 0.5. Therefore, this balance is created by using only a constant parameter. Considering the second question, [6] uses a random policy to select the winner action in the exploration phase and no other parameter such as Fitness or Strength is involve. In the next section, we propose an adaptive intelligent technique to develop an intelligent Exploration/Exploitation strategy.

5   Intelligent Exploration Method (IEM)

The main rational behind our proposed method is that a constant exploration rate seems to be not an appropriate EED strategy and we need to adapt it during the agent's life cycle. For example in the beginning of the learning procedure due to the lack of experience in the environment, acting with respect to previous experiences has no difference with random strategy. Therefore, in the beginning phase, existence or lack of exploration has no significant effect on online and long-term performance. However, in the middle phase, exploration helps us to find more information on the search space and may cause better long-term performance and in the final phase, exploration can help us to escape from local optima but also it may cause some kind of disturbance in agent's knowledge about the world and can reduce the agent's performance. It seems that adaptive changes in exploration probability can improve XCS's performance. Due to this hypothesis, we designed a system called IEM. It tries to propose a suitable exploration rate according to its information about the agent's performance and the environmental changes. IEM tries to distinguish the beginning, middle and final phases in order to propose the appropriate exploration probability for each phase. Figure 1, depicts XCS with IEM.

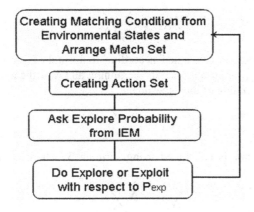

Fig. 1. XCS with IEM

5.1 IEM Internal Architecture

In this section, we describe the internal structure and overall architecture of IEM. As shown in Figure 2, IEM receives environment's state via its interface and chooses suitable control parameters with respect to its predefined rule base to apply to XCS.

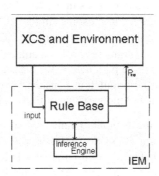

Fig. 2. IEM Architecture

IEM consists of different parts namely: input interface, rule base, inference engine and output interface. These parts are described in the following subsections.

5.2 The Input Parameters of IEM

The input parameters of IEM must indicate the current state of XCS with respect to its internal evolutionary process and its online performance. Our chosen input parameters are as follows:

- *PERF:* this factor is calculated using formula 1, where N_{ec} is number of trials with correct actions since the beginning and N_e is the total number of exploitation trials since the beginning of experiment.

$$PERF = N_{ec}\big/N_e \tag{1}$$

- N_{exr}: this factor is calculated according to formula 2; where N_r is the total number of explore trials since the beginning and N_t is the total number of trials since the beginning of the experiment.

$$N_{exr} = N_r\big/N_t \tag{2}$$

- N_{exp}: this factor is calculated using formula 3, where N_t and N_e are as the above.

$$N_{exp} = N_e\big/N_t \tag{3}$$

- *Age*: This parameter is used to distinguish XCS's beginning, middle and final phases and is calculated using formula 4, where N_t is as the above and T is the expected number of trials that XCS is going to accomplish.

$$Age = N_t\big/T \tag{4}$$

- F_{best}: Best fitness the chromosomes in the XCS's rule base.
- F_{mean}: Mean fitness of the chromosomes in the XCS's rule base.
- F_v: Fitness variance of the chromosomes in the XCS's rule base.
- D_{mean} and D_v: These parameters are calculated using Hamming Distance concept. To calculate these parameters, we measure all chromosome's distance with the best one in the population considering Hamming criterion and then we calculate the mean and the variance of these calculated distances.

Our motivation to choose these input parameters are as follows: First parameter determines the overall picture of XCS's performance. Second and third parameters show the state of the EED balance. Forth parameter is used to determine the current phase of XCS's learning procedure; and Fifth to ninth parameters are selected to determine the state of XCS's internal evolutionary process based on proposed parameters in [7].

5.3 IEM Output

IEM produces only one output; P_{exp} that is taken as the Exploration Probability by XCS. It means that XCS chooses a random action with probability of P_{exp} in current epoch; otherwise it selects the best action.

5.4 Rule Base of IEM

IEM is a Fuzzy Controller; so it uses a Fuzzy rule base. Due to the lack of learning ability of IEM, some of these rules are initiated using previous experiments and some

Table 1. Fuzzy rules, the relation between P_{exp} and *PERF*. This Table's structure is very simple, for example, indicated cell is interpreted as follows: *if PERF is low then P_{exp} must be high.*

PERF	P_{exp}
Low	**High**
Medium	Medium
High	Low

Table 2. Fuzzy rules, the relation between P_{exp}, *Nexp* and N_{exr}. This Table has two parts, first column shows N_{exp}'s Membership functions and first row shows the N_{exr}'s membership functions. For example, indicated cell is interpreted as follows: *if N_{exp} was medium and N_{exr} was low then P_{exp} must be high.*

N_{exp}	N_{exr}		
	Low	Medium	High
Low	Low	Low	Low
Medium	**High**	Medium	Low
High	High	Medium	Low

Table 3. Fuzzy rules, the relation between P_{exp} and D_{mean}. This Table is interpreted as Table 1.

D_{mean}	P_{exp}
Low	High
Medium	Medium
Normal	Low
High	Low

of them are drawn from [8], then they are tuned manually using many experiments with IEM and XCS. In Tables 1, 2 and 3, three types of such rules are shown. These rules are about relation between P_{exp}, N_{exp}, N_{exr}, D_{mean} and *PERF*.

Because of the fuzzy nature of IEM, each input and output variable must be associated with some membership functions. These membership functions are shown in Figure 3. It is notable that membership functions of the output parameter and the three first input parameters have been chosen experimentally, and others are delivered from [8].

5.5   A Life Cycle of IEM

In this section, we explain a sample life cycle of IEM to illustrate its overall architecture and its interaction with XCS. In all iterations, at first the input parameters are calculated using XCS's internal or environmental parameters. Then these parameters are fuzzified using Mamdani's techniques. After that, IEM rules with

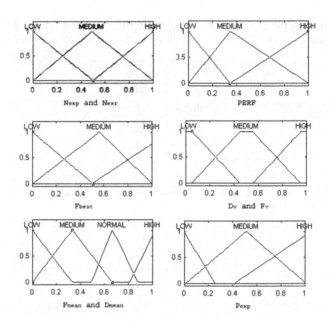

Fig. 3. Membership functions of input and output variables in IEM

trigger values greater than a predefined threshold (This threshold is set to 0.2 in this research experimentally) are selected for aggregation and defuzzification process. Our selected defuzzification method is Center of Gravity technique [9]. After defuzzification phase, the obtained P_{exp} is applied to XCS. XCS uses this value as the probability of exploring the environment.

6   Problems and Results

Our benchmark problems are chosen from known benchmark problems of single and multi step categories; Multiplexer family and Woods family. These problems are described here in brief, interested readers can refer to [2] for further reading.

Multiplexer Family: Boolean multiplexer functions are defined for binary strings of length $l = k + 2^k$. The function's value may be determined by treating the first k bits as an address that indexes into the remaining 2^k bits, and returning the indexed bit. For example, in the 6-multiplexer ($l=6$), the value for the input string 100010 is 1, since the "address", 10, indexes bit 2 of the remaining four bits. In disjunctive normal form, the 6-multiplexer is fairly complicated (the primes indicate negation):

$$F_6 = x_0'x_1'x_2 + x_0'x_1x_3 + x_0x_1'x_4 + x_0x_1x_5 \qquad (5)$$

To construct the payoff landscape, we associated two payoff values, 1000 and 0. Payoff 1000 was for the right answer and payoff 0 was for the wrong answer. There are more complicated instances of this problem such as MP11 (l=11) or MP20 (l=20). In this paper, we use MP11 and MP20 as our benchmark problems.

Woods Family: A maze consists of some squares which are arranged in a usually rectangular boundary. These squares are called cells. These cells are categorized into empty cells, which the agent can move through it and stays in it, blocked cells, that the agent cannot move through it and some cells with walls. In the latter type, the agent can stay in the cell but cannot move from this cell to the adjacent ones which have a wall in their common boundary with the current cell. Some other cells in the maze contain virtual food. The goal of the agent is to reach a food (G or F Cells) as fast as possible from any cell in the maze as the initial state. The agent can move in all eight directions. At each time step, the agent receives the environmental state which mentions the agents' adjacent cells using a coded string. Then it must choose a proper direction to move and carry out the corresponding action. If the action moves the agent to a food cell, the agent receives a reward r *(usually equal to 1000)* from the environment, unless no reward was given to the agent. As mentioned before, the task is to learn how to reach the food from any initial state using minimum number of steps. Our chosen maze is Woods2 which is depicted in Figure 4.

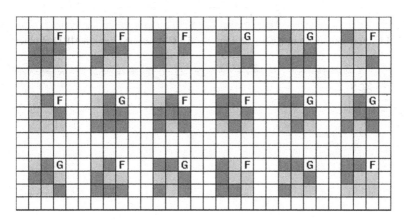

Fig. 4. Used instance of Woods2 Environment. This figure is downloaded from [10].

7 The Experimental Results

Our experiments are done separately by the XCS with IEM (we call it XCSI) and the original XCS based on [6]. For each problem, experiments are carried out 100 times separately and the final results are averaged over all of them. These results are shown in Figures 5 to 7. Vertical axis is system's performance for MP Family, which is equal to correct answers in last 50 exploit epochs, and average step to food for Woods family and the horizontal axis is *epoch-number/50*. All XCS parameters are initialized as in [6]; population size is 400 for MP11 and Woods2 and 1000 for MP20 for both XCS and XCSI.

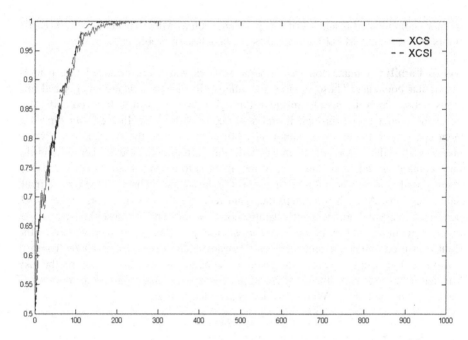

Fig. 5. Comparison of XCS (solid line) and XCSI (dashed line) in MP11 Problem

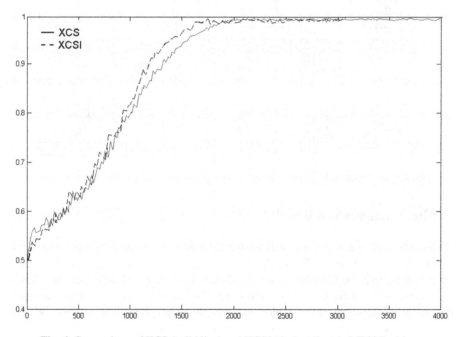

Fig. 6. Comparison of XCS (solid line) and XCSI (dashed line) in MP20 Problem

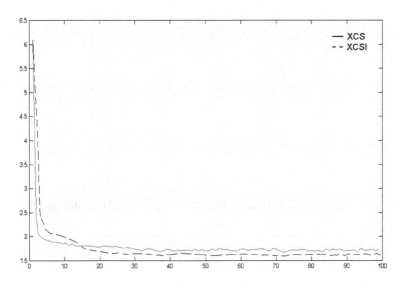

Fig. 7. Comparison of XCS (solid line) and XCSI (dashed line) in Woods2 Problem

8 Discussion

As the presented results show, XCSI achieves better performance than XCS in all benchmark problems. It is better to have a closer look at the IEM's operation during each run to describe this behavior. To do this, the best way is to inspect the proposed P_{exp} for each of these three problems.

In Figure 8 to 10, the proposed exploration rate by IEM rate is shown as vertical axis and the horizontal axis is *epoch-number/50*. Due to these Figures, the proposed exploration rate for these problems are low at the beginning and it grows higher with respect to the epoch number and the problem's complexity (as mentioned before, MP20 is more complex than MP11 and the Woods2 is more complex than the other two).

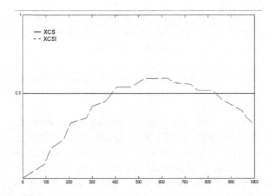

Fig. 8. The proposed Exploration Rate in XCS (Solid line) and XCSI (Dashed line) in MP11

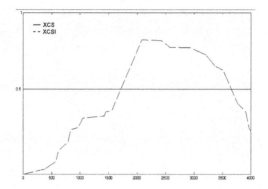

Fig. 9. The proposed Exploration Rate in XCS (Solid line) and XCSI (Dashed line) in MP20

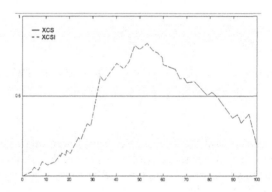

Fig. 10. The proposed Exploration Rate in XCS (Solid line) and XCSI (Dashed line) in Woods2

According to these Figures, we can explain the superiority of XCSI. This superiority is due to exploring more than XCS in the middle phase. In XCSI, this exploration epochs give the internal Genetic Algorithm more power to escape from local optima, better than XCS and less exploration at the beginning and final phase reduces needed epochs to reach the optimal performance. It is notable that IEM has no ability to learn from its environment and as mentioned before its rule base is set experimentally. It may work better with the ability to learn its rule base.

9   Conclusion

As we described before, it seems that constant exploration rate is not suitable for XCS and an adaptive one can improve XCS's performance. To realize this idea, we proposed an intelligent system called IEM. It is a fuzzy controller designed to propose the exploration rate for XCS. IEM is added to XCS and the new system, called XCSI, is applied to some benchmark problems. The obtained results demonstrate our intuition about usefulness of adaptive exploration rate for XCS performance. The main weakness of the proposed method is its static rule base. Another research is ongoing to add the learning capability to IEM.

References

[1] J.H. Holland, 1975, *Adaptation in Natural and Artificial Systems,* Ann Arbor: University of Michigan Press, Republished by the MIT press, 1992.
[2] S.W. Wilson, 1995, *Classifier Fitness Based on Accuracy,* Evolutionary Computation 3(2):149—175.
[3] S.W. Wilson, 1996, *Explore/Exploit strategies in autonomy,* In From animals to animats 4: Proceedings of the Fourth International Conference on Simulation of Adaptive Behavior (pp. 325-332), Cambridge MA: The MIT Press/Bradford Books.
[4] L. Rejeb and Z. Guessoum, 2005, *The Exploration-Exploitation Dilemma for Adaptive Agents,* Fifth European Workshop on Adaptive Agents and Multi-Agent Systems (AAMAS'05), to appear in Springer Lecture Note Series.
[5] Y. Chbany, F. Fouss, L. Yen, A. Pirotte and M. Saerens, 2005, *Managing the trade off between exploration and exploitation in reinforcement learning,* Technical report, Information System Unit, Universite cathaolique de louvain, Belgium.
[6] M. V. Butz, S. W. Wilson. 2002, *An Algorithmic Description of XCS,* Journal of Soft Computing, 6(3-4):144-153.
[7] J. H. Holmes, 1996, *Evolutionary-Assisted Discovery of Sentinel Features in Epidemiologic Surveillance,* PhD thesis, Drexel University.
[8] N. Parsa, 2005, *A Fuzzy Learning System to Adapt Genetic Algorithms,* MSc. Thesis, Iran University of Science and Technology.
[9] G. J. Klir, B. Yuan, 1995, *Fuzzy Sets and Fuzzy Logic: Theory and Applications,* Prentice Hall Press.
[10] http://www.cmp.uea.ac.uk/Research/kdd/projects.php?project=17, a web site by A.J. Bagnall and Z. Zatuchna.

Counter Example for Q-Bucket-Brigade Under Prediction Problem

Atsushi Wada[1,2], Keiki Takadama[3], and Katsunori Shimohara[4]

[1] National Institute of Information and Communication Technology
[2] ATR Cognitive Information Science Laboratories
2-2-2 Hikaridai, "Keihanna Science City" Kyoto 619-0288, Japan
[3] Department of Human Communication, The University of Electro-Communications
1-5-1 Chofugaoka, Chofushi, Tokyo 182-8585, Japan
[4] Faculty of Engineering, Doshisha University
1-3 Miyakodani, Tatara, Kyotanabe, Kyoto 610-0321, Japan

Abstract. Aiming at clarifying the convergence or divergence conditions for Learning Classifier System (LCS), this paper explores: (1) an extreme condition where the reinforcement process of LCS diverges; and (2) methods to avoid such divergence. Based on our previous work that showed equivalence between LCS's reinforcement process and Reinforcement Learning (RL) with Function approximation (FA) method, we present a counter example for LCS with the Q-bucket-brigade based on the 11-state star problem, a counter example originally proposed to show the divergence of Q-learning with linear FA. Furthermore, the empirical results applying the counter example to LCS verified the results predicted from the theory: (1) LCS with the Q-bucket-brigade diverged under prediction problems, where the action selection policy was fixed; and (2) such divergence was avoided by using the implicit-bucket-brigade or applying residual gradient algorithm to the Q-bucket-brigade.

1 Introduction

Learning Classifier Systems (LCSs) [1] are rule-based adaptive systems intended for a general framework to realize an intelligent behavior by combining two biologically inspired adaptive mechanisms – *learning* and *evolution* – with each essentially connecting to the fields of Reinforcement Learning (RL) and Evolutionary Computation (EC).

One of the main concern in LCS field has been *generalization* since XCS [2] was introduced, which became the currently mainstream model and was analyzed in theoretical aspect for its *accuracy-based* rule discovery process [3,4,5,6].

In such context, formal analysis of LCS's reinforcement process, together with generalization issue has become essential in this context. For a long time, the relationship between LCSs and Reinforcement Learning [7] has been regarded as one of the essential issue to be clarified. Comparison between the *bucket-brigade* algorithms in LCS and the concept of *Temporal Difference* (TD) in RL were carried out [8,9] due to the history of both sharing the concept of *reinforcement.*

X. Llorà et al. (Eds.): IWLCS 2003-2005, LNAI 4399, pp. 128–143, 2007.

However, when focusing on the generalization issue, few studies have dealt with LCSs' reinforcement process in company with the generalized representation of LCS's rule condition to be rigidly compared with TD update of RL with generalization.

Towards our goal to build the foundations of LCS seamlessly connected to the basis of RL, this paper addresses the issue of the convergence proof for LCS's reinforcement process. We have approached to this issue by focusing on Function Approximation method, a common generalization technique in RL[1], to be compared with LCS's reinforcement process with rule condition generalization. In [13,14], we revealed that the reinforcement process of ZCS[15] with the Q-bucket-brigade is equivalent with Q-learning with Function Approximation (FA) method[16] within the class of the linear approximation. Also, a disappointing results was derived that currently there exists no convergence proof for Q-learning with linear FA, while some counter example exist that leads the learning to diverge. This problem motivated us to propose ZCS with *residual gradient algorithm*[17], a RL technique that can introduce convergence to Q-learning with linear FA[18] [2].

In this paper, we proceed to further steps exploring: (1) an extreme condition where the reinforcement process of LCS diverges; and (2) the methods to avoid such divergence. For this objective, we present a counter example for LCS with the Q-bucket-brigade based on the 11-state star problem, which is originally proposed as the counter example for Q-learning with linear FA, and present an empirical results by applying it to Reinforcement learning-based XCS (RXCS), a LCS based on XCS but modified to be consistent with Q-learning with FA.

The rest of this paper is organized as follows. Section 2 introduces the current state of the convergence proofs for RL methods. Section 3 introduces RXCS. Section 5 proposes the 11-state star problem, the counter example for LCS and Section 6 gives the experimental results of applying the 11-state star problem to RXCS. Finally, Section 7 includes our discussions and conclusion.

2   State of Convergence Proofs for Reinforcement Learning

In this section, we first explain the properties of RL methods identifying the types of RL methods, which affects the availability of the convergence proofs. And next, based on the difference between these properties, the current state of the convergence proofs for RL methods is introduced.

---

[1] Issues relating LCS with Function approxmation method for RL has been attracting increasing attention. For example, several studies are presented from performance improvement aspect [10] and from high representation capability aspect [11,12].

[2] The approach applying residual gradient algorithm to LCS was also presented in [10], which contributed to further improvement of XCSG, a variant of XCS, to build accurately estimated payoff landscapes, while the main concern is rather different from ours, which examine closely a convergence or a divergence condition for LCS's reinforcement process.

2.1 Properties of Reinforcement Learning

Prediction and control problems. RL has two aspects regarding its learning: (1) *policy evaluation*, which estimates the action values, which are often *Q-values*, for an arbitrary policy π^3; and (2) *policy improvement*, which improves the current policy π to a better policy π' referring to the current action values. In *prediction problems*, the policy is fixed through the learning and the action values for that policy is estimated. In *control problems*, policy evaluation and policy improvement are performed at the same time.

On-policy and off-policy methods. *On-policy* methods, such as Sarsa, estimate the action values of the policy which controls the actual action selection. On the other hand, *off-policy* methods, such as Q-learning estimate a policy different from the policy controlling the actual action selection.

Classes of the approximated action value function. Several representations for designing the generalized action value function have been proposed for RL methods, such as state-aggregation, tile-coding, and highly sophisticated representations such as RBF networks and Neural Networks are applicable. These representations can be categorized into the following classes by their mathematical properties: (I) tabular; (II) state-aggregation; (III) linear approximation; and, (IV) non-linear approximation.

2.2 Convergence Proofs

Referring to RL literature, the current state of the convergence proofs for RL methods under the prediction and the control problems can be described as Figs. 1 (a) and (b), respectively. In both the figures, the white regions, the light gray regions and the thick gray regions each represent the current state: the convergence proof is available, the convergence proof is not available but proved to converge near optimal and oscillates, and the convergence proof is not available. Classes (I), (II), and (III) denote different types of function approximations: tabular, state-aggregation and linear approximation, respectively.

Under the prediction problems, on-policy methods are proved to converge within the class of linear approximation including both the tabular and the state aggregation, which is illustrated as the left column in Fig. 1 (a) [19,20,21,22,23,24]. However, in the case of off-policy methods, these proofs only covers the class of state aggregation. Within the class of linear approximation in general, off-policy methods are known to have the risk of divergence[17].

Under the control problems, the range where convergence proofs are available shrinks to the class of state aggregation for both the case of on-policy methods and off-policy methods, which is illustrated in Fig. 1 (b) [25,26,27,28]. In the case of the class of linear approximation, on-policy methods are proved to oscillate near optimal[24] and off-policy methods have the risk of divergence[17].

---

[3] In RL literature, *policy* π_{xa} is defined as a set of probability for all the possible combinations of taking a possible action a at a possible state x.

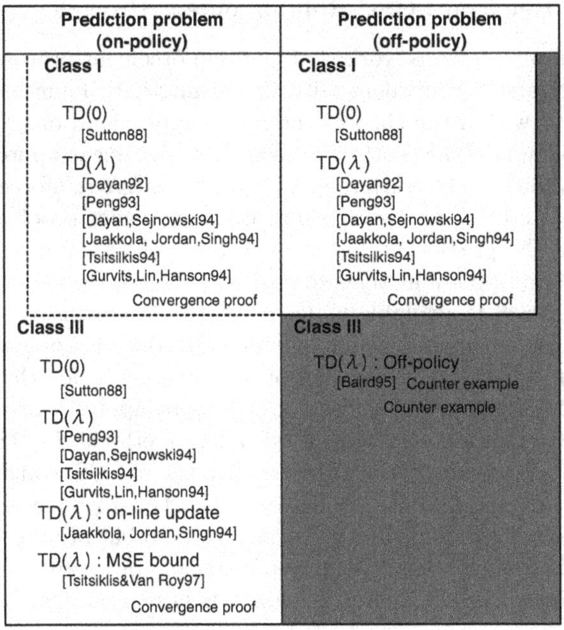

Prediction problem (on-policy)	Prediction problem (off-policy)
Class I TD(0) [Sutton88] TD(λ) [Dayan92] [Peng93] [Dayan,Sejnowski94] [Jaakkola, Jordan,Singh94] [Tsitsilkis94] [Gurvits,Lin,Hanson94] Convergence proof	**Class I** TD(0) [Sutton88] TD(λ) [Dayan92] [Peng93] [Dayan,Sejnowski94] [Jaakkola, Jordan,Singh94] [Tsitsilkis94] [Gurvits,Lin,Hanson94] Convergence proof
Class III TD(0) [Sutton88] TD(λ) [Peng93] [Dayan,Sejnowski94] [Tsitsilkis94] [Gurvits,Lin,Hanson94] TD(λ) : on-line update [Jaakkola, Jordan,Singh94] TD(λ) : MSE bound [Tsitsiklis&Van Roy97] Convergence proof	**Class III** TD(λ) : Off-policy [Baird95] Counter example Counter example

(a) Prediction problems

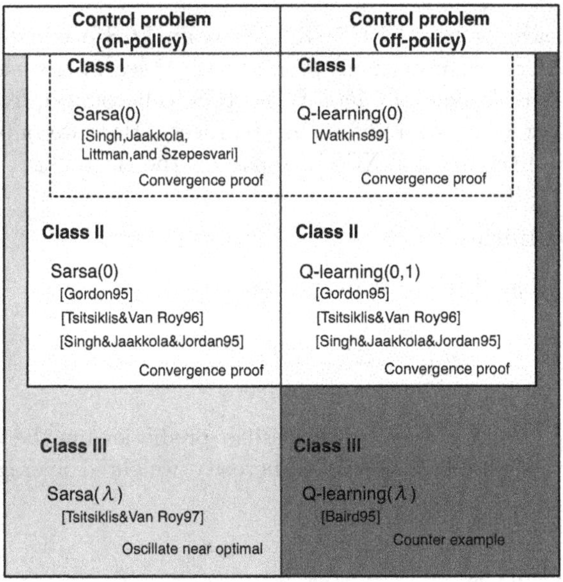

Control problem (on-policy)	Control problem (off-policy)
Class I Sarsa(0) [Singh,Jaakkola, Littman,and Szepesvari] Convergence proof	**Class I** Q-learning(0) [Watkins89] Convergence proof
Class II Sarsa(0) [Gordon95] [Tsitsiklis&Van Roy96] [Singh&Jaakkola&Jordan95] Convergence proof	**Class II** Q-learning(0,1) [Gordon95] [Tsitsiklis&Van Roy96] [Singh&Jaakkola&Jordan95] Convergence proof
Class III Sarsa(λ) [Tsitsiklis&Van Roy97] Oscillate near optimal	**Class III** Q-learning(λ) [Baird95] Counter example

(b) Control Problems

Fig. 1. The state of convergence proofs under prediction problems and control problems

2.3   Relation Between LCSs' Reinforcement Process

We presented in our previous work that the reinforcement process of ZCS with the Q-bucket-brigade is equivalent with Q-learning with Function Approximation (FA) method within the class of the linear approximation. Thus, the categories for off-policy methods with the class (III), the linear approximation will tell us the availability of the convergence proofs for ZCS's reinforcement process. From Figs. 1 (a) and (b), we can see that no convergence proof is available for such under both the prediction and the control problem. But we can also say that if we modify off-policy method to on-policy method within the class (III), the convergence proof is available under prediction problems and the proof for oscillating near the optimal is available under the control problems.

Another approach to avoid the risk of the divergence for the categories of off-policy method within the FA class of (III) is to apply *residual gradient algorithm*[17] to the update equation of off-policy methods. In [18], we applied residual gradient algorithm to the Q-bucket-brigade of ZCS, which resulted in a LCS with off-policy update but avoided its risk of the divergence.

Note that these discussions are not specific to ZCS but also applicable to any LCSs having RL with FA equivalent reinforcement processes, including Reinforcement learning based XCS (RXCS) which is presented in the next section.

3   Reinforcement Learning Based XCS (RXCS)

Reinforcement learning based XCS (RXCS) is an LCS originally proposed in [14] comparing XCS's reinforcement process with Q-learning with linear FA. In this study, RXCS is designed by modifying XCS to become equivalent with RL with linear FA regarding its reinforcement process. Here we only describe the modifications from the original XCS. See [29] for the entire algorithm of XCS.

3.1   Payoff Definition

The payoff definition of RXCS is defined as:

$$P(a_i) = \frac{\sum_{cl_k \in [\mathrm{M}]|_{a_i}} p_k \times num_k}{\sum_{cl_k \in [\mathrm{M}]|_{a_i}} num_k}, \tag{1}$$

where the fitness weighted average of classifier predictions in the XCS's original payoff definition is modified to the numerosity weighted average of classifier predictions.

3.2   Update Process

The update equation of the classifier prediction is defined as:

$$p_j \leftarrow p_j + \beta(P - P_{-1})\frac{num_j}{\sum_{cl_k \in [\mathrm{A}_{-1}]} num_k}, \tag{2}$$

where P_{-1} is a numerosity-weighted prediction for the classifiers included in the previous action set $[A_{-1}]$ defined as:

$$P_{-1} = \frac{\sum_{cl_k \in [A_{-1}]} p_k \times num_k}{\sum_{cl_k \in [A_{-1}]} num_k}. \tag{3}$$

Q-Bucket-Brigade and Implicit-Bucket-Brigade. In original XCS, the target value P for the update in Equation 2 is defined:

$$P \leftarrow r + \gamma \max_a P(a), \tag{4}$$

which is defined as a sum of the current reward and the discounted payoff value for the current greedy action $a^* = \arg\max_a P(a)$. This type of update, namely, the *Q-bucket-brigade* is originally introduced for ZCS in [15] with another alternative update named the *implicit-bucket-brigade*, which can be described by modifying Equation 4 as:

$$P \leftarrow r + \gamma P(a), \tag{5}$$

where the max operator is removed from the update equation for the Q-bucket-brigade.

Residual-Bucket-Brigade. Following the same process that we applied residual gradient algorithm to ZCS in [18], RXCS with residual gradient algorithm can be obtained. This is simply realized by adding the update process for the classifiers in the greedy action set $[M]|_{a^*}$ defined as:

$$p_j \leftarrow p_j - \gamma \beta (P - P_{-1}) \frac{num_j}{\sum_{cl_k \in [A_{-1}]} num_k}, \tag{6}$$

which works complementary with the ordinary update process for classifiers in the previous action set $[A_{-1}]$ defined as Equation 6. Here, we name this update process, the *residual-bucket-brigade* for convenience.

3.3   Convergence Proofs

As RXCS is designed to be equivalent with RL with linear FA, the discussion in 2.3 is also applicable to RXCS, which means that: (i) RXCS with the Q-bucket-brigade is within the category of off-policy method within the FA class of (III) the linear approximation, which might have the risk of the divergence of the learning; and (ii) such risk can be avoided in RXCS with the implicit-bucket-brigade or the residual-bucket-brigade.

4   Counter Example for Off-Policy Update

In this section, a counter example for RXCS with the Q-bucket-brigade is proposed, which is based on the 11-state star problem originally presented in [30].

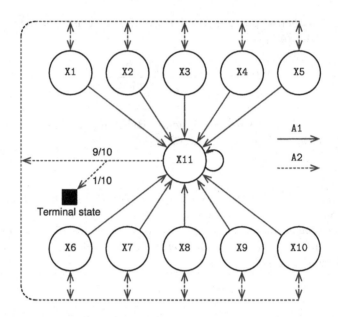

Fig. 2. The state transition diagram for the 11-state star problem, which is originally proposed by Baird as a counterexample for Q-learning with linear FA

4.1   The 11-State Star Problem

The 11-state star problem is originally proposed by Baird as a counter example for Q-learning with linear FA, whose state transitions are described as Fig. 2. The circles in the figure denotes the states $\{X1, \ldots, X11\}$. Every transition receives zero reward, and each state has two actions, the action A1 represented by a solid line, and the other action A2 represented by a dotted line. In all states, the action A1 invoke the transition to state X11. And the action A2 invoke the transition to one of the randomly chosen states within X1 through X10 except for the case when the current state is X11, where the transition follows the same rule with probability 9/10 but otherwise transits to the terminal state with probability 1/10. The discount factor γ is set to 0.9.

The approximated action value function is designed as described in Table 1, where the value of each state is given by the single approximation parameter or the linear combination of two approximation parameters. The function-approximation system is simply a lookup table, except for one additional approximation parameter $\theta(0)$ giving generalization. Note that the coefficient of the parameter $\theta(0)$ in the action value for (X11, A1) is twice as large as that of other action values for A1. This difference in the coefficient regarding the center state X11 is known to cause the monotonic increase of the value of the approximation parameter and thus derives the divergence of the learning.

Table 1. Action value function for the 11-state star problem

State	Action values for A1	Action values for A2
X1	$\theta(0) + 2\theta(1)$	$\theta(12)$
X2	$\theta(0) + 2\theta(2)$	$\theta(13)$
X3	$\theta(0) + 2\theta(3)$	$\theta(14)$
X4	$\theta(0) + 2\theta(4)$	$\theta(15)$
X5	$\theta(0) + 2\theta(5)$	$\theta(16)$
X6	$\theta(0) + 2\theta(6)$	$\theta(17)$
X7	$\theta(0) + 2\theta(7)$	$\theta(18)$
X8	$\theta(0) + 2\theta(8)$	$\theta(19)$
X9	$\theta(0) + 2\theta(9)$	$\theta(20)$
X10	$\theta(0) + 2\theta(10)$	$\theta(21)$
X11	$2\theta(0) + \theta(11)$	$\theta(22)$

4.2   Representation for LCS

To apply the 11-state star problem to LCS, the following steps are required: (1) represent the states and the actions using LCS's representation; and (2) represent the approximated action value function as a classifier population. Here, we adopt the ternary representation, the most common representation for LCS.

For the former step, a simple conversion rules are designed that convert: (a) the states {X1, ..., X11} into a corresponding 4-bits binary strings {0000, ..., 1011}; and (b) the actions A1 and A2 into single bits 0 and 1, respectively.

For the latter step, we propose a classifier population design that represents the approximated action value function for the 11-state star problem. Table 2 shows how the population is composed. For each state-action pair, there exists a corresponding classifier that identically matches that state-action pair. The classifier 0 is the only exception that matches with any states and having the action part of 0, that is, the action A1. In the case of RXCS, this composition of the classifier population represents the approximated action value function described as Table 3. From the table, we can see that the essential property of the 11-state problem is successfully expressed, where the coefficient of p_0, the prediction of the classifier 0 in the action value for (X11, A1) is twice as large as that of other action values for A1. This is due to the payoff definition of RXCS, where the predictions of classifiers in the action set is weighted by the value of each classifier's numerosity and averaged.

5   Simulation Experiments

In this section, we apply the 11-state star problem to RXCS with the Q-bucket-brigade, the implicit-bucket-brigade and the residual-bucket-brigade. These three cases are tested for the condition representing a prediction problem with a fixed and a decaying learning rate, whose detail is explained as follows:

Table 2. The design of classifier population for representing the 11-state star problem

ID	Matching states	Action	Numerosity	$condition_j$: $action_j$
0	$\{X1, X2, \ldots, X11\}$	A1	2	#### : 0
1	$\{X1\}$	A1	4	0001 : 0
2	$\{X2\}$	A1	4	0010 : 0
3	$\{X3\}$	A1	4	0011 : 0
4	$\{X4\}$	A1	4	0100 : 0
5	$\{X5\}$	A1	4	0101 : 0
6	$\{X6\}$	A1	4	0110 : 0
7	$\{X7\}$	A1	4	0111 : 0
8	$\{X8\}$	A1	4	1000 : 0
9	$\{X9\}$	A1	4	1001 : 0
10	$\{X10\}$	A1	4	1010 : 0
11	$\{X11\}$	A1	1	1011 : 0
12	$\{X1\}$	A2	1	0001 : 1
13	$\{X2\}$	A2	1	0010 : 1
14	$\{X3\}$	A2	1	0011 : 1
15	$\{X4\}$	A2	1	0100 : 1
16	$\{X5\}$	A2	1	0101 : 1
17	$\{X6\}$	A2	1	0110 : 1
18	$\{X7\}$	A2	1	0111 : 1
19	$\{X8\}$	A2	1	1000 : 1
20	$\{X9\}$	A2	1	1001 : 1
21	$\{X10\}$	A2	1	1010 : 1
22	$\{X11\}$	A2	1	1011 : 1

Table 3. Action value function

State	Action values for A1	Action values for A2
X1	$(2p_0 + 4p_1)/6$	p_{12}
X2	$(2p_0 + 4p_2)/6$	p_{13}
X3	$(2p_0 + 4p_3)/6$	p_{14}
X4	$(2p_0 + 4p_4)/6$	p_{15}
X5	$(2p_0 + 4p_5)/6$	p_{16}
X6	$(2p_0 + 4p_6)/6$	p_{17}
X7	$(2p_0 + 4p_7)/6$	p_{18}
X8	$(2p_0 + 4p_8)/6$	p_{19}
X9	$(2p_0 + 4p_9)/6$	p_{20}
X10	$(2p_0 + 4p_{10})/6$	p_{21}
X11	$(2p_0 + p_{11})/3$	p_{22}

Action selection policy: The action selection policy is fixed to satisfy the condition for prediction problems. The probability of taking the actions A1 and A2 are fixed to 1/10 and 9/10, respectively.

Fixed and decaying learning rate: In both the cases of adopting the fixed and the decaying learning rate, the learning rate β is initially set to 0.01.

In the case of adopting the decaying learning rate, β is decreased by the decaying coefficient of $1/n$, where n is initially set to 1 but incremented by 1 in every 1000 steps[4].

The common conditions used for the experiments are as follows. The discount factor γ is set to 0.9. RXCS is suppressed with its rule discovery process of GA invocation, covering, and deletion. In all cases, ten simulations are performed with each, including a total of 100,000 episodes.

5.1 Empirical Results

Figures 3, 4 and 5 presents the results for RXCS with the Q-bucket-brigade, the implicit-bucket-brigade and the residual-bucket-brigade, respectively. In all the figures, the graphs on the upper-hand side, labeled (a) represents the cases for the fixed learning rate and the graphs on the lower-hand side, labeled (b) represents the cases for the decaying learning rate. In all the graphs, the x-axis, the y-axis and the z-axis measures the number of episodes, the identification number of the classifier and the value of the corresponding classifier prediction.

The Q-bucket-brigade with both the fixed and the decaying learning rate showed the monotonical increase of the classifier predictions as shown in Figures 3 (a) and (b), while the the implicit-bucket-brigade and the residual-bucket-brigade converged to the correct value of 0 in both the cases of the fixed and the decaying learning rate.

6   Discussion and Conclusion

So far, the convergence regarding the reinforcement process of LCS is discussed from both the aspects of theory and practice. From the theoretical aspect, we referred to the convergence proofs of RL and clarified that: (1) RXCS with the Q-bucket-brigade is within the category of off-policy method within the FA class of (III) the linear approximation, which might have the risk of the divergence of the learning; and (2) such risk can be avoided in RXCS with the implicit-bucket-brigade or the residual-bucket-brigade. For an experimental aspect, we presented LCS version of the 11-state star problem, the counter example for off-policy RL methods with linear FA. The empirical results applying this counter example to RXCS verified that the properties of each bucket-brigade algorithms predicted from the theory applies to actual learning dynamics as follows: (1) RXCS with the Q-bucket-brigade diverged under prediction problems, where the action selection policy was fixed; and (2) such divergence was avoided by using the implicit-bucket-brigade or applying residual gradient algorithm to the Q-bucket-brigade.

---

[4] This decaying condition satisfies the general convergence condition required for β_t, the learning rate at the total steps t within an episode as follows: $\sum_t \beta_t \to \infty$, $\sum_t \beta_t^2 \to 0$.

PARAMETER VALUES

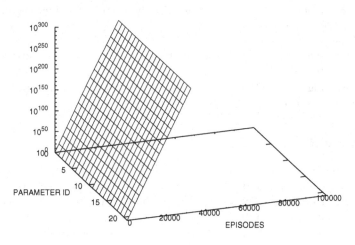

(a) Fixed learning rate.

PARAMETER VALUES

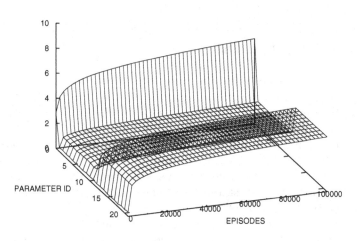

(b) Decaying learning rate.

Fig. 3. The dynamics of the prediction value for each classifier in the classifier population of RXCS with the Q-bucket-brigade under prediction problems

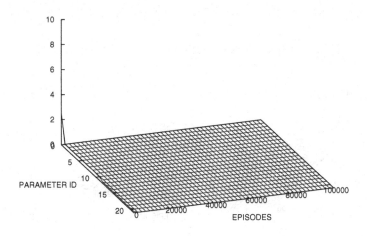

(a) Fixed learning rate.

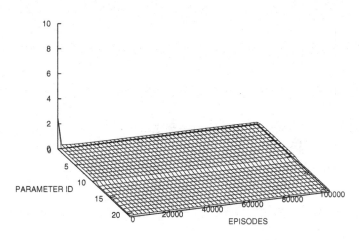

(b) Decaying learning rate.

Fig. 4. The dynamics of the prediction value for each classifier in the classifier population of RXCS with the implicit-bucket-brigade under prediction problems

PARAMETER VALUES

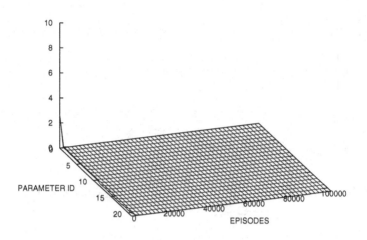

(a) Fixed learning rate.

PARAMETER VALUES

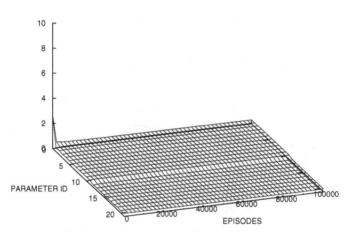

(b) Decaying learning rate.

Fig. 5. The dynamics of the prediction value for each classifier in the classifier population of RXCS with the residual-bucket-brigade under prediction problems

The conditions for LCSs presented here are quite strict from practical application point of view. However, our approach keeps a rigid consistency between LCS's reinforcement process and RL's theory, which showed that one of the essence from RL's theory, convergence proofs for learning, can be applied to LCS and successful for predicting the properties of learning dynamics such as the divergence of Q-bucket-brigade under a specifically designed ill conditions.

Here, a natural question occurs whether such divergence can be observed for LCS under normal conditions. We attempt an additional experiments applying control problems to RXCS by replacing the fixed action policy with ϵ-greedy action selection with the exploration rate *epsilon* set to 0.1. In all the cases for implicit, Q, and residual bucket-brigades, the learning converged and made no difference between the three algorithms.

In RL fields, at the moment, the divergence or the convergence under such condition is still a difficult issue, which can be categorized as a class of control problem with asynchronous update applying linear approximations, even for the most simplest case of using gradient descent based update [5], and much strong analysis that can apply to this domain is awaited.

The other question is supposed regarding the lack of evolutionary process in LCS, the rule discovery process, throughout our analysis. As the original theories for RL methods referred do not suppose any dynamical change in the design of approximated action value function, any genetic operation to modify the classifier population in LCS breaks the consistency with RL with function approximation. Although, at least, the results presented in this paper applies to a situation for an LCS such as the last phase of the learning where the rule discovery is already succeeded and suppressed for the reinforcement process to stabilize.

In addition, from XCS's accuracy-based rule discovery point of view, it is possible to expect that the divergence of an action value due to the corresponding overestimated or underestimated classifier predictions can be eliminated under the selection pressure towards accuracy. Such phenomenon might put light on the accuracy-based rule discovery not only from the generalization issue but also from the divergence avoiding issue.

Towards the final goal of developing a unified theory of LCS, the influence of evolutionary process in detail must be interpreted from RL perspective, which might be able to integrate analyses for reinforcement and evolutionary processes of LCSs.

Acknowledgments

This research was partly conducted for 'Research on Human Communication' with funding from the National Institute of Information and Communications Technology, and supported in part by a Grant-in-Aid for Scientific Research (B) (No. 17360424) of Ministry of Education, Culture, Sports, Science and Technology (MEXT).

---

[5] In the case of synchronous update, where the update delta value for each action values are aggregated during a single episode and applied in the end of the episode at once, a rigorous analysis is presented by [31].

References

1. Holland, J.H.: Escaping brittleness: the possibilities of general-purpose. Machine Learning, an artificial intelligence approach **2** (1986) 593–623
2. Wilson, S.W.: Classifier fitness based on accuracy. Evolutionary Computation **3** (1995) 149–175
3. Kovacs, T.: Evolving optimal populations with xcs classifier systems. Technical Report CSRP-96-17, University of Birmingham, School of Computer Science (1996)
4. Butz, M.V., Pelikan, M.: Analyzing the evolutionary pressures in XCS. In L. Spector, e.a., ed.: Proceedings of the Genetic and Evolutionary Computation Conference (GECCO-2001), San Francisco, CA, Morgan, Kaufmann (2001) 935–942
5. Butz, M.V., Goldberg, D.E., Lanzi, P.L.: Bounding learning time in XCS. In K. Deb, e.a., ed.: Proceedings of the Genetic and Evolutionary Computation Conference (GECCO-2004), Springer (2004) 739–750
6. Butz, M.V., Kovacs, T., Lanzi, P.L., Wilson, S.W.: Toward a theory of generalization and learning in XCS. IEEE Transactions on Evolutionary Computation **8** (2004) 28–46
7. Sutton, R., Barto, A.: An introduction to reinforcement learning. MIT Press, Cambridge, MA. (1998)
8. Dorigo, M., Bersini, H.: A comparison of Q-learning and classifier systems. In Cliff, D., Husbands, P., Meyer, J.A., Wilson, S.W., eds.: Proceedings of From Animals to Animats, Third International Conference on Simulation of Adaptive Behavior, Cambridge, MA., MIT Press (1994) 248–255
9. Lanzi, P.L.: Learning classifier systems from a reinforcement learning perspective. Soft Computing **6** (2002) 162–170
10. Butz, M.V., Lanzi, P.L., Goldberg, D.E.: Gradient descent methods in learning classifier systems: Improving xcs performance in multistep problems. IEEE Transactions on Evolutionary Computation **9** (2005) 452–473
11. Booker, L.: Adaptive value function approximations in classifier systems. In: The Eighth International Workshop on Learning Classifier Systems (IWLCS2005). (2005) 90–91
12. O'Hara, T., Bull, L.: A memetic accuracy-based neural learning classifier system. In: Proceedings of the IEEE Congress on Evolutionary Computation. (2005) 2040–2045
13. Wada, A., Takadama, K., Shimohara, K., Katai, O.: Comparison between Q-learning and ZCS Learning Classifier System: From aspect of function approximation. In: The 8th Conference on Intelligent Autonomous Systems,. (2004.)
14. Wada, A., Takadama, K., Shimohara, K., Katai, O.: Learning classifier system equivalent with reinforcement learning with function approximation. In: The Eighth International Workshop on Learning Classifier Systems (IWLCS2005). (2005) 24–29
15. Wilson, S.W.: ZCS: A zeroth level classifier system. Evolutionary Computation **2** (1994) 1–18
16. Sutton, R.S.: Generalization in reinforcement learning: Successful examples using sparse coarse coding. In Touretzky, D.S., Mozer, M.C., Hasselmo, M.E., eds.: Advances in Neural Information Processing Systems. Volume 8., Cambridge, MA., The MIT Press (1996) 1038–1044
17. Baird, L.C.: Residual algorithms: Reinforcement learning with function approximation. In Prieditis, A., Russell, S.J., eds.: Machine Learning, Proceedings of the Twelfth International Conference on Machine Learning (ICML1995), San Francisco, CA, Morgan Kaufmann (1995) 30–37

18. Wada, A., Takadama, K., Shimohara, K., Katai, O.: Learning Classifier Systems with Convergence and Generalization. In: Foundations on Learning Classifier Systems. Springer, London, UK (2005) 285–304
19. Dayan, P., Sejnowski, T.J.: TD(λ) converges with probability 1. Machine Learning **14** (1994) 295–301
20. Jaakkola, T.S., Jorda, M.I., Singh, S.P.: On the convergence of stochastic iterative dynamic programming algorithms. IEEE Transactions on Automatic Control **6** (1994) 1185–1201
21. Peng, J., Williams, R.J.: On the convergence of stochastic iterative dynamic programming algorithms. Adaptive Behavior **1** (1993) 437–454
22. Sutton, R.S.: Learning to predict by the methods of temporal differences. Machine Learning **3** (1988) 9–44
23. Tsitsiklis, J.N.: Asynchronous stochastic approximation and q-learning. Machine Learning **16** (1994) 185–202
24. Tsitsiklis, J.N., Roy, B.V.: An analysis of temporal-difference learning with function approximation. IEEE Transactions on Automatic Control **42** (1997) 674–690
25. Gordon, G.J.: Stable function approximation in dynamic programming. In Prieditis, A., Russell, S., eds.: Proceedings of the Twelfth International Conference on Machine Learning, San Francisco, CA, Morgan Kaufmann (1995) 261–268
26. Singh, S.P., Jaakkola, T., Jordan, M.I.: Reinforcement learning with soft state aggregation. In Tesauro, G., Touretzky, D., Leen, T., eds.: Advances in Neural Information Processing Systems 7, Cambridge, MA., The MIT Press (1995) 361–368
27. Singh, S.P., Jaakkola, T., Littman, M.L., Szepesvári, C.: Convergence results for single-step on-policy reinforcement-learning algorithms. Machine Learning **38** (2000) 287–308
28. Watkins, J.C.H.: Learning from delayed rewards. PhD thesis, Cambridge University (1989)
29. Butz, M.V., Wilson, S.W.: An Algorithmic Description of XCS. In: Advances in Learning Classifier Systems, Forth International Workshop, IWLCS 2001. Springer-Verlag, Berlin, Germany (2002) 253–272
30. Baird, L.C.: Reinforcement Learning Through Gradient Descent. PhD thesis, Carnegie Mellon University, Pittsburgh, PA (1999)
31. Merke, A., Schoknecht, R.: Convergence of synchronous reinforcement learning with linear function approximation. In: ICML '04: Proceedings of the twenty-first international conference on Machine learning, New York, NY, USA, ACM Press (2004) 75

An Experimental Comparison Between ATNoSFERES and ACS

Samuel Landau[1], Olivier Sigaud[2], Sébastien Picault[3], and Pierre Gérard[4]

[1] TAO/INRIA-Futurs, LRI, Univ. de Paris-Sud, 91 405 Orsay, France
Samuel.Landau@lri.fr
[2] AnimatLab, LIP6, Univ. Pierre et Marie Curie, 75 015 Paris, France
Olivier.Sigaud@lip6.fr
[3] SMAC, LIFL, Univ. de Lille I, 59 655 Villeneuve d'Ascq, France
Sebastien.Picault@lifl.fr
[4] ADAge, LIPN, Univ. de Paris-Nord, 93 430 Villetaneuse, France
Pierre.Gerard@lipn.univ-paris13.fr

Abstract. After two papers comparing ATNoSFERES with XCSM, a Learning Classifier System with internal states, this paper is devoted to a comparison between ATNoSFERES and ACS (an Anticipatory Learning Classifier System). As previously, we focus on the way perceptual aliazing problems encountered in non-Markov environments are solved with both kinds of systems. We shortly present ATNoSFERES, a framework based on an indirect encoding Genetic Algorithm which builds finite-state automata controllers, and we compare it with ACS through two benchmark experiments. The comparison shows that the difference in performance between both system depends on the environment. This raises a discussion of the adequacy of both adaptive mechanisms to particular subclasses of non-Markov problems. Furthermore, since ACS converges much faster than ATNoSFERES, we discuss the need to introduce learning capabilities in our model. As a conclusion, we advocate for the need of more experimental comparisons between different systems in the Learning Classifier System community.

Keywords: Evolutionary Algorithms, Perceptual Aliazing, Augmented Transition Networks.

1 Introduction

Most Learning Classifier Systems (LCS) [1] are used to control agents involved in a sensori-motor loop with their environment. Such agents perceive situations through their sensors as vectors of several attributes, each attribute representing a perceived feature of the environment. As pointed out by Lanzi [2], LCS are adaptive architectures based on *Reinforcement Learning* (RL) techniques [3], but endowed with generalization capabilities. Thanks to a LCS, an agent can *learn* the optimal policy – *i.e.* which action to perform in every situation, in order

X. Llorà et al. (Eds.): IWLCS 2003-2005, LNAI 4399, pp. 144–160, 2007.
© Springer-Verlag Berlin Heidelberg 2007

to maximize a reward obtained in the environment. The policy is defined by a set of rules – or classifiers – specifying an action according to some *conditions* concerning the perceived situation.

Standard RL algorithms are generally used in situations where the state of the agent-environment interaction is always known without ambiguity. But in real world environments, it often happens that agents perceive the same situation in several different states, eventually requiring different optimal actions, giving rise to the so called *"perceptual aliazing"* problem. In such a case, the environment is said *non-Markov*, and agents cannot perform optimally if their decision at a given time step only depends on their perceptions at the same time step.

There are several attempts to apply LCSs to non-Markov problems, relying on different approaches to the problem. For instance, in XCSM [4] added explicit internal states to the classical (condition, action) pair of the classifiers used in XCS [5]. From XCS again, [6] proposed in CXCS a rule-chaining mechanism able to build a bridge over ambiguous situations. ACS, an Anticipatory LCS (ALCS), uses a similar rule-chaining mechanism to solve non-Markov problems.

In two recent papers [7,8], we have presented a new framework, "ATNoS-FERES" [9], also used to automatically design the behavior of agents and able to cope with non-Markov environments. ATNoSFERES relies on an evolutionary approach instead of classical reinforcement learning techniques, but we have shown in [7] that the resulting graph-based representation was semantically very similar to the LCS representation, giving rise to a detailed comparison between both classes of systems. In particular, we have shown that two important advantages of the graph-based representation were its minimality and its readability. As a result, the structure of the controller gives a lot of information about the structure of the problem faced by the system. In these papers, ATNoSFERES was compared with XCSM on the well-known Maze10 environment and then on a new environment called 12-Candlesticks.

In the present paper, we provide a new comparison between ATNoSFERES and another LCS, ACS. We rely on a study from [10] to compare the performance of both systems on two distinct environments. Our comparison reveals new features of the interaction of LCSs with non-Markov problems.

In the next section, we summarize the features and properties of the ATNoS-FERES model, and we highlight the formal similarity between ATNoSFERES and LCS representations. In section 3, we briefly present the different approaches used in LCSs to cope with non-Markov problems. Then we actually compare ATNoSFERES with ACS in section 4. This new study reveals that some problems found difficult with ACS appear easier with ATNoSFERES and *vice versa*. We discuss this point in section 5. Finally, we draw lessons from the fact that ATNoSFERES converges slower than ACS to conclude that we should include on-line learning mechanisms in our model, and we highlight the need of more experimental comparisons between classes of Learning Classifier Systems now that the field is getting more mature.

2   The ATNoSFERES Model and Learning Classifier Systems

2.1   Graph-Based Expression of Behaviors

The architecture provided by the ATNoSFERES model [9,11] involves an ATN[1] graph [12] which is basically an oriented, labeled graph with a Start (or initial) node and an End (or final) node (see figure 7). Nodes represent states while edges represent transitions of an automaton.

Like LCSs, ATNoSFERES binds conditions expressed as a set of attributes to actions, and is endowed with the ability to generalize conditions by ignoring some attributes. But in ATNoSFERES, the conditions and actions are used in a graph structure that provides internal states.

The graph describing the behaviors is built from a genotype by adding nodes and edges to a basic structure containing only the *Start* and *End* nodes. The graph-building process was described in [7,8] and will not be detailed here again. For the self-consistency of the paper, we just have to mention that the process is separated into two steps:

1. The bitstring (genotype) is translated into a sequence of tokens.
2. The tokens are interpreted as instructions of a robust programming language, dedicated to graph building.

Since any sequence of tokens is meaningful, the graph-building language is highly robust to any variations affecting the genotype, thus there is no specific syntactical nor semantical constraint on the genetic operators. In addition, the sequence of tokens is to some extent order-independent and a given graph can be produced from very different genotypes, which guarantees a degeneracy property.

2.2   ATNoSFERES Model and Learning Classifier Systems

As explained in more details in [7] and illustrated in figure 1, an ATN such as those evolved by ATNoSFERES can be translated into a list of classifiers. The nodes of the ATN play the role of internal states and endow ATNoSFERES with the ability to deal with perceptual aliazing. The edges of the ATN are characterized by several informations which can also be represented in classifiers: the source and destination nodes of the edge correspond to internal states; the conditions associated to the edges correspond to the conditions of the classifiers and the actions associated to the edges correspond to the actions of the classifiers.

3   Background: LCSs and Non-markov Problems

Dealing with simple `Condition-Action` classifiers does not endow an agent with the ability to behave optimally in perceptually aliazed problems. In such problems, it may happen that the current perception does not provide enough information to always choose the optimal action: as soon as the agent perceives

---

[1] ATN stands for "Augmented Transition Networks".

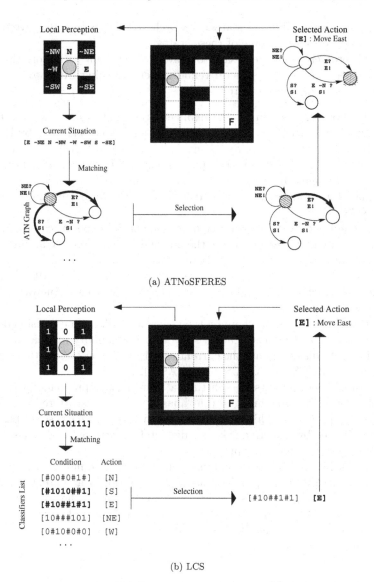

(a) ATNoSFERES

(b) LCS

Fig. 1. The sensori-motor loop with ATNoSFERES and a standard LCS. The agent perceives the presence/absence (resp. 1/0) of blocks in each of the eight surrounding cells and must decide towards which of the eight adjacent cells it should move. In AT-NoSFERES, from its current location, the agent perceives [E ¬NE N ¬NW ¬W ¬SW S ¬SE] (token E is true when the east cell is empty). From the current state (node) of its graph, two edges (in bold) are eligible, since the condition parts of their label match the perceptions. One is selected either deterministically or not, then its action part (move east) is performed and the current state is updated. In a LCS case, the agent perceives [01010111] (starting north and rotating clockwise). Within the list of classifiers characterizing it, the LCS first selects those matching the current situation. Then, it selects one of the matching classifiers and the corresponding action is performed.

the same situation in different states, it will choose the same action even if this action is inappropriate in some of these states.

For such problems, it is necessary to provide the system with more than just current perceptions. In the general reinforcement learning framework, several kinds of solutions have been tested.

- The first one consists in adding explicit internal states to the perceptions involved in the decisions of the system. This approach was used by Holland in his early LCSs thanks to an internal message list [13]. But both [14] and [15] reported unsatisfactory performance of Holland's system on non-Markov problems. In the context of more recent LCS research, the explicit internal state solution was adopted by [16] in ZCSM and by [17] in XCSM and XCSMH.
- The second one, memory window management, is a special case of explicit internal state management where the internal state consists in an immediate memory of the past of length k. Some systems use a fixed size window (see [18] for a review) while others use a variable size window (*e.g.* [19]). The next solution, rule-chaining, can be seen as an alternative view of the variable size window mechanism.
- The third one consists in chaining the decisions, making one decision depend on the decisions previously taken, so as to use a memory of what was done previously to disambiguate the current situation. Among LCSs, this solution was used in ZCCS [20], CXCS [6] and ACS [21].
- The fourth one consists in splitting a non-Markov problem into several Markov problems, making sure that aliased states are scattered among different sub-problems. This solution has been investigated first by [22], and then improved by [23]. To our knowledge, no LCS actually uses this solution, despite its very interesting properties.
- The last solution consists in building a finite state automaton corresponding to the structure of the problem, as [24] or [25] do, in a context where the structure of the problem is known in advance. This is the solution chosen in ATNoSFERES, using a Pittsburg style evolutionary algorithm, but in a context where the agents do not know anything about the structure of the problem before starting.

4   Experimental Comparison with ACS

4.1   ACS

In previous papers, we have compared ATNoSFERES with XCSM on two non-Markov problems. In order to go deeper into the comparison between the abilities of ATNoSFERES and LCSs to cope with the perceptual aliazing problem, we present in this section a comparison with another system, ACS.

The Anticipatory Classifier System has been developed by Stolzmann [26]. It differs from classical Learning Classifier Systems by adding to the perception-action rules an "effect part" that represents a perceptual anticipation of the consequences of the action upon the environment. ACS relies on an Anticipatory Learning Process (ALP) [26] and has been successfully applied to both Markov and non-Markov environments.

The main feature of ACS with respect to XCS-like LCSs relies in the fact that their use of anticipation make it possible to design some efficient heuristics that are believed to make the system converge faster, though no explicit performance comparison has been published yet. Gérard and Sigaud have proposed two ALCSs similar to ACS, namely YACS [27] and MACS [28], that have been shown to be faster than ACS, but are limited to Markov and deterministic environments.

In ACS, in order to deal with non-Markov environments, it was chosen to use a rule-chaining mechanism like in CXCS [6]. In that case, the effect part of a classifier consisting in a behavioral sequence is intended to represent the perceptual consequence of the sequence of actions. As it is the case with CXCS, this feature makes ACS able to deal efficiently with non-Markov environments [21].

In order to build such a behavioral sequence, a new parameter was added to ACS, namely "BS_{max}". BS_{max} represents the maximal length of the behavioral sequences that ACS may build. Its value must be decided before starting any run.

4.2   Experimental Setup

We tried to reproduce an experimental setup as close as possible to that used in [4] with the Maze10 environment and ACS in E1 and E2 environments, taking into account the specificities of our model. This setup has been applied to all the experiments presented in this paper.

Perception/Action abilities and Tokens. The agents used for the experiments are able to perceive the presence/absence of walls or the presence of food in the eight adjacent cells of the grid, these three perceptions being mutually exclusive. They can move in adjacent cells (the move will be effective if the cell is empty or contains food). Thus, the genetic code includes 24 condition tokens, 8 action tokens, 7 stack manipulation tokens and 4 node creation/connection tokens. We used 7 bits encoding to define the tokens ($2^7 = 128$ tokens, which means that some tokens are encoded twice or more).

In [8], we demonstrated that the performances of ATNoSFERES could be increased by using a new token, *selfConnect*, endowing our model with the ability to build easily self-connecting edges from a node to itself. This new token has been used in all the experiments presented below.

Course of Experiments. Each experiment involves the following steps:

1. Initialize the population with $N = 300$ agents with random bitstrings.
2. For each generation, build the graph of each agent and evaluate it in the environment.

3. Select the 20 % best individuals of the population and produce new ones by crossing over the parents. The system performs probabilistic mutations (with a 1% rate) and insertions or deletions of codons (with a 0.5% rate) on the bitstring of the offspring.
4. Iterate the process in 2 with the new generation.

Fitness function. Each individual is evaluated by putting it into the environment, starting on a blank cell in the grid, and letting it try to find the food within a limited amount of time (the limit is 20 time steps in all experiments described below). The agent can perceive the food, and it can perform only one action per time step; when this action is incompatible with the environment (*e.g.* go west when the west cell contains an obstacle), it is simply discarded (the agent loses one time step and stays on the same cell).

The fitness of the agent for each run is the remaining time if the food has been found within the time limit. Thus, the selection pressure encourages short paths to food. For one generation, each agent is evaluated one time starting on each empty cell, then its total fitness for this generation is the sum of the fitnesses computed for each run. Each agent is reevaluated at each generation in order to average its fitness over generations. This is necessary because of the non-deterministic aspects of the automata.

Indeed, there are several potential sources of non-determinism in our automata. The first one is due to the fact that several arcs might be eligible from the current node in the current situation. In that case, we can either choose one arc randomly, giving rise to a non-deterministic behavior, or assign fixed priorities (by order of creation, for instance) to arcs, so as to keep the automata deterministic. In all the experiments presented here, we have chosen the deterministic stance, after having checked that we obtain better performance with such a choice.

But there are still two sources of non-determinism in our automata. In a situation where no arc is eligible, or when an edge to cross does not carry any action label, one action is chosen randomly. Thus an automaton will be fully deterministic only in the case where one arc can be elected in any encountered situation, and if all such arcs bear an action to perform. This explains the need to average the performance over several runs.

4.3   Experimental Environments

The experiments described below take place in two non-Markov environments (E1 and E2, see figure 4.3) that have been used in [10] to study how ACS deals with non-Markov problems. E1 presents 20 aliazed situations (among the 44 free cells) which are perceived as 9 distinct situations. E2 presents 36 aliazed situations (among 48 free cells), which are perceived as 5 distinct situations.

On figure 3, we show the number of steps an optimal agent among several may need to reach food from each starting cell, given that its perception is limited (an omniscient agent could perform even better).

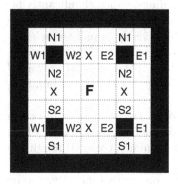

(a) E1 environment

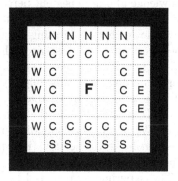

(b) E2 environment

Fig. 2. E1 and **E2** environments. **F** (food) is the goal. Other marked cells represent aliazed situations (identical letters imply the same perception).

4.4   Comparison with ACS

Before comparing, we have to emphasize a major difference between the way ACS and ATNoSFERES deal with these environments. This difference regards the implicit selection of possible movements. In ACS experiments, as they are described in [10, § 4.1 and 4.2], the only movements tested in each free position are transitions towards surrounding free cells (for example, if the cell to the north contains an obstacle, the move to the north is not considered as a possible move, thus it is not tested). This constitutes a kind of prior domain-dependent

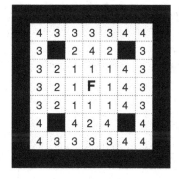

(a) E1 environment

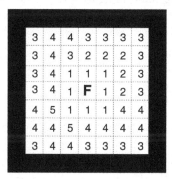

(b) E2 environment

Fig. 3. One optimal policy for E1 (resp. E2), represented by the number of steps needed to reach food from each Start cell. Other equivalent policies can be obtained at least by applying all possible rotations and symmetries to all the numbers given. In E1, the optimal average number of steps to food is 2.8181 steps. In E2, it is 2.9792 steps.

knowledge about consistent perceptions-actions bindings, which significantly biases the learning process by reducing the number of classifiers to test. In [29], we have shown that prohibiting the use of this bias can severely impair some learning algorithms. For instance, McCallum's U-Tree algorithm [19] which works well in non-Markov mazes such as those studied here if the agent is prevented from bumping into walls, might grow an infinitely deep tree if it keeps bumping into the same wall in an aliased situation.

In ATNoSFERES, on the contrary, any move token can be used as an action label. When the corresponding movement is impossible, the agent stays where it is and loses a time step (it is penalized only in an indirect way, through the fitness function).

The experiments reported here were carried out on various initial genotype sizes. In E1, the genotypes that have been tested are between 40 and 150 tokens long (with step 10), as in E2. Using these different sizes was necessary because we do not know in advance the minimum size required to produce an efficient automaton.

The original population genotype sizes may drift during an evolution, since some genetic operators insert or delete parts of the genotype randomly. Each experiment is stopped after 10,000 generations, and 10 experiments have been performed in each experimental situation.

4.5   Results

Figure 4 gives the respective fitness values obtained by the best automata in E1 and E2 experiments, depending on initial lengths of the genotypes. Each cross in the figures represent the performance of the best automaton obtained after 10,000 generations in one run. Thus there are ten crosses for each initial length. From figure 4 (a), it can be seen that in E1, ATNoSFERES easily reaches the performance of ACS in the case where $BS_{max} = 1$, but hardly reaches

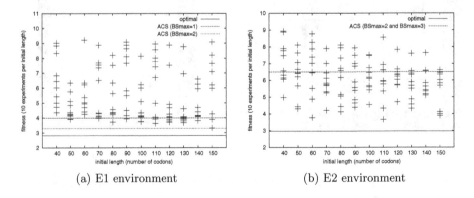

(a) E1 environment   (b) E2 environment

Fig. 4. Minimal average time to reach food in E1 (resp. E2) experiments with "deterministic" automata as a function of the initial length of the bitstring

the performance of ACS with $BS_{max} = 2$, which is very close to the optimal performance.

In E2, the performance obtained with ATNoSFERES is significantly better than the one obtained with ACS with $BS_{max} = 2$ and $BS_{max} = 3$. Indeed, ATNoSFERES is about twice closer to the optimum performance.

In order to check whether ATNoSFERES could reach an even higher performance in E1, we took the best run on figure 4 (a) and ran it up to 100,000 generations. The best performance was slightly improved again, reaching 3.2 (it was 3.3 after 10,000 runs).

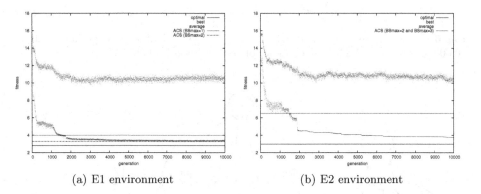

(a) E1 environment   (b) E2 environment

Fig. 5. Best fitness evolution in E1 (resp. E2) experiment as a function of generations; the shape and smoothness of the curve are representative for all E1 (resp. E2) evolutions. The thickness of the curves (particularly manifest in E1) is due to the indeterministic behavior of agents. In E2, it seems that the pressure towards deterministic behavior is stronger.

Figure 5 gives the evolution of the best fitnesses, respectively in E1 and E2 environments. It appears clearly that gradual improvements occur in both environments.

4.6 Representative Solutions

E1 environment. We present on figure 8 the best automaton obtained in E1 experiments after 100,000 generations, on figure 7 the best automaton obtained after 10,000 generations, and on figure 6 a more representative automaton obtained after 10,000 generations. From these figures it is clear that the most common solutions found are nearly reactive. The graph of the more common automata contains a single node (in addition to the Start and End node that always exist in ATNoSFERES graphs), which means that a reactive behavior already performs well in E1. The results show that this kind of behavior is produced in most cases and gets high fitness values, more easily than solutions involving internal states.

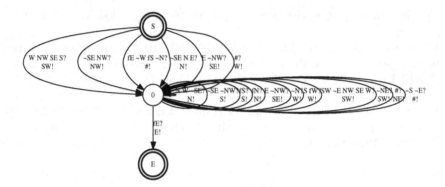

Fig. 6. A representative automaton found with ATNoSFERES in E1 experiment (after 10,000 generations). Its average number of steps to food is about 3.8.

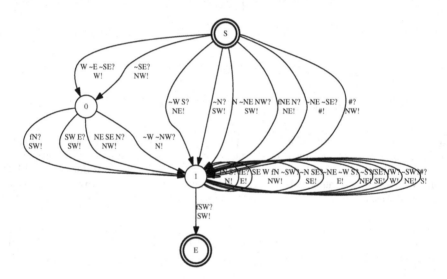

Fig. 7. The best automaton found with ATNoSFERES in E1 experiment (after 10,000 generations). Its average number of steps to food is about 3.3.

However, the automaton depicted on figure 7 shows that adding one node can already improve significantly the global performance.

The main difference between the best automaton obtained after 10,000 generations and the one obtained after 100,000 generations is that the latter contains several additional arcs. In particular, the agent will more often take into account the presence of food (label f on the edges) in its immediate surrounding to reach it immediately.

Indeed, we can see on figure 9 (a) that in several situations where the food is visible the agent needs more than one step to reach it, though a more efficient behavior is obvious. ATNoSFERES has a lot of difficulties in finding these

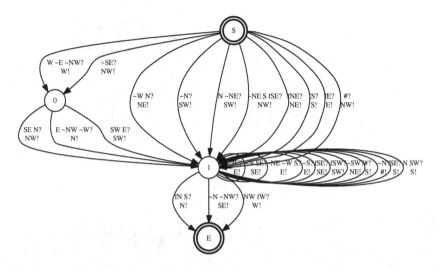

Fig. 8. The best automaton found with ATNoSFERES in E1 experiment (after 100,000 generations). Its average number of steps to food is about 3.2.

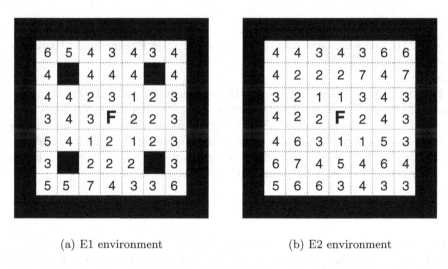

(a) E1 environment   (b) E2 environment

Fig. 9. Best policy found with ATNoSFERES in E1 (resp. E2) in 10,000 generations, represented by the number of steps needed to reach food from each Start cell (see figure 3 for optimal policy)

reactive rules that a reinforcement learning algorithm combining exploration and exploitation would find immediately.

However, even if these additional arcs could improve the performance a bit more, this would not be enough to reach the true optimal performance. A carefully hand-crafted optimal automaton needs much more internal states than the ones shown in this section.

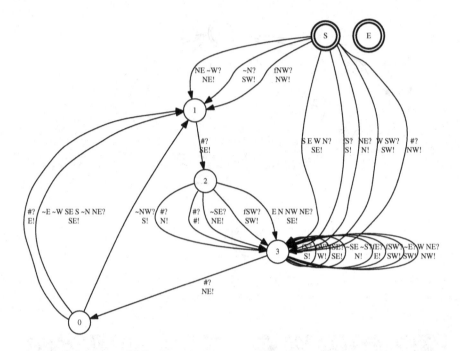

Fig. 10. The best automaton found with ATNoSFERES in E2 experiment. Its average number of steps to food is about 3.8.

E2 environment. Figure 10 gives the best automaton found in E2 environment. From this figure it is immediately clear that a good automaton in E2 needs more nodes than it is the case in E1. This seems to imply that reactive and nearly reactive behaviors perform much worse in E2 than in E1. This fact, in addition to the fact that ATNoSFERES clearly outperforms ACS on E2 while it is less the case in E1, will be at the heart of the discussion that follows.

5   Discussion

The experimental study presented in the previous section reveals that different subclasses of non-Markov problems should be distinguished more accurately. Indeed, some problems, like E1, are actually non-Markov, but in such a way that reactive behaviors can still perform well on such problems.

In E1, our study has shown that through an evolutionary process, it is easy to gradually grow a set of *ad hoc* rules (which are to some extent independent from each other), even more if the agent is tested from each cell: thus, an agent can start with a few rules that are efficient for a few cells, and evolve from one generation to another rules that are useful for additional cells. From such a reactive solution, built by the accumulation of small changes, it is unlikely to develop internal states to deal with a few particular cases, since it requires at the same time additional nodes, linked with consistent edges, conditions and actions.

We meet again the *structural cost* mentioned in [7]: "simple", incremental good solutions are preferred to structurally complex optima.

On the contrary, other problems, like E2, should be said "highly non-Markov", since reactive policies perform very poorly on such problems. In E2, there is no hope that a reactive behavior could lead to the food in a reasonable amount of time, due to the location and the nature of aliazed situations.

Our comparative study has revealed that ACS performs very well on the first subclass of problems and more poorly on the second, while ATNoSFERES performs consistently on both subclasses.

Now we should ask ourselves why this is so. On first thoughts, one might consider that the maximal length of sequences in ACS plays a major role in the phenomenon. One could expect that setting BS_{max} to more than 3 in E2 should fix the problem. A closer examination, however, reveals that this is not so.

In [10], the authors show that setting BS_{max} to 3 is enough to let ACS build a completely reliable model of E2, under the form of (`situation`, `action`, `next situation`) classifiers. This explains why they did not try $BS_{max} = 4$ or more.

But the performance concern and the model reliability concern are not strictly correlated. Regarding the convergence to stable reward performance, [10] emphasize that increasing the maximum length of the behavioral sequence "does not improve the 'steps to food' performances", *i.e.* a "good" behavioral solution can be exploited without having built an exhaustive representation of the environment.

One reason explaining that building longer action sequences would not improve the performance comes from the fact that these sequences specify a blind series of actions to perform without interruption and without checking between its beginning and its end the situation perceived in the environment by the agent. These sequences can improve the performance of the agent when they let it jump over ambiguous situations, but they have two main drawbacks:

- first, they do not help the agent when it is starting from an ambiguous situation, since at the first time step the agent benefits from no memory to help disambiguating its situation;
- second, once a sequence is elected, the agent will at least perform the number of actions specified in the sequence.

Since the number of steps to the food given by the optimal policy in E1 and E2 is generally less than 4, it is very unlikely that letting the agent perform sequences of 4 actions or more will help reaching the optimal performance.

Even worse, if an agent starts from an ambiguous situation and then follows a long sequence of actions, this sequence will delay the time at which the agent can discover its actual location and then follow an optimal path to the food.

Indeed, our experience with ATNoSFERES in small environments like E1 and E2 is that the main issue for the agent consists in discovering as fast as possible where it is from an initially ambiguous situation and then follow the shortest path to the goal. Maybe the situation about the use of sequences would be different in much bigger environments, but we will not treat this issue here.

Finally, we must compare the number of elementary runs necessary to reach a good performance with ACS and ATNoSFERES. In the experiments reported in [10], ACS needs about 60,000 steps (resp. 120,000 steps) to build an exhaustive internal model of E1 (resp. E2) given a convenient length of the behavioral sequence used as action part in ACS. With ATNoSFERES, about 1500 generations of 300 individuals are necessary to obtain a performance similar to that of ACS with $BS_{max} = 1$ in E1 and $BS_{max} = 2$ or 3 in E2, which makes about 450,000 runs of 6 to 15 steps on average. Thus it is clear that ATNoSFERES still needs several orders of magnitude more steps than ACS to converge.

This can be easily explained by the fact that ATNoSFERES evolves automata thanks to a blind GA process while ACS relies on a reinforcement learning algorithm which extracts information about the environment from its experiences. From this comparison, it is clear that an area for a major improvement of ATNoSFERES consists in endowing it with reinforcement learning capabilities. This is our immediate agenda for future work.

A source of inspiration in that direction comes from the SAMUEL system [30]. Like ATNoSFERES, SAMUEL is a *Pittsburg* style system based on a single chromosome GA, but it also includes lamarckian operators that endow it with basic learning capabilities. As a result, as claimed by the author, "Samuel represents an integration of the major genetic approaches to machine learning, the Michigan approach and the Pittsburg approach". Most of the operators used in SAMUEL can be transposed in ATNoSFERES, the main difference being that ATNoSFERES does not provide a high level symbolic representation and that SAMUEL does not include any mechanism to solve perceptual aliasing problems.

6   Conclusion and Future Work

In this paper, we have applied ATNoSFERES to non-Markov environments that have been investigated with ACS. Our experiments confirm that ATNoSFERES encounters more difficulties in producing an optimal behavior in some environments where reactive solutions are highly valuable than in environments that are more difficult for ACS.

Such a result suggests that the difficulties of different non-Markov problems with different hidden-state structure such as E1 and E2 should be distinguished in more details than is usually done. Along that line, we believe that, thanks to the information ATNoSFERES provides on the structure of different problems, it can be seen as a tool that may help understanding which kind of system will perform best in which kind of environment and why.

Finally, we would like to highlight the fact that the comparative studies we provided with ATNoSFERES both in this paper and in [7] and [8] should be generalized in the LCS community. Previously, we have compared ATNoSFERES with XCSM on some environments qualitatively, without comparing both systems performances. Here we have compared ATNoSFERES with ACS quantitatively on other environments, relying on the experiments presented on the available literature. Since XCSM and ACS have not been tested on the same environments, a precise comparison of their respective performance has never been published yet. A lot of work deserves to be done to provide more global comparisons between several systems and classes of systems. We strongly believe that such comparisons would greatly enhance the understanding of the current state of the art in the LCS research community.

References

1. Holland, J.H.: Adaptation in Natural and Artificial Systems: An Introductory Analysis with Applications to Biology, Control, and Artificial Intelligence. University of Michigan Press, Ann Arbor, MI (1975)
2. Lanzi, P.L.: Learning Classifier Systems from a Reinforcement Learning Perspective. Technical report, Dip. di Elettronica e Informazione, Politecnico di Milano (2000)
3. Sutton, R.S., Barto, A.G.: Reinforcement Learning, an introduction. MIT Press, Cambridge, MA (1998)
4. Lanzi, P.L.: An Analysis of the Memory Mechanism of XCSM. [32]
5. Wilson, S.W.: Classifier Fitness Based on Accuracy. Evolutionary Computation **3**(2) (1995) 149–175
6. Tomlinson, A., Bull, L.: CXCS. In Lanzi, P.L., Stolzmann, W., Wilson, S.W., eds.: Learning Classifier Systems: from Foundations to Applications. Springer Verlag, Heidelberg (2000) 194–208
7. Landau, S., Picault, S., Sigaud, O., Gérard, P.: A comparison between ATNoSFERES and XCSM. In: GECCO 2002: Proceedings of the Genetic and Evolutionary Computation Conference, New York, Morgan Kaufmann Publishers (2002) 926–933
8. Landau, S., Picault, S., Sigaud, O., Gérard, P.: Further Comparison between ATNoSFERES and XCSM. [31]
9. Landau, S., Picault, S.: ATNoSFERES: a Model for Evolutive Agent Behaviors. In: Proceedings of the AISB'01 Symposium on Adaptive Agents and Multi-Agent Systems. (2001)
10. Métivier, M., Lattaud, C.: Anticipatory Classifier System using Behavioral Sequences in Non-Markov Environments. [31] 143–163
11. Picault, S., Landau, S.: Ethogenetics and the Evolutionary Design of Agent Behaviors. In Callaos, N., Esquivel, S., Burge, J., eds.: Proceedings of the 5th World Multi-Conference on Systemics, Cybernetics and Informatics (SCI'01). Volume III. (2001) 528–533
12. Woods, W.A.: Transition Networks Grammars for Natural Language Analysis. Communications of the Association for the Computational Machinery **13**(10) (1970) 591–606
13. Holland, J.H., Reitman, J.S.: Cognitive Systems based on adaptive algorithms. Pattern Directed Inference Systems **7**(2) (1978) 125–149
14. Robertson, G.G., Riolo, R.L.: A tale of two classifier systems. Machine Learning **3** (1988) 139–159
15. Smith, R.E.: Memory exploitation in learning classifier systems. Evolutionary Computation **2**(3) (1994) 199–220
16. Cliff, D., Ross, S.: Adding memory to ZCS. Adaptive Behavior **3**(2) (1994) 101–150
17. Lanzi, P.L., Wilson, S.W.: Toward optimal classifier system performance in non-markov environments. Evolutionary Computation **8**(4) (2000) 393–418

18. Lin, L.J., Mitchell, T.M.: Memory approaches to reinforcement learning in non-markovian domains. Technical Report CMU-CS-92-138, Carnegie Mellon University, School of Computer Science (1992)
19. McCallum, R.A.: Reinforcement Learning with Selective Perception and Hidden State. PhD thesis, University of Rochester, Rochester, NY (1995)
20. Tomlinson, A., Bull, L.: A zeroth level corporate classifier system. In Lanzi, P.L., Stolzmann, W., Wilson, S.W., eds.: Learning Classifier Systems. From Foundations to Applications. Volume 1813 of Lecture Notes in Artificial Intelligence., Berlin, Springer-Verlag (2000) 306–313
21. Stolzmann, W.: Latent Learning in Khepera Robots with Anticipatory Classifier Systems. In Wu, A.S., ed.: Proceedings of the 1999 Genetic and Evolutionary Computation Conference (GECCO'99). (1999) 290–297
22. Wiering, M., Schmidhuber, J.: HQ-Learning. Adaptive Behavior 6(2) (1997) 219–246
23. Sun, R., Sessions, C.: Multi-agent reinforcement learning with bidding for segmenting action sequences. In Meyer, J.A., Wilson, S.W., Berthoz, A., Roitblat, H., Floreano, D., eds.: From Animals to Animats 6: Proceedings of the Sixth International Conference on Simulation of Adaptive Behavior, Paris, MIT Press (2000) 317–324
24. Meuleau, N., Peshkin, L., Kim, K.E., Kaelbling, L.P.: Learning finite-state controllers for partially observable environments. In: Fifteenth Conference on Uncertainty in Artificial Intelligence. AAAI. (1999) 427–436
25. Hansen, E.A.: Finite Memory Control of Partially Observable Systems. PhD thesis, University of Massachusetts, Amherst, MA (1998)
26. Stolzmann, W.: Anticipatory Classifier Systems. [32] 658–664
27. Gérard, P., Stolzmann, W., Sigaud, O.: YACS: a new Learning Classifier System with Anticipation. Journal of Soft Computing (2001)
28. Gérard, P., Meyer, J.A., Sigaud, O.: Combining latent learning with dynamic programming. European Journal of Operation Research **to appear** (2003)
29. Sigaud, O., Gérard, P.: Contribution au problème de la sélection de l'action en environnement partiellement observable. In Drogoul, A., Meyer, J.A., eds.: Intelligence Artificielle Située. Hermès, Paris (1999) 129–146 (In French).
30. Grefenstette, J.J.: Lamarckian learning in multi-agent environments. In Belew, R., Booker, L., eds.: Proceedings of the Fourth International Conference on Genetic Algorithms, San Mateo, CA, Morgan Kaufmann (1991) 303–310
31. Stolzmann, W., Lanzi, P.L., Wilson, S.W., eds.: Proceedings of the International Workshop on Learning Classifier Systems (IWLCS'02). LNAI, Granada, Springer-Verlag (2002)
32. Koza, J.R., Banzhaf, W., Chellapilla, K., Deb, K., Dorigo, M., Fogel, D.B., Garzon, M.H., Goldberg, D.E., Iba, H., Riolo, R., eds.: Proceedings of the Third Annual Conference on Genetic Programming, University of Wisconsin, Madison, Wisconsin, USA, Morgan Kaufmann (1998)

The Class Imbalance Problem in UCS Classifier System: A Preliminary Study

Albert Orriols-Puig and Ester Bernadó-Mansilla

Enginyeria i Arquitectura La Salle
Universitat Ramon Llull
Quatre Camins, 2. 08022, Barcelona, Spain
{aorriols,esterb}@salleURL.edu

Abstract. The class imbalance problem has been said recently to hinder the performance of learning systems. In fact, many of them are designed with the assumption of well-balanced datasets. But this commitment is not always true, since it is very common to find higher presence of one of the classes in real classification problems. The aim of this paper is to make a preliminary analysis on the effect of the class imbalance problem in learning classifier systems. Particularly we focus our study on UCS, a supervised version of XCS classifier system. We analyze UCS's behavior on unbalanced datasets and find that UCS is sensitive to high levels of class imbalance. We study strategies for dealing with class imbalances, acting either at the sampling level or at the classifier system's level.

1   Introduction

Learning Classifiers Systems (LCSs) [11,12] are rule-based systems that have been demonstrated to be highly competitive in classification problems with respect to other machine learning methods. Nowadays, XCS [28,27], an evolutionary online learning system, is one of the best representatives of LCSs.

The performance of XCS on real classification problems has been tested extensively in many contributions [18,19,3,5,17]. In addition, some analyses on the factors that make a problem hard for XCS have been made [15], and some theories have been formulated [5]. This work focuses on one of the complexity factors which is said to hinder the performance of standard learning methods: the *class imbalance* problem.

The class imbalance problem corresponds to classification domains for which one class is represented by a larger number of instances than other classes. The problem is of great importance since it appears in a large number of real domains, such as fraud detection [9], text classification [6], and medical diagnosis [20]. Traditional machine learning approaches may be biased towards the majority class and thus, may predict poorly the minority class examples. Recently, the machine learning community have paid increasing attention to this problem and how it affects the learning performance of some well-known classifier systems such as *C5.0*, *MPL*, and *support vector machines* [22,23,10]. In the LCS's framework, some approaches have been proposed to deal with class imbalances

X. Llorà et al. (Eds.): IWLCS 2003-2005, LNAI 4399, pp. 161–180, 2007.

in epidemiological data [13]. The aim of this paper is to enhance the analysis on class imbalances into the LCS's framework, and debate whether this problem affects LCSs, to what degree, and, if it is necessary, study different methods to overcome the difficulties.

Our analysis is centered on Michigan-style learning classifier systems. We choose UCS [7] as the test classifier system for our analysis, with the expectation that our results and conclusions can also be extended to XCS and other similar LCSs. UCS is a version of XCS that learns under a supervised learning scheme. In order to isolate the class imbalance problem and control its degree of complexity, we designed two artificial domains. We study UCS's behavior on these problems and identify factors of complexity when the class imbalance is high, which makes us to analyze different approaches to deal with these difficulties.

The remainder of this paper is organized as follows. Section 2 describes the UCS classifier system, focusing on the differences with XCS. Section 3 gives the details on the domain generation. In section 4, UCS is trained in the designed problems, and the class imbalance effects are analyzed. Section 5 describes the main approaches for dealing with the class imbalance problem, and sections 6 and 7 analyze these approaches under UCS's framework. Finally, we summarize our main conclusions, give limitations of the current study, and provide directions for further work.

2   Description of UCS

UCS [18,7] is a Michigan-style classifier system derived from XCS [28,27]. The main difference is that UCS was designed under a supervised learning scheme, while XCS follows a reinforcement learning scheme. In the following, we give a brief description of UCS, emphasizing the main differences with XCS. For more details, the reader is referred to [7].

2.1   Representation

UCS evolves a population of [P] classifiers. Each classifier has a rule of type *condition* → *class*, as in XCS, and a set of parameters estimating the quality of the rule.

The main parameters of a rule are: a) the rule's accuracy acc, b) the fitness F, c) the experience exp, d) the niche size ns, f) the last time of the GA activation ts, and g) the numerosity num.

2.2   Performance Component

UCS learns incrementally according to a supervised learning scheme. During *learning*, examples are provided to the system. Each example comes with its attributes $x = (x_1,... x_n)$ and its corresponding class c. Then, the system creates a match set [M] consisting of those classifiers whose conditions match the input example. From [M], the classifiers that correctly predict the class c form the correct set [C]. The remaining classifiers belong to [!C]. If [C] is empty, the covering operator is activated, creating a new classifier with a generalized condition matching x and class c.

In *exploit* or *test* mode, an input x is presented and UCS must predict its associated class. In this case, the match set [M] is formed, and the system selects the best class from the vote (weighted by fitness) of all classifiers present in [M].

2.3   Parameter Updates

In learning mode, the classifier parameters are updated. First of all, the classifier's accuracy is updated:

$$acc = \frac{number\ of\ correct\ classifications}{experience}$$

Fitness is calculated as a function of accuracy:

$$F = (acc)^{\nu}$$

where ν is a parameter set by the user. A typical value is 10. Thus, accuracy accumulates the number of correct classifications that each classifier has done, and fitness scales exponentially with accuracy.

The experience of a classifier exp is updated every time a classifier participates in a match set. The niche set size ns stores the average number of classifiers in [C]; it is updated each time the classifier belongs to a correct set.

2.4   Discovery Component

In UCS, the genetic algorithm (GA) is used as the search mechanism in a similar way to that in XCS. The GA is applied to [C] instead of all the population. It selects two parents from [C] with a probability proportional to fitness and copies them. Then, the copies are recombined and mutated with probabilities χ and μ respectively. The resulting offspring are introduced into the population. First, each offspring is checked for subsumption. In an offspring can not be subsumed, it is inserted in the population, deleting potentially poor classifiers if the population is full. The deletion probability is computed in the same way as in XCS (see [14]).

3   Dataset Design

In order to isolate the class imbalance problem from other factors that affect UCS's performance [15], two artificial domains were generated. Each one tries to highlight different traits of the system. These are the checkerboard problem (denoted as *chk*) and the position problem (denoted as *pos*). They are described in the following.

3.1   Chk Domain Generation

The *chk* problem is based on the balanced checkerboard problem, used as a benchmark in [8,26]. It has two real attributes (x and y) that can take values from zero to one ($x, y \in [0, 1]$). Instances are grouped in two non-overlapping classes, drawing a checkerboard in the feature space.

The complexity of the problem can be varied along three different dimensions (similarly to [23]): the degree of *concept complexity* (c), the *dataset size* (s), and the *imbalance level* between the two classes (i). *Concept complexity* defines the number of class boundaries, pointed as a complexity factor in XCS [8]. The *dataset size* is the size of the balanced dataset. The *imbalance level* determines the ratio between the number of minority class instances and the number of majority class instances.

The generation process creates a balanced two-class domain, and then proceeds to unbalance it by removing some of the minority class instances. The original balanced problem is defined by the dataset size s, and the concept complexity c, which defines c^2 alternating squares. We randomly drew points into the feature space so that each checkerboard square received s/c^2 instances. For the balanced dataset, $i=0$.

An imbalance degree i corresponds to the case where the minority class has $1/2^i$th of its normally entitled points, while the majority class maintains the same points as in the original dataset. This means that the ratio between the minority class instances and the majority class instances is $1/2^i$. Given s, c, and an imbalance level i, each square of the majority class has s/c^2 instances, while each square of the minority class has $s/(c^2 \cdot 2^i)$ instances. For example, for $c = 4$, $s = 4096$, and $i = 3$, each square of the majority class is represented by 256 examples, and each square of the minority class is represented by 32 examples. The domain generation unbalances the dataset iteratively. For $i = 1$, it takes the balanced dataset and removes half of the instances from the minority class. For $i = 2$, it takes the dataset obtained in the previous step and again removes half of the minority class instances, and so on.

3.2   Pos Domain Generation

The *pos* problem [7] has multiple classes and different proportions of examples per class. Given a binary input x of fixed length l, the output class corresponds to the position of the leftmost one-valued bit. If there is not any one-valued bit, the class is zero. The length of the string l determines both the complexity of the problem and the imbalance level.

In the position problem, the most specific rules are activated very sparsely and thus they have very few opportunities to reproduce. Our motivation to include such an extreme domain in the current analysis, rather than trying to solve this particular problem, is to validate our findings in the checkerboard domain.

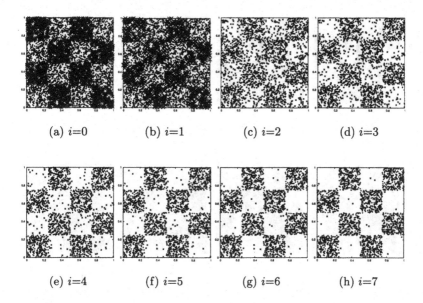

(a) $i=0$ (b) $i=1$ (c) $i=2$ (d) $i=3$

(e) $i=4$ (f) $i=5$ (g) $i=6$ (h) $i=7$

Fig. 1. Training datasets for the *chk* problem, generated with parameters $s=4096$, $c=4$, and imbalance levels from 0 to 7

4   UCS's Behavior on Unbalanced Datasets

In this section we analyze UCS's performance with unbalanced datasets, using the artificial problems described in the last section.

4.1   *Chk* Problem

We ran UCS in the checkerboard problem with a fixed dataset size $s=4096$ and concept complexity $c=4$, which corresponds to sixteen alternating squares. We varied the imbalance level from $i=0$ to $i=7$. For $i=0$, the dataset is balanced, with 2048 instances per class. For increasing i values we took out half of the instances of the minority class. Thus, the last configuration ($i=7$) corresponds to 256 instances per square belonging to the majority class, and only two instances for each square of the minority class. Figure 1 depicts all the datasets generated, showing the location of each training point in the feature space.

Since there is not overlapping among instances of different classes, the minority class instances should not be dealt as noise. Even in the most unbalanced dataset (figure 1(h)), all the minority class instances are isolated from the regions containing the majority class instances, and this should be enough to let the system discover the minority class regions. However, it is reasonable to expect that some regions will be more difficult to evolve, depending on the distance of the training points of the minority class to the training points of the majority class.

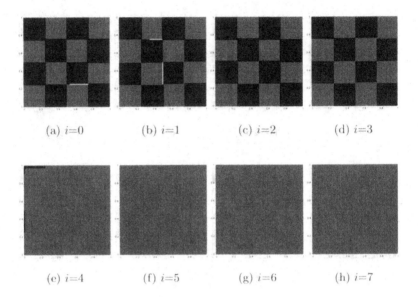

Fig. 2. Boundaries evolved by UCS in the *chk* problem with imbalance levels from 0 to 7. Plotted in black are the regions belonging to the minority class and plotted in gray are the regions of the majority class.

UCS was trained with each of the datasets for 200,000 learning iterations, with the following parameter settings (see [21] for the notation): N=400[1], α=0.1, β=0.2, δ=0.1, ν=10, θ_{del}=20, θ_{sub} = 20, acc_0=0.99, χ=0.8, μ=0.04, θ_{GA}=25, GASub = true, [A]Sub=false. Specify was enabled with parameters N_{SP} =20 and P_{SP} =0.5 (see [16]).

Figure 2 shows the classification boundaries obtained by UCS in all the datasets. Figure 2(a) corresponds to the balanced dataset. As expected, UCS is able to discover the optimal ruleset. A similar behavior is shown for imbalance levels i={1,2,3}, as seen in figures 2(b), 2(c), and 2(d) respectively. In these cases, the class imbalance does not prevent UCS from evolving the correct boundaries.

The problem arises with imbalance levels equal or greater than 4. Figure 2(e) shows that the system is not able to classify correctly any of the minority class squares. Looking at the training dataset, shown in figure 1(e), it seems feasible that UCS could learn the minority class regions since the number of instances representing these regions define a distinguished space in the eyes of a human beholder. In addition, any model evolved with higher imbalance levels is not able to discover any minority class region, as shown in figures 2(f), 2(g), and 2(h). This abrupt change in the UCS's behavior led us to make a deeper analysis.

Table 1 shows the rules evolved by UCS in the *chk* problem for imbalance level i=4. The table shows, to our surprise, that UCS evolved accurate and maximally general classifiers that cover the minority class regions. Actually, the

---

[1] The value was set to sixteen times the optimal population size, as suggested in [7].

Table 1. Most numerous rules evolved by UCS in the *chk* problem with imbalance level *i*=4, sorted by class and numerosity. Columns show respectively: the rule number, the condition and class of the classifier, where 0 is the majority class and 1 the minority class, the accuracy (acc), fitness (F), and numerosity (num).

id	condition	class	acc	F	num
1	[0.509, 0.750] [0.259, 0.492]	: 1	1.00	1.00	39
2	[0.000, 0.231] [0.252, 0.492]	: 1	1.00	1.00	38
3	[0.000, 0.248] [0.755, 1.000]	: 1	1.00	1.00	35
4	[0.761, 1.000] [0.000, 0.249]	: 1	1.00	1.00	34
5	[0.255, 0.498] [0.520, 0.730]	: 1	1.00	1.00	33
6	[0.751, 1.000] [0.514, 0.737]	: 1	1.00	1.00	31
7	[0.259, 0.498] [0.000, 0.244]	: 1	1.00	1.00	27
8	[0.501, 0.743] [0.751, 1.000]	: 1	1.00	1.00	18
9	[0.500, 0.743] [0.751, 1.000]	: 1	1.00	1.00	9
10	[0.751, 1.000] [0.531, 0.737]	: 1	1.00	1.00	8
	...				
18	[0.509, 0.750] [0.246, 0.492]	: 1	0.64	0.01	1
19	[0.000, 1.000] [0.000, 1.000]	: 0	0.94	0.54	20
20	[0.000, 1.000] [0.000, 0.990]	: 0	0.94	0.54	13
21	[0.012, 1.000] [0.000, 0.990]	: 0	0.94	0.54	10
	...				
64	[0.012, 1.000] [0.038, 0.973]	: 0	0.94	0.54	1

eight most numerous rules are those that cover the eight minority class regions, and all of them are accurate. Besides these rules, the table also shows some less numerous rules predicting the majority class. These are overgeneral rules; they cover inaccurately almost all the feature space. From these results two questions arise. Why does UCS evolve these overgeneral rules? And why the system does not properly use the specific rules to classify the minority class regions instead of these overgeneral rules?

To explain UCS's tendency to evolve these overgeneral rules, other populations evolved with lower imbalance levels were checked. We found that all populations evolved with an imbalance level higher than 1 contained the most general rule ($at_1 \in [0, 1]$ and $at_2 \in [0, 1]$), as shown in figure 2 for imbalance level *i*=3.

We hypothesize that the generalization pressure produced by the GA induces the creation of these overgeneral rules. Once created, these rules are activated in nearly any action set, because of its overgeneral condition. In balanced or low-unbalanced datasets, these overgeneral rules tend to have low accuracy and consequently, low fitness. For example, the most general rule has a 0.50 of accuracy in a balanced dataset. Thus, overgeneral rules with low fitness tend to have low probabilities of participating in reproductive events and finally, they are removed from the population. The problem comes out with high imbalance levels, where overgeneral rules have a high tendency to be maintained in the population. The reason is that the data distribution does not allow to penalize the classifier's accuracy so much, as long as the minority class instances are sampled in a lower frequency. So, the higher the imbalance level, the more accurate an

Table 2. Most numerous rules evolved by UCS in the *chk* problem with imbalance level $i=3$, sorted by class and numerosity. Columns show respectively: the rule number, the condition and class of the classifier, where 0 is the majority class and 1 the minority class, the accuracy (acc), fitness (F), and numerosity (num).

id	condition		class	acc	F	num
1	[0.251, 0.498]	[0.000, 0.244]	: 1	1.00	1.00	39
2	[0.501, 0.751]	[0.760, 1.000]	: 1	1.00	1.00	37
3	[0.000, 0.246]	[0.259, 0.500]	: 1	1.00	1.00	36
4	[0.259, 0.499]	[0.504, 0.751]	: 1	1.00	1.00	33
5	[0.506, 0.746]	[0.263, 0.498]	: 1	1.00	1.00	30
6	[0.751, 1.000]	[0.502, 0.749]	: 1	1.00	1.00	29
7	[0.752, 1.000]	[0.000, 0.240]	: 1	1.00	1.00	27
8	[0.000, 0.246]	[0.759, 1.000]	: 1	1.00	1.00	20
		...				
25	[0.000, 0.233]	[0.584, 1.000]	: 1	0.13	0.00	1
26	[0.000, 1.000]	[0.000, 1.000]	: 0	0.89	0.31	13
27	[0.010, 1.000]	[0.000, 1.000]	: 0	0.89	0.31	12
		...				
60	[0.051, 1.000]	[0.017, 0.926]	: 0	0.89	0.31	1

overgeneral classifier is considered (and also the higher fitness it has). This effect is clearly seen in tables 2 and 1. Observe that for $i=3$ (table 2), overgeneral rules have accuracies of 0.89. Similar overgeneral rules for $i=4$ (table 1) have 0.94 of accuracy. Consequently, in high imbalance levels overgeneral rules tend to have higher accuracies, presenting more opportunities to be selected by the GA, and also lower probabilities of being removed by the deletion procedure.

After analyzing why overgeneral rules are created and maintained in the population as the imbalance level increases, let's consider why the system does not predict the minority class even though it evolved the appropriate rules. For $i=3$, UCS was able to predict the minority class regions but not for $i=4$, although apparently the populations evolved were similar. For $i=3$, there are several numerous and overgeneral rules, but their vote in the prediction array is not enough to overcome the vote of the accurate rules predicting the minority class. Therefore, UCS is able to predict accurately the minority class. The problem arises at a certain point in the imbalance level that makes the vote of the overgeneral rules higher than the vote of the accurate rules. In our datasets, this happens at imbalance level 4. The population evolved in this dataset consists of 64 macroclassifiers, 46 of them predicting the majority class and only 18 predicting the minority class. Taking into account the classifiers' numerosities, there are more than 100 microclassifiers covering all the feature space with the majority class, and only 32 microclassifiers in average for each of the minority class squares. When an instance belonging to a minority class region is shown to UCS, the prediction vote for the majority class is greater, and this makes UCS to classify this instance wrongly.

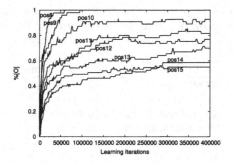

Fig. 3. Percentage of optimal population evolved by UCS in the *pos* problem. Each curve corresponds to the average of five seeds.

4.2 *Pos* Problem

UCS was trained with the *pos* problem with condition lengths l from 8 to 15. We ran UCS for 400000 iterations with the the following parameter settings: $N=25\cdot(l+1)$, $\alpha=0.1$, $\beta=0.2$, $\delta=0.1$, $\nu=10$, $\theta_{del}=20$, $\theta_{sub}=20$, $acc_0=0.99$, $\chi=0.8$, $\mu=0.04$, $\theta_{GA}=25$, GASub = true, [A]Sub=false, Specify=true, $N_{SP}=20$, P_{SP} =0.5. The experiments were averaged over five different runs.

To analyze UCS's behavior on this problem, we show the curves of performance of UCS during training. We consider here the percentage of optimal classifiers (%[O]) [15] achieved by UCS along the iterations. We use this metric instead of accuracy, because accuracy is biased towards the majority classes. Since we aim to evaluate the system's capability to evolve all the classes, we use a measure that gives equal importance to each of the classes independently of the *a priori* probabilities of each class. Alternatively, a measure of cost per class could be used.

Figure 3 depicts the percentage of optimal population achieved during training. Each curve represents a different level of complexity in the *pos* problem, ranging from $l=8$ until $l=15$. It shows that UCS has difficulties in learning all the optimal classifiers as the condition length grows. Table 3 shows an example of the population evolved by UCS for the *pos* problem at $l=12$. Note that the system can discover the most general optimal rules, being not able to discover the three most specific ones. This behavior is also observed in other evolved populations. As expected, more general rules have higher numerosities. This behavior is attributed to the fact that specific rules activate less often than more general ones, and thus they have fewer reproductive opportunities. Therefore, in a problem with class imbalances the system has more opportunities to learn rules that cover the majority class than those that cover the minority class [7].

5 How to Deal with Class Imbalances

This section reviews different strategies for dealing with the class imbalance problem. The first two are general methodologies applicable to any type of classifier

Table 3. Most numerous rules evolved by UCS in the *pos* problem for l=12, sorted by numerosity. Columns show respectively: the rule number, the condition and class of the classifier, the accuracy (acc), fitness (F), and numerosity (num).

condition	class	acc	F	num
1###########	:12	1.00	1.000	56
0001#########	: 9	1.00	1.000	49
01##########	:11	1.00	1.000	46
001#########	:10	1.00	1.000	43
00001########	: 8	1.00	1.000	32
000001#######	: 7	1.00	1.000	24
0000001######	: 6	1.00	1.000	16
00000001#####	: 5	1.00	1.000	11
000000001####	: 4	1.00	1.000	10
0000000001###	: 3	1.00	1.000	5
00#00#00000#	: 1	0.20	0.000	4
0000000#01##	: 3	0.70	0.028	2
0000000000#0	: 2	0.66	0.017	2

since they are based on dataset resampling. The aim of resampling is to balance the *a priori* probabilities of the classes. The last methods are specially designed for UCS, although they can also be adapted to other classifier schemes.

Oversampling. This method consists of oversampling the examples of the minority class until their number is equal to the number of instances in the majority class [22]. Two variants can be considered. *Random oversampling* resamples at random the minority class examples. *Focused resampling* oversamples mainly those instances closer to class boundaries.

Undersampling. Undersampling consists of eliminating some of the majority class instances until we reach the same number of majority class instances as minority class instances [22]. Two classic schemes are *random undersampling*, which removes at random majority class instances, and *focused resampling*, which removes only those instances further away from class boundaries.

Adaptive sampling. This method, initially proposed in [4], is inspired in *oversampling* and in the way *boosting* [25] works. It proposes to maintain a weight for each dataset instance (initially set to 1), which indicates the probability that the sample process selects it. Weights are updated incrementally when the system makes a prediction under exploit mode. Depending on whether the system has made a correct prediction or not on a given example, the weight for that example will be decreased or increased by a factor α (in the experiments we fixed $\alpha = 0.1$).

Class-sensitive accuracy. *Class-sensitive accuracy* modifies the way in which UCS computes accuracy so that each class is considered equally important regardless of the number of instances representing each class. UCS was slightly modified to compute the experience of a classifier per each class. The proportion of examples covered per each class is taken in account to calculate the classifier's fitness, counterbalancing the bias produced by class imbalances. See [1] for further details.

Selected techniques. We chose to analyze three main representative approaches: random oversampling, adaptive sampling, and class-sensitive accuracy with weighted experience. Undersampling was not considered for being too extreme in the case of highly imbalanced datasets. Under our point of view, undersampling majority class instances to the same degree as minority class instances may produce a loss of valuable information and may change class boundaries unnecessarily. The problem may degenerate into a problem of sparsity, for which classifier schemes in general are expected to show poor generalization capability. Next section compares each of the selected strategies under the checkerboard problem.

6   UCS in the *Chk* Problem

We run UCS in the checkerboard problem for imbalance levels from $i=0$ to $i=7$ using the three aforementioned strategies. We use the same parameter settings as those in section 4, adding the new parameter θ_{acc} for the case of class-sensitive computation. θ_{acc} is set to 50 to protect the fitness decrease of young classifiers.

6.1   Random Oversampling

Figure 4 shows the boundaries evolved by UCS under random oversampling. Note that UCS was able to evolve some boundaries for the minority class examples even for the highest imbalance levels. In many cases, these boundaries do not reach the real boundaries of the original balanced dataset. But this result is reasonable since the distribution of training points has changed with respect to the original dataset.

Under oversampling, UCS works as the problem was a well balanced dataset, because the proportion of minority class examples has been adjusted a priori. UCS sees a dataset with the same number examples per class, but with some gaps in the feature space that are not represented by any example. These gaps are mostly covered by rules predicting the majority class rather than by minority class rules. In fact, what happens is that rules from both classes tend to expand as much as possible into these gaps until they reach points belonging to the opposite classes. That is, rules tend to expand as long as they are accurate. Thus, there are overlapping rules belonging to different classes in the regions that are not covered by any example. When we test UCS in these regions, the majority class rules have higher numerosities and their vote into the prediction array is higher. The reason why majority class rules have higher numerosities is that their boundaries are less complex, so in many cases a single rule suffices for all the region. This rule tends to cover all the examples of a majority class square and benefits from long experience and numerosity. On the contrary, minority

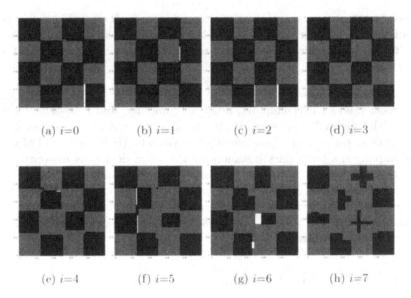

(a) i=0   (b) i=1   (c) i=2   (d) i=3

(e) i=4   (f) i=5   (g) i=6   (h) i=7

Fig. 4. Class boundaries evolved by UCS with random oversampling in the chk problem with imbalance levels from 0 to 7

class regions are more complex, and rules covering these regions tend to have less experience and numerosity.

6.2   Adaptive Sampling

Figure 5 shows the results obtained by UCS under the adaptive sampling strategy. UCS evolved part of the squares belonging to the minority class. For i=4 and i=5, the boundaries evolved by UCS almost approximate the real boundaries of the original problem. In these cases, adaptive sampling allowed UCS to evolve fairly good approximations with respect to the original results shown in figure 2. For higher imbalance levels, UCS found more difficulties in finding good approximations for the minority class squares. In these cases, the result achieved under the adaptive sampling strategy is worse than that achieved by UCS under oversampling (see figure 4).

We tried to use a more disruptive function for the weight computation of the adaptive sampling strategy but we found no improvements. On the contrary, trying to use a higher α parameter so that weights could be further increased if instances were poorly classified led to oscillations in the weights and difficulties to stabilize the boundaries evolved.

Analyzing the behavior of UCS under these two strategies (not detailed for brevity), we found that under adaptive sampling there is less generalization pressure towards the minority class rules than with oversampling. The reason is that, with adaptive sampling, once all instances are well classified, weights stabilize and then, all instances are sampled as the original a priori probabilities. Under oversampling, minority class instances are always sampled at the same

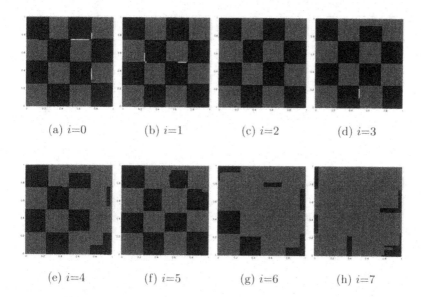

(a) $i=0$ (b) $i=1$ (c) $i=2$ (d) $i=3$

(e) $i=4$ (f) $i=5$ (g) $i=6$ (h) $i=7$

Fig. 5. Class boundaries evolved by UCS with adaptive sampling in the chk problem with imbalance levels from 0 to 7

a priori probability as majority class instances, keeping the same generalization pressure towards both classes. This may justify why, under adaptive sampling, UCS finds more difficulties in the generalization of rules covering the minority class instances, especially for the highest imbalance levels.

6.3   Class-Sensitive Accuracy

Figure 6 shows the results of UCS under class-sensitive accuracy. Note that the boundaries evolved in all imbalance levels are better at discovering minority class regions than those evolved by raw UCS. However, for the lowest imbalance levels (i.e., $i=[1\text{-}3]$), there is a little tendency to leave some blank spaces near the class boundaries. These gaps predominantly belong to the minority class regions. The reason is that rules covering minority class regions easily get inaccurate when they overpass slightly into the majority class regions. Rules classifying the majority class squares and getting inside minority class squares have less probability to cover minority instances (because they are less frequent) so that their accuracy is not penalized as much. This gap effect in the class boundaries was even stronger for class-sensitive accuracy without the weighted experience modification, as shown in [1]. Note that for the balanced dataset, i.e., $i=0$, the gap effect is also present although with few incidence. These gaps also appeared slightly under oversampling and adaptive sampling, although in the latter to a lower extent.

For imbalance levels $i=4$ and $i=5$, UCS with class-sensitive accuracy clearly improves raw UCS in terms of the boundaries evolved. Furthermore, the

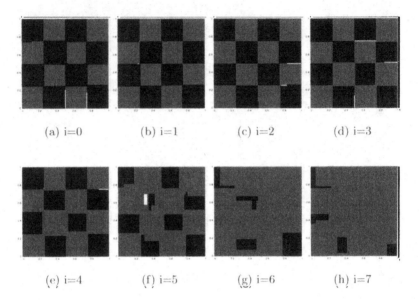

Fig. 6. Boundaries evolved by UCS with *class-sensitive accuracy* for 0 to 7. Black regions are those classified as minority class. Grey regions are regions classified as majority class. White regions are non-covered domain regions.

tendency of evolving overgeneral rules is restrained. Table 4 depicts the most numerous rules evolved by UCS for imbalance level $i=4$. Note that overgeneral rules are not evolved. Instead there are maximally general rules covering each of the alternating squares.

Finally, figures 6(g) and 6(h) show the models evolved with imbalance levels $i=6$ and $i=7$. As the imbalance level increases, the system finds it harder to evolve the minority class regions. For the highest class imbalances, UCS can only draw partially four of the minority class regions. Looking at the evolved population, not shown for brevity, we confirm that the problem is not attributable to the evolution of overgeneral rules but to the fact that the imbalance ratio is so high (1:128) that it could be considered as an sparsity problem. There are so few representatives of the minority class regions that we may debate whether these points are representative of a sparse class region or whether they belong to noisy cases. In the latter case, we would acknowledge that UCS should not find any distinctive region.

7   Contrasting Results with *Pos* Problem

In last section, we isolated and analyzed the imbalance class problem under the *chk* domain. Now, we analyze the *pos* problem, which combines jointly different complexity factors. Increasing the condition length increases not only its imbalance level but also other identified complexity factors such as the number of classes, the size of the training dataset and the condition length itself. Our

Table 4. Most numerous rules evolved by UCS with class-sensitive accuracy in the *chk* problem for imbalance level $i=4$, sorted by numerosity. Columns show respectively: the rule number, the condition and class of the classifier, where 0 is the majority class and 1 the minority class, the accuracy for each class (acc_0 and acc_1), fitness (F), and numerosity (num).

id	condition		class	acc_0	acc_1	F	num
1	[0.000 - 0.200]	[0.492 - 0.756]	:0	1	-	1.00	34
2	[0.754 - 1.000]	[0.757 - 1.000]	:0	1	-	1.00	32
3	[0.000 - 0.253]	[0.000 - 0.242]	:0	1	-	1.00	26
4	[0.730 - 1.000]	[0.260 - 0.527]	:0	1	-	1.00	15
5	[0.491 - 0.759]	[0.480 - 0.753]	:0	1	-	1.00	12
6	[0.000 - 0.250]	[0.770 - 1.000]	:1	-	1	1.00	20
7	[0.519 - 0.748]	[0.260 - 0.483]	:1	-	1	1.00	19
8	[0.751 - 1.000]	[0.000 - 0.247]	:1	-	1	1.00	17
9	[0.257 - 0.454]	[0.510 - 0.722]	:1	-	1	1.00	15
10	[0.000 - 0.246]	[0.253 - 0.460]	:1	-	1	1.00	15
11	[0.763 - 1.000]	[0.526 - 0.740]	:1	-	1	1.00	13
12	[0.482 - 0.786]	[0.000 - 0.241]	:0	1	0	0.39	19
13	[0.264 - 0.565]	[0.699 - 1.000]	:0	1	0	0.87	18
14	[0.114 - 0.547]	[0.156 - 0.529]	:0	1	0	0.66	15
		...					
69	[0.156 - 0.547]	[0.201 - 0.507]	:0	1	0	0.43	1

purpose here is to analyze the three strategies under a more difficult problem, as a previous step to the analysis of real-world problems which will be left as a future work.

7.1 Oversampling

UCS was run under oversampling with the same parameter settings as in the original *pos* problem (see section 4.2). Figure 7 shows the percentage of optimal population achieved by UCS in the *pos* problem for condition-lengths l from 8 to 15. Curves are averages of five runs. The figure shows high oscillations in the learning curves of UCS. Note that the learning curves have worsened significantly with respect to the original results with UCS (as shown in figure 3).

Making a deeper analysis, some harmful traits of oversampling, which were not observed in the *chk* problem, come out. In the *pos* problem, changing the a priori probabilities of examples makes accurate generalizations very hard to be evolved. UCS learns accurate and maximally general rules by the presence of the appropriate examples and counter-examples. While the presence of numerous examples favor the generalization of rules, counter-examples set the limit for these generalizations. If rules overgeneralize, the presence of counter-examples makes the rule inaccurate. Therefore rules generalize as long as they cover all the examples of the same class and cover no counter-examples. In the *pos* problem, we oversample minority class examples. Thus, the system gets a higher number of examples for the minority class rules, but on the contrary receives few

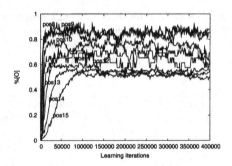

Fig. 7. Percentage of optimal population evolved by UCS under oversampling in the *pos* problem for l=8 until l=15

proportion of counter-examples for these rules. The result is that UCS tends to overgeneralize the rules covering the minority class examples. And the discovery of specific rules for the minority class examples remains unsolved.

We wonder why this effect did not arise in the *chk* problem. The reason is that the *chk* problem has originally the same generalization for each of the rules. Oversampling makes each rule to receive the same proportion of examples and counter-examples. So it is easier to find accurate generalizations.

The results of oversampling on the *position* problem suggest that this method could be harmful depending on the topology of the problem. So this is a method that should be applied with caution in real-world problems, at least for classifier schemes using similar learning patterns to those of UCS.

7.2   Adaptive Sampling

Figure 8 shows the percentage of optimal population achieved by UCS under adaptive sampling. Curves are averages of five runs. See that the oversampling effect does not appear here. If a rule is inaccurate because it does not classify properly an example, the probability of sampling that example is increased. Thus, this example will serve as a counter-example for overgeneral rules, and as an example to help discover rules covering it accurately. Note that the learning curves have improved with respect to the original problem (figure 3), although there is still a high difficulty in discovering the optimal populations for the highest complex levels.

7.3   Class-Sensitive Accuracy

Figure 9 shows the percentage of optimal population evolved by UCS with class-sensitive accuracy. The figure does not reveal significant improvements with respect to raw UCS, as shown in figure 3. The problem here is that UCS is receiving very few instances of the most specific classes, i.e., it receives exactly the same instances as in the original problem. For example, for l=15, UCS receives only one instance of class 0 each 32768 instances. Thus, even though UCS weighs

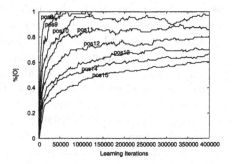

Fig. 8. Percentage of optimal population evolved by UCS under adaptive sampling in the *pos* problem from $l=8$ to $l=15$

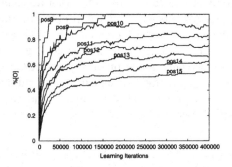

Fig. 9. Percentage of optimal population evolved by UCS with class-sensitive accuracy in the *pos* problem, from $l=8$ to $l=15$

the contribution of each class equally, if the minority class instances come very sparsely, rules covering them have few reproductive events. We find it interesting to analyze a hybrid strategy between adaptive sampling and class-sensitive accuracy so that the benefits of each approach could be combined.

8   Conclusions

We analyzed the class imbalance problem in UCS classifier system. We found that UCS has a bias towards the majority class, especially for high degrees of class imbalances. Isolating the class imbalance problem by means of artificially designed problems, we were able to explain this bias in terms of the population evolved by UCS.

In the checkerboard problem, we identified the presence of overgeneral rules predicting the majority class which covered almost all the feature space. For a given imbalance level, these rules overcame the vote of the most specific ones and thus, UCS predicted all instances as belonging to the majority class. Different strategies were analyzed to prevent the evolution of these overgeneral rules. We found that all tested strategies (oversampling, adaptive sampling, and class-sensitive accuracy) prevented UCS from evolving these overgeneral rules, and class boundaries —for both the minority and majority classes— were approximated fairly better than with the original setting. However, the analysis on the position problem revealed many inconveniences in the oversampling strategy which make UCS's learning unstable. This leads us to discard this method for real-world datasets.

The study would be much enhanced with the analysis of the class imbalance problem on other LCSs. Preliminary experiments made with two other evolutionary learning classifier systems, *GAssist* [2] and HIDER [24], showed that they are even more sensitive to the class imbalance problem. Moreover, the analysis of the class imbalance problem in other classifier schemes such as nearest neighbors, support vector machines and C4.5, and their comparison with LCSs, could give higher understanding on how imbalance class problems affect classifier schemes, and whether they affect LCSs to a higher degree than others.

Also, we would like to extend this analysis to other artificial problems, as well as to real-world datasets. We suspect that the proposed strategies may be very sensitive in noisy problems, i.e, in problems having misclassified instances. The combination of noisy instances and strategies for dealing with class imbalances may worsen significantly the generalization of learners, resulting in too overfitted boundaries. As this is a feature often present in real-world datasets, this should be analyzed in detail as a previous step to understand the behavior of these strategies on real-world datasets.

Acknowledgements

The authors thank S.W. Wilson and two anonymous reviewers for useful comments on a previous version of the paper. Also thanks to the support of *Enginyeria i Arquitectura La Salle*, Ramon Llull University, as well as the support of *Ministerio de Ciencia y Tecnología* under project TIN2005-08386-C05-04, and *Generalitat de Catalunya* under Grants 2005FI-00252, 2002SGR-00155, and 2005SGR-00302.

References

1. A. Orriols Puig and E. Bernadó Mansilla. The Class Imbalance Problem in UCS Classifier System: Fitness Adaptation. In *Congress on Evolutionary Computation*, volume 1, pages 604–611, Edinburgh, UK, 2-5 September 2005. IEEE.
2. J. Bacardit. *Pittsburgh genetic-based machine learning in the data mining era: representations, generalization and run-time*. PhD thesis, Department of Computer Science. Enginyeria i Arquitectura la Salle, Ramon Llull University, Barcelona, 2004.
3. J. Bacardit and M.V. Butz. Data Mining in Learning Classifier Systems: Comparing XCS with GAssist. In *Seventh International Workshop on Learning Classifier Systems (IWLCS-2004)*, 2004.
4. E. Bernadó Mansilla. Complejidad del Aprendizaje y Muestreo de Ejemplos en Sistemas Clasificadores. In *Tercer Congreso Espaol de Metaheursticas, Algoritmos Evolutivos y Bioinspirados (MAEB'2004)*, pages 203–210, 2004.

5. M.V. Butz. *Rule-based Evolutionary Online Learning Systems: Learning Bounds, Classification and Prediction.* PhD thesis, Illinois Genetic Algorithms Laboratory (IlliGAL) - University of Illinois at Urbana Champaign, Urbana Champaign 117, 2004.

6. N. Chawla, K. Bowyer, L Hall, and W. Kegelmeyer. SMOTE: Synthetic Minority Over-sampling Technique. *Journal of Artificial Intelligence Research,* 16:321–357, 2002.

7. E. Bernadó Mansilla and J.M. Garrell Guiu. Accuracy-Based Learning Classifier Systems: Models, Analysis and Applications to Classification Tasks. *Evolutionary Computation,* 11(3):209–238, 2003.

8. E. Bernadó Mansilla and T.K. Ho. Domain of Competence of XCS Classifier System in Complexity Measurement Space. *IEEE Transactions on Evolutionary Computation,* 9(1):82–104, 2005.

9. R.E. Fawcett and F. Provost. Adaptive Fraud Detection. *Data Mining and Knowledge Discovery,* 3(1):291–316, 1997.

10. G.M. Weiss. *The effect of small disjunct and class distribution on decision tree learning.* PhD thesis, Graduate School - New Brunswick. Rutgers, The State University of New Jersey, New Brunswick, New Jersey, 2003.

11. J.H. Holland. *Adaptation in Natural and Artificial Systems.* The University of Michigan Press, 1975.

12. J.H. Holland. Adaptation. In R. Rosen and F. Snell, editors, *Progress in Theoretical Biology,* volume 4, pages 263–293. New York: Academic Press, 1976.

13. J.H. Holmes. Differential negative reinforcement improves classifier system learning rate in two-class problems with unequal base rates. In *Genetic Programming 1998: Proceedings of the Third Annual Conference,* pages 635–642. Morgan Kaufmann, 1998.

14. T. Kovacs. Deletion Schemes for Classifier Systems. In Wolfgang Banzhaf, Jason Daida, A.E. Eiben, Max H. Garzon, Vasant Honavar, Mark Jakiela, and Robert E Smith, editors, *Proceedings of the Genetic and Evolutionary Computation Conference, (GECCO-99),* pages 329–336. Morgan Kaufmann, 1999.

15. T. Kovacs and M. Kerber. What makes a problem hard for XCS. In *Lanzi, P. L., Stolzmann, W., & Wilson, S. W. (Eds.), Advances in Learning Classifier Systems: Third International Workshop, IWLCS,* pages 80–99. Springer-Verlag, 2000.

16. P.L. Lanzi. A study of the generalization capabilities of XCS. In Thomas Bäck, editor, *Proc. of the Seventh Int. Conf. on Genetic Algorithms,* pages 418–425, San Francisco, CA, 1997. Morgan Kaufmann.

17. M. V. Butz, D. E. Golberg, and P. L. Lanzi. Bounding Learning Time in XCS. In *Proceedings of the Genetic and Evolutionary Computation Conference (GECCO'2004),* Berlin, 2004. Springer Verlag.

18. E. Bernadó Mansilla. *Contributions to Genetic Based Classifier Systems.* PhD thesis, Enginyeria i Arquitectura la Salle, Ramon Llull University, Barcelona, 2002.

19. E. Bernadó Mansilla, F.X. Llorà Fàbrega, and J.M. Garrell Guiu. XCS and GALE: a Comparative Study of Two Learning Classifier Systems on Data Mining. In P.L. Lanzi, W. Stolzmann, and S.W. Wilson, editors, *Advances in Learning Classifier Systems, 4th International Workshop,* volume 2321 of *Lecture Notes in Artificial Intelligence,* pages 115–132. Springer, 2002.

20. P.M. Murphy and D.W. Aha. UCI Repository of machine learning databases, University of California at Irvine, Department of Information and Computer Science, 1994.

21. M.V. Butz and S.W. Wilson. An algorithmic description of XCS. In P.L. Lanzi, W. Stolzmann, and S.W. Wilson, editors, *Advances in Learning Classifier Systems: Proceedings of the Third International Workshop*, volume 1996 of *Lecture Notes in Artificial Intelligence*, pages 253–272. Springer, 2001.

22. N. Japkowicz and S. Stephen. ‘The Class Imbalance Problem: Significance and Strategies. In *2000 International Conference on Artificial Intelligence (IC-AI'2000)*, volume 1, pages 111–117, 2000.

23. N. Japkowicz and S. Stephen. The Class Imbalance Problem: A Systematic Study. *Intelligent Data Analisis*, 6(5):429–450, November 2002.

24. R. Giráldez, J.S. Aguilar-Ruiz, and J.E. Riquelme. Evolutionary Learning of Hierarchical Decision Rules. *IEEE Transactions on Systems, Man, and Cybernetics*, 33(2):324.331, 2003.

25. R.E. Schapire. A Brief Introduction to Boosting. In *Sixteenth International Joint Conference on Artificial Intelligence*, 1999.

26. C. Stone and L. Bull. For Real! XCS with Continuous-Valued Inputs. *Evolutionary Computation*, 11(3):299–336, 2003.

27. S.W. Wilson. Generalization in the XCS Classifier System. In J.R. Koza, W. Banzhaf, K. Chellapilla, K. Deb, M. Dorigo, D. Fogel, M.H. Garzon, D.E. Goldberg, H. Iba, and R. Riolo, editors, *Genetic Programming: Proceedings of the Third Annual Conference*, pages 665–674. Morgan Kaufmann, 1998.

28. S.W. Wilson. Classifier Fitness Based on Accuracy. *Evolutionary Computation*, 3(2):149–175, 1995.

Three Methods for
Covering Missing Input Data in XCS

John H. Holmes, Jennifer A. Sager, and Warren B. Bilker

Center for Clinical Epidemiology and Biostatistics
University of Pennsylvania School of Medicine
Philadelphia, PA 19104 USA
jholmes@cceb.med.upenn.edu
wbilker@cceb.upenn.edu
Department of Computer Science
University of New Mexico
Albuquerque, NM 87131, USA
sagerj@cs.unm.edu

Abstract. Missing data pose a potential threat to learning and classification in that they may compromise the ability of a system to develop robust, generalized models of the environment in which they operate. This investigation reports on the effects of three approaches to covering these data using an XCS-style learning classifier system. Using fabricated datasets representing a wide range of missing value densities, it was found that missing data do not appear to adversely affect LCS learning and classification performance. Furthermore, three types of missing value covering were found to exhibit similar efficiency on these data, with respect to convergence rate and classification accuracy.

1 Introduction

Learning Classifier Systems (LCS) are used for a number of functions, including agent control and data mining. All of the environments in which LCS operate are potentially plagued by the problem of incomplete, or *missing*, data. Missing data arise from a number of different scenarios. In databases, fields may have values that are missing because they weren't collected, or they were lost or corrupted in some way during processing. In real-time autonomous agent environments, data may be missing due to the malfunctioning of a sensor. In any case, missing data can cause substantial inaccuracies due to their frequency, their distribution, or their association with other features that are important to learning and classification. As a result, the problem of missing data has attracted substantial attention in the data mining and machine learning communities [1, 2, 10, 11, 15, 16].

Although one study [12] has investigated the use of a genetic algorithm in analyzing clinical data with missing values, and one other [9] has investigated their use in spectral estimation, the effects of missing data on LCS learning and classification performance have been described only by Holmes and Bilker [8]. They found that missing data adversely affect learning and classification performance in a stimulus-response LCS based on the NEWBOOLE [3] paradigm, and this effect is

X. Llorà et al. (Eds.): IWLCS 2003-2005, LNAI 4399, pp. 181–192, 2007.

positively correlated with increasing fractions of missing data. However, no work to date has investigated the effects of missing data on XCS-type LCS, nor on the use of real-time, system-level imputation for dealing with these data.

This paper reports on an investigation into the comparative efficacy of three methods of covering missing data and their effects on the learning and classification performance of an XCS-type LCS, EpiXCS, when it is applied to a simulated datasets with highly controlled numbers of missing values. This investigation focuses on the use of LCS in a simulated data mining task, rather than one in agent-based environments. However, the results of this investigation are applicable to a variety of settings wherever missing data are present in the environment.

1.1 Types of Missing Data

The values of fields in a database can be considered as "responses" to a query, such that for a field such as gender, the value for any given record (or row) in the database reflects a response to the question "What is the gender of [what or whom is represented by the record]?" within the response domain {MALE, FEMALE}. Responses can be *actual*, that is, valid responses within the domain, or they can be *missing*, such that a response value does not exist for that field in the record.   Note the important distinction between missing data and erroneous data: missing data are not responsive, while erroneous data are responsive, but not within the response domain for the field.

Missing responses, or more generally, *missing data*, are typically categorized into one of three types, depending on the pattern of the response [5, 13] on a given field, x, and the other fields, represented collectively as y, in the database. The first type of missing data is characterized by responses to x that are statistically independent of responses to x or y. That is, the probability of a missing value for x is independent of the value of x, as well as of the values of the variables y. This type of missing data is referred to as *missing completely at random* (MCAR). An example of MCAR data would be where the value for gender is randomly missing for some cases, but the "missingness" of gender for any particular case is unrelated to the value of y, as well as the true, but unknown, value of x itself.

A second type of missing data occurs when the probability of a response to x is dependent on the response to y (or, more simply, the value of y). Data such as these are *missing at random* (MAR). An example of MAR data would be where the value for gender is missing when the value of y is at a certain value, or more specifically, if the probability of a missing value for gender is highest when another field, such as race, is equal to Asian. In this case, the missing values for gender are MAR. While the probability of a missing value for gender is essentially random, there is an implicit dependency on race which lessens the degree of randomness of response to the gender field.   Thus, it can be seen that MAR data are qualitatively less desirable, and potentially more problematic than MCAR, in analyses and possibly classification.

The last type of missing data is *not missing at random* (NMAR), and these pose the greatest threat to data analysis. NMAR data are found where the probability of a response to x is dependent on the value of x or a set of data which have not been measured. An example of NMAR data would be where the probability of a missing value for gender is highest when gender is male.   NMAR data are not ignorable in a statistical analysis, due to the possibility of extreme bias that may be introduced by them.

In traditional statistical analyses, missing data may be dealt with by using a variety of procedures loosely grouped under the rubric of *imputation*, which calls for the replacement of missing data with statistically plausible values created by means of one of numerous algorithmic approaches [13]. This is the only viable option when there is a large fraction of missing data. However, for cases where the fraction of missing data is small, it may be reasonable to omit cases with missing data only for MCAR. For MAR or NMAR data, omitting these cases will result in uncorrected bias, so even where the fraction of missing data is small, imputation should be considered in analysis, and it is reasonable to assume that it should be considered in using LCS for classification.

This paper reports on the application of three covering methods MCAR data. The rationale for restricting the investigation to MCAR data is that this type of missing data is the most common of the three, even if it is not the most potentially deleterious.

1.2 Covering in XCS

In the XCS paradigm, covering occurs when the number of actions represented in a Match Set [M] fail to reach a predetermined threshold (specified as θ_{mna} in [4]). One way in which this will occur is when an input, σ, cannot be matched by any classifier in the population. In this situation, a copy of σ will be created and inserted into the population after adding "wild-card" or "don't care" values for specific features at some probability. Covering occurs perhaps most frequently during the early training phase when starting with an empty classifier population, although it can occur later, especially if the dimensionality of the environment (such as a training set) is high.

There are two potential problems associated with this approach to covering. The first is that it does not account for missing data in σ during the process of generating [M]. Specifically, a σ such as 010??0, where the ?s represent missing values for the fourth and fifth features in a six-feature σ, would be evaluated for matching only on the first three and the sixth feature. Thus a way for dealing with missing values in σ when matching it against the classifiers in the population is needed. Even if this problem were addressed, a second problem is that there is a danger of overgeneralization of the covering classifier. For example, if there are large numbers of missing values in a given σ that is being covered, and these missing values are represented by #s, the covering classifier could become saturated with #s as they are injected into it at some probability $p_\#$. Potentially, this problem could be addressed by incorporating a heuristic that ensures against overgeneralization during the covering procedure, but this seems somewhat arbitrary. Other approaches that reflect statistical imputation and thus are more grounded in analytic practice, could provide more robust ways to deal with missing input data during matching and covering. Three such approaches are discussed in the next section.

1.3 Alternative Approaches: Covering in EpiXCS

Missing values in input (training or testing) data are addressed in EpiXCS during the creation of the Match Sets ([M]) by one of three approaches. Each of these approaches assumes that missing values are specifically encoded as such. It doesn't matter to EpiXCS which characters are used to encode missing values as long as they do not conflict with the natural coding for a given feature. That is, if a dichotomous feature is normally coded 0 or 1, then a different value must be used to indicate a missing value for that feature, such as 9 or "*".

Approach 1: Wild-to-Wild. The first approach to handling missing values in a σ is *Wild-to-Wild*, in which any classifiers in the population that match on the *specific* features of an input case are added to [M]. The features of population classifiers that correspond to features with missing values in σ are considered matches as well. Thus, an input case consisting of six features, 001??0 (where ?'s are missing values and treated as "wild cards," or *wilds*), will match (among others) 00#110, ##1010, or 001110 (where #s are "don't cares" or "wilds") in the population. In this approach, "wilds" exist in both the input data (as missing values) and in the classifier population (as "don't cares"). Thus, if matching classifiers exist in the population at that time step, no covering will need to occur. However, if no matching classifiers exist in the population the input case will be added to the classifier population as 001##0, where the missing values have now been replaced by the "#" symbol. The "Wild-to-Wild" approach is perhaps the most intuitively obvious way of covering σs with missing data.

Approach 2: Population Average. The *Population-Average* approach uses a measure of central tendency for covering missing values in σ. In this approach, the population mean (for continuously-coded features) or median (for categorically-coded features) will be used as the covering value. For example, consider a σ that consists of six numeric features:

$$2 \quad 2.0 \quad ? \quad 38.0 \quad 19 \quad 4.54$$

where ? is a missing value for the third (categorically-coded) feature, and the population mode for the third feature at that time step is 394. In this case, the feature will be replaced with that value, so that the new (covering) classifier to be inserted into the population will be:

$$2 \quad 2.0 \quad 394 \quad 38.0 \quad 19 \quad 4.54$$

Approach 3: Random Assignment. The *random assignment* approach replaces a feature's missing value with a value that is randomly selected within the range for that feature. For example, if the range for the third feature in the preceding example is 45 to 400, based on the classifiers in the population at a given time step, the missing value would be replaced with a value randomly selected within this range. Categorically-coded features are preserved to the extent that missing data are not replaced with real values. A variant on this approach, which uses a random number selected within the range of the standard deviation is also available in EpiXCS, but not used in this investigation, which focuses on dichotomous, rather than continuous data.

2 Methods

2.1 Data

Generation of the baseline datasets. This investigation used datasets that were created with the DataGen [14] simulation dataset generator. This software facilitates the creation of datasets for use in testing data mining algorithms and is freely available on the Web. Twenty-five baseline datasets were created, each containing 500 records consisting of 10 dichotomously coded ($\in \{0,1\}$) predictor features and one dichotomously coded ($\in \{0,1\}$) class feature. No missing values were incorporated into the baseline datasets, and although each dataset contained the same number of features, each was unique in that significant differences existed between the datasets with respect to the distribution of the predictor features ($p >> 0.05$) and in the association of each predictor with the class features ($p >> 0.05$).

The baseline datasets were created in such a way as to incorporate noise at a rate of 20%; thus, over the 5,000 feature-record pairs in each dataset, there were 1,000 features that conflicted or contradicted a putative association with the class feature. This was done to ensure that the dataset was sufficiently difficult in terms of learning and classification. In addition to incorporating noise, the user of DataGen has the capability of specifying the number of expected conjuncts per rule; the higher the number, the more complex the relationships between the predictor features and the class. For this investigation, the maximum number of conjuncts per rule was set at six. After examining the resulting conjuncts, one of the 10 predictor features was found to be prevalent in most, or all, of the rules. This feature was used to hold missing values, and thus corresponds to x that is discussed in Section 1.1. The baseline datasets are described in the Appendix.

Generation of datasets with missing values. From the baseline datasets, separate versions were created to simulate 30 increasing proportions, or *densities*, of missing data, ranging from 2.5% to 75%, in 2.5% intervals. The density of missing data was determined as a proportion of the possible feature-record pairs that result from multiplying the number of possible candidate features by the number of records (500). In each of the datasets, only one feature was replaced with a missing value. The actual number of records that contained a missing value changed, depending on the missing value density. For example, at 5% density, there were a total of 25 (500*0.05) records with missing values, all in $\phi\varepsilon\alpha\tau\upsilon\rho\varepsilon$ x. In summary, separate datasets were created at 30 missing value densities for each of the 25 baseline datasets, for a total of 750 datasets; with the addition of the 25 baseline datasets, there were 775 datasets in all. Each of these datasets provided separate pools of data from which training and testing cases were drawn, as described below.

Creation of training and testing sets. This investigation focused on a supervised learning task, requiring the creation of training and testing sets.Once created, the 775 datasets were partitioned recursively into training and testing sets by randomly selecting records without replacement at a sampling fraction of 0.50. Thus, each training and testing set contaned 250 mutually exclusive records. Care was taken to sample the records so as to preserve the original class distribution, which was 50% positive and 50% negative cases.

2.2 EpiXCS

System description. EpiXCS is an XCS version of EpiCS [7], a stimulus-response LCS employing the NEWBOOLE model [3]. It was developed to apply the XCS paradigm to the unique challenges of classification and knowledge discovery in epidemiologic data. EpiXCS uses the XCS class library implemented by Lanzi (http://xcslib.sourceforge.net/), and implements several additional features that tailor the XCS paradigm to the demands of epidemiologic data and users who are not familiar with learning classifier systems. The distinctive features of EpiXCS include a graphical knowledge discovery workbench for parameterization and rule visualization, facilities for handling missing input data, multi-threaded and batch processing, and a methodology for determining risk as a classification metric. EpiXCS uses a variety of test characteristic-based metrics, such as area under the receiver operating characteristic curve and positive and negative predictive values as a means for driving the performance and reinforcement components. Binary, categorical, ordinal, and real data formats are all acceptable, even in the same dataset.

2.3 Metrics and Analytic Issues

Several metrics were used to evaluate the learning and classification performance of EpiXCS in this investigation.

Learning metrics. First, the *area under the receiver operating characteristic curve* (AUC) was used to evaluate evolving classification accuracy during learning and accuracy on classifying novel data. The AUC is preferable to the traditional "accuracy" metric (usually expressed as "percent correct"), as it is not sensitive to imbalanced class distributions such as is found in the simulation data used in this investigation [6]. In addition, the AUC represents, as a single metric, the relationship between the true positive and false positive rates, thereby taking into account the different types of error that can be measured in a two-choice decision problem.

Second, *convergence rate* was evaluated by means of a metric, λ, created specifically for this purpose. This metric was calculated as follows:

$$\lambda = \left(\frac{AUC_{Shoulder}}{Shoulder} \right) 1000 \qquad (1)$$

Shoulder is the iteration at which 95% of the maximum AUC obtained during training is first attained, and $AUC_{Shoulder}$ is the AUC obtained at the shoulder. Thus, the higher the value of λ, the faster the system reaches convergence on the training data. As the first AUC is not measured until the 100th iteration, and the maximum AUC measurable is 1.0, the maximum value of λ is 10.0. The minimum ξ is 0.0.

Third, the ability of the trained EpiXCS system to classify previously unseen cases of similar genre to the training cases was assessed. This was done by comparing the AUCs obtained at testing across the range of missing value densities. In addition to classification accuracy, as measured by the AUC, it is important to assess the extent to which novel data is unclassifiable, and therefore doesn't factor in to the calculation of the AUC. A metric designed for this purpose, the *Indeterminant Rate* (IR), was used to quantify the proportion of testing cases that could not be classified on testing:

$$\text{Indeterminant Rate} = \frac{\text{Number of testing cases not classifiable}}{\text{Total number of testing cases}} \qquad (2)$$

These metrics were used in a variety of statistical analyses. To evaluate the effects of missing data on learning performance, the κs were correlated by Spearman's rho (ρ) the nonparametric equivalent of Pearson's r. The nonparametric test was chosen because the independent variable in the correlation analyses, missing value density, is ordered-categorical. The λs were compared, using the baseline dataset as the reference, across the range of the missing value densities.

Classification metrics. Classification performance was evaluated by AUC, adjusted by the IR, as described above. In addition, the sensitivity, specificity, and positive and negative predictive values were calculated as described below and derived from the confusion matrix in Figure 1.

	Gold standard (class value in data)	
As classified by EpiXCS	Positive	Negative
Positive	A	B
Negative	C	D

Fig. 1. 2x2 confusion matrix for a two-choice decision problem. The columns represent the "gold standard," or the classifications as they exist in the data. The rows represent the classification. A=True positives; B=False positives; C=False negatives; D=True negatives.

$$\text{Sensitivity} = \text{True positive rate} = \frac{A}{A+C} \tag{3}$$

$$\text{Specificity} = \text{True negative rate} = \frac{D}{B+D} \tag{4}$$

$$\text{Positive predictive value} = \frac{A}{A+B} \tag{5}$$

$$\text{Negative predictive value} = \frac{D}{C+D} \tag{6}$$

Sensitivity and specificity are prior probabilities: they indicate the accuracy of positive or negative classifications, respectively, made by EpiXCS in the past, prior to the current classification. Thus, one would want to know the sensitivity and specificity in deciding whether or not to use EpiXCS for a given decision task. The predictive values are posterior probabilities, and indicate the positive or negative classification accuracy of EpiXCS on a decision task just performed. The predictive values reflect the confidence one might place in a classification made by EpiXCS.

2.4 Experimental Procedure

Training. EpiXCS was trained over 2,500 iterations, comprising a *training epoch*. At each iteration, the system was presented with a single training case, σ. As training cases were drawn randomly from the training set with replacement, it could be assumed that the system would be exposed to all such cases with equal probability over the course of the 2,500 iterations of the training epoch. At the 0th and each 100th iteration thereafter, the convergence of EpiXCS was evaluated by presenting the taxon of every case in the training set, in sequence, to the system for classification. As these iterations constituted a test of the training set, the reinforcement component and the genetic algorithm were disabled on these occasions. The decision advocated by EpiXCS for a given training case was compared to the known classification of the training case. The decision type was classified in one of four categories: true positive, true negative, false positive, and false negative, and tallied for each classifier. From the four decision classifications, the AUC and IR were calculated and written to a file for analysis.

Testing. After the completion of the designated number of iterations of the training epoch, EpiXCS entered the testing epoch, in which the final learning state of the system was evaluated using every case in the testing set, each presented only once in sequence. As in the interim evaluation phase, the reinforcement component and the genetic algorithm, were disabled during the testing phase. At the completion of the testing phase, the AUC and IR were calculated and written to a file for analysis, as was done during the interim evaluations. The entire cycle of training and testing comprised a single *trial*; a total of 20 trials were performed for this investigation for each of the 775 datasets.

Parameterization. EpiXCS was parameterized as described in Butz and Wilson [4], except that the population size was set to 500, which was found empirically to be optimal. In addition, the baseline prediction error (ε_0) was evaluated over a range from 0.5 to 4.0 to determine if the randomness of the data affected learning performance; the default of 1.0 was found to be optimal. Both action set and genetic algorithm subsumption were performed.

3 Results

3.1 Effects of Missing Data on Learning Performance

The convergence rate (λ) of EpiXCS was remarkably stable across all missing value densities using MCAR data. No variance was noted in progressing from low to high densities, indicating that the system is not affected by even high proportions of missing input data during learning. In addition, relatively little variation was found

Table 1. Convergence rate (λ) for each covering method. Values averaged over the 20 runs, and then the 25 datasets at each density.

Covering Method	Mean λ	SD λ
Mode	5.45	2.77
Random assignment	5.49	2.73
Wild-to-Wild	5.44	2.69

between the three covering methods, as shown in Table 1. In addition, no correlation was found between λ and missing value density.

3.2 Effects of Missing Data on Classification Performance

The evaluation of the effect of missing data on classification performance focused on comparing the various test characteristics obtained on the testing set with missing value density, separately for each type of covering. These characteristics included sensitivity, specificity, AUC, IR, and positive and predictive values. Virtually no effect of missing data density was observed on classification performance; the mean values for these metrics were virtually identical across the range of densities. Slight differences in the mean values for these metrics were noted between the three covering methods. These differences were not significant, and are shown in Table 2.

Table 2. Test characteristics indicating classification performance on testing data. Values averaged over the 20 runs, and then the 25 datasets at each density. Standard deviation represented in parentheses.

	Mode	Random	Wild-to-Wild
Area under the curve	0.970 (0.021)	0.969 (0.023)	0.969 (0.022)
Sensitivity	0.967 (0.029)	0.966 (0.032)	0.966 (0.031)
Specificity	0.973 (0.026)	0.973 (0.027)	0.973 (0.027)
Positive predictive value	0.973 (0.026)	0.972 (0.026)	0.973 (0.026)
Negative predictive value	0.969 (0.027)	0.968 (0.029)	0.968 (0.027)
Indeterminant Rate	0.0 (0.0)	0.0 (0.0)	0.0 (0.0)

Correlation of classification accuracy with missing value density. No correlation was found between classfication accuracy, using any of the above test characteristics, and missing value density.

4 Discussion

Table 1 clearly demonstrates that neither missing value density nor covering method affected convergence rate, either positively or negatively. This indicates, at least on the simple MCAR datasets used in this investigation, that EpiXCS, and indeed XCS in general, is insensitive to even large amounts of missing data during the training phase in supervised learning environments. However, it is not yet clear what would happen when MAR, or particularly NMAR, data need to be covered. Nor is it clear that this level of learning performance would be seen in larger, more complicated datasets, consisting of large numbers of features with mixtures of the three types of missing data, especially where the feature set y consists of larger numbers of features that may include complicated interactions.

Table 2 demonstrates a similar phenomenon: classification accuracy is essentially not affected by MCAR-type missing data, across a wide range of missing value densities. Some slight differences were observed in the evaluation metrics using the different covering methods, but it is not clear that this should dictate the use of one covering method over another. In fact, this study indicates that much more investigation is needed into the properties of the three covering methods, and in the face of a wide variety of contrived as well as real datasets.

Limitations of this study. While there is much in this investigation to suggest that EpiXCS is insensitive to missing data in terms of convergence rate and classification accuracy, there are several ways to confirm these conclusions. First, only one feature in the data was used as a candidate for assigning missing data.. It would be interesting to extend the patterns of missing data to sets of features that included more than one each, such that x (and/or y, when extended to MAR and NMAR data) would have many features contained within them. It should be noted, however, that doing so would substantially increase the complexity of the analysis, due to the possibility for interactions, so these would have to be handled carefully in creating the datasets.

Second, as noted previously, this study used small datasets. While these provide the basic groundwork for further investigation, more work needs to be done in extending this work to larger and real-world data that contain a variety of features with missing data as well as missing data types.

5 Conclusions

This investigation is the first report into the effects of covering missing data on the learning and classification performance in an XCS-based learning classifier system. EpiXCS is insensitive to the missing data used in this study, but this is by no means the end of the story. Even in the face of the results presented here, researchers would be wise to exercise caution when employing LCS in any environment that may contain missing data.

A future task, in addition to researching the effects of covering a wider range of missing value densities and patterns in a variety of datasets and dataset sizes, is to study the effects of pre-processing imputation on LCS performance. Imputation performed by value assignment or multivariate methods ("multiple imputation") is typically accomplished by processing the data prior to statistical modeling, and there is no reason why these procedures couldn't be used before modeling with XCS. In a real sense, covering is a form of real-time, system-based imputation, but it is highly non-traditional in the statistical and machine learning worlds, where missing data are imputed ("covered") even prior to exposure to the system. Thus, an interesting question remains: are standard methods of imputation better than the covering methods described here, or are they superfluous? Either way, the answer to this question has serious implications for the use of LCS in a variety of environments and domains, including maze learning and knowledge discovery, to name two.

References

1. Anand S.S., Bell D.A., Hughes J.G.: EDM: a general framework for data mining based on evidence theory. Data & Knowledge Engineering (1996) 18(3):189-223.
2. Bing, L., Ke, W., Lai-Fun, M., Xin-Zhi, Q.: Using decision tree induction for discovering holes in data. PRICAI'98: Topics in Artificial Intelligence. 5th Pacific Rim International Conference on Artificial Intelligence. Springer-Verlag, Berlin (1998), 182-93.

3. Bonelli, P., Parodi, A., Sen, S., Wilson, S.: NEWBOOLE: A fast GBML system, in: Porter, B. and Mooney, R. (eds.), Machine Learning: Proceedings of the Seventh International Conference. Morgan Kaufmann, San Mateo, CA (1990), 153-159.

4. Butz MV and Wilson SW: An algorithmic description of XCS. In: Lanzi, P. L., Stolzmann, W., and Wilson, S. W. (eds.):, Advances in Learning Classifier Systems. Third International Workshop, Lecture Notes in Artificial Intelligence (LNAI-1996). Berlin: Springer-Verlag (2001), 211-230.

5. Fengzhan, T., Hongwei, Z., Yuchang, L., Chunyi, S.: Incremental learning of Bayesian networks with hidden variables. Proceedings 2001 IEEE International Conference on Data Mining. IEEE Computing. Society, Los Alamitos, CA (2001), 651-2.

6. Holmes J.H.: Quantitative methods for evaluating learning classifier system performance In forced two-choice decision tasks. In: Wu, A. (ed.) Proceedings of the Second International Workshop on Learning Classifier Systems (IWLCS99). Morgan Kaufmann, San Francisco (1999), 250-257.

7. Holmes JH, Durbin DR, Winston FK: The Learning Classifier System: An evolutionary computation approach to knowledge discovery in epidemiologic surveillance. Artificial Intelligence in Medicine (2000) 19(1):53-74.

8. Holmes JH and Bilker WB: The effect of missing data on learning classifier system classification and prediction performance. Advances in Learning Classifier Systems. Lecture Notes in Artificial Intelligence. Lanzi PL, Stolzmann W, and Wilson SW (eds.). Berlin, Springer Verlag, Vol. 2661: 46-60 (2003).

9. Jui-Chung. H., Bor-Sen, C., Wen-Sheng, H., Li-Mei, C.: Spectral estimation under nature missing data. 2001 IEEE International Conference on Acoustics, Speech, and Signal Processing. Proceedings IEEE, Piscataway, NJ (2001), 3061-4.

10. Kryszkiewicz, M.: Association rules in incomplete databases. Methodologies for Knowledge Discovery and Data Mining. Third Pacific-Asia Conference, PAKDD-99. Springer-Verlag, Berlin (1999), 84-93.

11. Kryszkiewicz, M. and Rybinski, H.: Incomplete database issues for representative association rules. Foundations of Intelligent Systems. 11th International Symposium, ISMIS'99. Springer-Verlag, Berlin (1999), 583-91.

12. Laurikkala J., Juhola M., Lammi S., Viikki K.: Comparison of genetic algorithms and other classification methods in the diagnosis of female urinary incontinence. Methods of Information in Medicine (1999), 38(2):125-131.

13. Little R.J.A. and Rubin, D.B.: Statistical Analysis with Missing Data. John Wiley and Sons, New York (1986).

14. Melli, G.: http://www.datasetgenerator.com/

15. Ng, V. and Lee J.: Quantitative association rules over incomplete data. SMC'98 Conference Proceedings. 1998 IEEE International Conference on Systems, Man, and Cybernetics, IEEE, New York (1998), 2821-6.

16. Sarle W.S.: Prediction with missing inputs. Joint Conference on Intelligent Systems 1999 (JCIS'98). Association for Intelligent Machinery (1998), 399-402.

Appendix

Variable-by-variable description of the 25 baseline datasets created by the DataGen generator. Cell values are modes for each predictor variable, for each dataset. The class distribution for each dataset was 50% Class 1 and 50% Class 2.

Dataset	Variable									
	1	2	3	4	5	6	7	8	9	10
1	1	1	0	1	1	1	0	0	0	1
2	1	1	0	0	1	1	0	0	1	1
3	0	0	0	1	1	0	0	0	0	0
4	1	1	0	0	1	0	0	1	0	0
5	1	0	1	0	0	0	1	1	1	0
6	0	0	0	1	0	1	0	0	1	1
7	0	0	0	0	1	0	0	0	0	0
8	1	0	1	0	1	0	0	1	0	1
9	1	0	1	1	1	0	0	1	1	0
10	0	1	1	1	1	1	1	1	0	0
11	1	0	1	0	1	0	0	1	1	0
12	0	0	1	1	0	1	1	1	0	1
13	1	0	0	0	0	0	1	1	0	1
14	1	1	1	1	1	1	0	1	1	1
15	0	1	0	1	0	0	0	1	0	0
16	1	1	0	0	1	1	0	1	1	0
17	1	1	1	0	0	0	1	1	1	1
18	0	1	1	0	0	0	0	1	1	0
19	1	1	1	1	1	1	0	0	0	1
20	0	1	1	1	0	0	0	1	0	0
21	0	1	1	0	1	1	1	1	1	1
22	0	0	0	1	1	1	1	1	0	0
23	0	0	1	1	1	1	1	1	1	1
24	1	0	1	0	1	1	1	1	1	0
25	0	0	0	1	0	0	0	0	0	0

A Hyper-Heuristic Framework with XCS: Learning to Create Novel Problem-Solving Algorithms Constructed from Simpler Algorithmic Ingredients

Javier G. Marín-Blázquez[1] and Sonia Schulenburg[2]

[1] Ramón y Cajal Program Researcher
Department of Information and Communications Engineering (DIIC)
Facultad de Informática, Campus de Espinardo
Universidad de Murcia, 30100 Murcia, Spain
jgmarin@um.es
[2] Royal Society of Edinburgh Fellow
Centre of Intelligent Systems and their Applications (CISA)
Appleton Tower, Room 3.14, Crichton Street
Edinburgh University, Edinburgh, EH8 9LE, United Kingdom
sonia.schulenburg@ed.ac.uk

Abstract. Evolutionary Algorithms (EAs) have been successfully reported by academics in a wide variety of commercial areas. However, from a commercial point of view, the story appears somewhat different; the number of success stories does not appear to be as significant as those reported by academics. For instance, Heuristic Algorithms (HA) are still very widely used to tackle practical problems in operations research, where many of these are NP-hard and exhaustive search is often computationally intractable. There are a number of logical reasons why practitioners do not embark so easily in the development and use of EAs. This work is concerned with a new line of research based on bringing together these two approaches in a harmonious way. The idea is that instead of using an EA to learn the solution of a specific problem, use it to find an algorithm, i.e. a solution process that can solve well a large family of problems by making use of familiar heuristics. The work of the authors is novel in two ways: within the Learning Classifier Systems (LCS) current body of research, it represents the first attempt to tackle the Bin Packing problem (BP), a different kind of problem to those already studied by the LCS community, and from the Hyper-Heuristics (HH) framework, it represents the first use of LCS as the learning paradigm. Several reward schema based on single or multiple step environments are studied in this paper, tested on a very large set of BP problems and a small set of widely used HAs. Results of the approach are encouraging, showing outperformance over all HAs used individually and over previously reported work by the authors, including non-LCS (a GA based approach used for the same BP set of problems) and LCS (using single step environments). Several findings and future lines of work are also outlined.

X. Llorà et al. (Eds.): IWLCS 2003-2005, LNAI 4399, pp. 193–218, 2007.
© Springer-Verlag Berlin Heidelberg 2007

1 Introduction

Among the main criticisms of bio-inspired algorithms, and in more general terms of stochastic based problem solving techniques, is the fact that they involve some randomness throughout several stages of the method. In addition to this, they generally offer no guarantees of solution quality; the user may have to do many runs to sample the quality of possible solutions. Many algorithms also have a sizeable set of parameters that need to be set, a process that can require considerable skill and experience. The delivered solution may also be fragile, in the sense that there is little continuity between problem specification and EA solution: if you change the problem only slightly, the solution found by re-running the EA changes drastically. Even renowned academical texts [31] can be very cautious about them, and users may justifiably prefer to use simpler deterministic approaches even if those approaches generally produce poorer results.

While evolutionary algorithms have reached maturity in recent years, it is fair to say that the impact of commercial applications deployed has not met the expectations set some years ago. Many of the problems are NP-hard [15] and exhaustive search, often computationally intractable, is not a viable option. Also, many state-of-the-art developments on search are too problem-specific or too knowledge-intensive to be implemented in cheap, easy-to-use computer systems. As a result, users often employ simpler heuristics which are easy to understand and implement, even at the expense of poor performances. Commercially, there is not enough interest in solving optimisation problems to optimality, or even close to optimality. Instead the interest seems to be in obtaining "good enough - soon enough - cheap enough" kind of solutions. It is for these reasons that HAs are very widely used to tackle practical problems in operations research. They are also simple, easy to understand and inspire confidence. Besides, many optimisation problems have simple heuristic methods to solve them, with the additional benefit that these heuristics may have associated theoretical best and worst performance limits.

Hyper-heuristics attempt to avoid some of the drawbacks of bio-inspired and stochastic methods, while exploiting the trust that the heuristic algorithms inspire. The idea is to use the evolutionary methods to search for a novel problem-solving algorithm rather than to solve individual problems. The novel algorithm is to be constructed from simple heuristic ingredients and should, ideally, be fast and deterministic and with good worst-case performance across a range of problems. The general framework that is considered is that of algorithms that use a simplified problem state description and associate a simple heuristic with different states. That is, repeatedly: determine the current state of the partially-solved problem; look up which heuristic to use to extend the solution a little; apply it -until the problem has been fully solved.

In the following sections, an example of using hyper-heuristic methods to tackle one-dimensional bin-packing problems is described. XCS [37] is used to learn a set of rules which associate characteristics of the current state of a problem with specific heuristics being used. The set of rules is used to solve problems as mentioned: given the initial problem characteristics P, a heuristic H is chosen to pack a bin, thus gradually altering the characteristics of the problem that remains to be solved. At each step, a rule appropriate to the current problem state P' is chosen, and the process repeats until all items have been packed. This means that different widely used heuristics can be used to pack one bin, as opposed to only using one.

XCS represents an elegant and simple way to try to fabricate a composite algorithm, and the interest lies in assessing competitiveness of its performance. It has already been presented in [30,33], that even when the model was trained using only a few problems, it generalised and also performed well on a large number of unseen problems. That was a useful step towards the concept of using EAs to generate strong solution processes, rather than merely using them to find good individual solutions. Later work [27,29,28], using alternate EAs (a genetic algorithm) and applied to a different kind of problem (timetabling) provided further evidence of the benefits of the hyper-heuristics approach.

In [30] an XCS was applied using exclusively single-step environments, meaning that rewards were available only after each action had taken place. Here the approach is extended to multi-step environments (preliminary results were presented at [33], and analysed under two different reward schemes. In the first one, rewards are assigned on the basis of **actions** (after one action is the case of single-step), and in the second one, on the basis of **states**. Note that a state can have a large number of actions, and chains of various lenghts can be selected for both types of rewards. These approaches are tested using a large set of benchmark BP problems and a small set of eight heuristics, consisting of widely used algorithms, and combinations of these devised by the authors.

The work reported here attempts to find an algorithm that iteratively builds a solution, however it is important to note that this is atypical in the field. It is more common for an algorithm to start with a complete candidate solution and then search for improvements by making modifications, eg by some kind of controlled neighbourhood search. Hyper-heuristics could also be applied to try to discover a good, fast algorithm of that type, with tightly-bounded performance costs; this might be a topic for further research and outside the scope of the present work.

The rest of this paper is organised as follows: in section 2 the idea of hyper-heuristic is further developed and justified. Section 3 explains the different styles of LCS and the reasons of choosing the classifier used in this work, the Extended Classifier System (XCS). Section 4 introduces one-dimensional bin-packing problems and some of their features. In section 5 the different heuristics used to solve the bin-packing problem are presented, followed by descriptions of how the state of the problem solving procedure is represented. Both, heuristics and state correspond, respectively, to the actions and the messages (context) for the XCS. Section 6 reports the experiments performed, including set-up and results obtained. Finally, in section 7 conclusions are drawn and further lines of research are suggested.

2   The Idea of Hyper-Heuristics

Despite the aforementioned mistrust from general practitioners with regards to evolutionary techniques, their use is often justified simply by results. Evolutionary algorithms can be excellent for searching very large spaces, at least when there is some reason to believe that there are 'building blocks' to be found. A 'building block' is a fragment, in the chosen representation, such that chromosomes which contain it tend to have higher fitness than those which don't. EAs bring building blocks together by chance recombination, and building blocks which are not present in the population at all may still be generated by mutation. Some EAs allow, as well, reinforcement learning, to be applied when some measure of reward can be granted to actions that produced good results. This type of learning procedure is especially useful in tasks where the solution to be found is a sequence of actions. In particular, learning classifier systems have repeatedly shown capabilities to deal with this kind of problems.

Hyper-Heuristics represents a step towards a new way of using EAs that may solve some problems of acceptability, mentioned above, for potential real-world use. The basic idea of HH is as follows: instead of using an EA to discover *a solution to a specific problem*, the EA is used to try to fabricate *a solution process applicable to many problem instances* built from simple, well-understood heuristics. Such a solution process might consist of using a certain heuristic initially, but after a while the nature of the remainder of the task may be such that a different heuristic becomes more appropriate to use. Once such solution process is discovered, it can be provided to users that can apply it as many times, and to as many different problems, as desired. The problems will be solved using the reassuring and simple heuristics that users are familiar with. A good general overview of hyper-heuristics and their applications can be found in [5].

For example, in [35] an early version of this idea was used to tackle large exam timetabling problems by choosing two heuristics and associated parameters, together with a test for when to switch from using the first to using the second. This was motivated by the unsurprising observation that different academic institutions have very different constraints. One institution might have some very large exams, limited exam seating and many smaller exams, so that the important task early on is to pack those large exams together as far as possible in order to have plenty of space to deal with placing the many smaller exams. Another institution might not have very large exams, but instead the exams can be clustered such that there are very few inter-cluster constraints, and exam clusters can therefore be viewed as relatively independent sub-problems, for which one might naturally choose some other heuristic that placed little emphasis on packing large exams.

The key idea in hyper-heuristics is to use members of a set of known and reasonably understood heuristics to transform the state of a problem. The key observation is a simple one: the strength of a heuristic often lies in its ability to make some good decisions on the route to fabricating an excellent solution. Why not, therefore, try to associate each heuristic with the problem conditions under which it flourishes and hence apply different heuristics to different parts or phases of the solution process?

The alert reader will immediately notice an objection to this whole idea. Good decisions are not necessarily easily recognisable in isolation. It is a sequence of decisions that builds a solution, and so there can be considerable epistasis involved - that is, a non-linear interdependence between the parts. However, many general search procedures such as evolutionary algorithms and, in particular, classifier systems, can cope with a considerable degree of epistasis, so the objection is not necessarily fatal.

Therefore, a possible framework for a hyper-heuristic algorithm:

1. Start with a set H of heuristic ingredients, each of which is applicable to a problem state and transform it to a new problem state.
2. Let the initial problem state be S_0
3. If the problem state is S_i then find the ingredient that is in some sense most suitable for transforming that state. Apply it, to get a new state of the problem S_{i+1}
4. If the problem is solved, stop. Otherwise go back to 3.

There could be many variants of this, for example, in which the set H varies as the algorithm runs or in which suitability estimates are updated across the iterations, or in which the size of a single state transformation varies because the heuristic ingredients are dynamically parameterised. There is very considerable scope for research here.

3 Learning Classifier Systems

This section is intended for readers not familiar with LCS in general. For those who are, it might be best to go to the following section.

3.1 Introduction to LCS

As defined in [17], a "classifier system is a machine learning system that learns syntactically simple string rules (called classifiers) to guide its performance in an arbitrary environment", where machine learning primarily refers to systems which acquire and improve knowledge by using input information.

In LCS, the general idea is to learn concepts through decision rules that account for positive examples in order to "predict a classification of previously unseen examples, or suggest (possibly more than one) classifications of partially specified descriptors" [26]. The learning takes place by adjusting certain values associated to the rules according to the environmental feedback they receive and by discovering new and better rules. Considering that chromosomes are good ways of representing knowledge as well as good candidates to be manipulated by genetic operators, the GA community quickly responded with two distinctive approaches that were labelled according to the Universities where they were developed: Pittsburgh (or 'Pitt approach', led by De Jong and his students) and Michigan (mainly led by Holland, Reitman and Booker).

In the Pittsburgh-style classifier approach (also known as population of rule-sets approach), the competition occurs between rule-sets. The idea here is "to represent an entire rule set as a string (an individual), maintain a population of candidate rule sets, and use selection and genetic operators to produce new generations of rule sets" [10]. So, the population is composed of multiple rule-sets competing to mate other rule-sets and exchanging individual rules with the hope to combine rules over many generations to form an effective rule-set.

In the Michigan-style approach (population of rules approach), it is pointed out in [10] that "the members of the population are individual rules and a rule set is represented by the entire population (e.g., see [18,3])". In this view, individuals compete via fitness for reproductive rights. The competition of rules takes place within the set of rules, and highly fit individuals have the opportunity to match with other highly fit individuals with the hope of increasing survival characteristics of their progeny.

The main difference between these two approaches, apart from their State of origin, resides in the nature of the members of the population created: either single rules, or sets of rules.

3.2   Choosing the Appropriate Classifier

The classifier used in this work is of the Michigan type. As mentioned above, learning classifier systems of the Michigan type evolve a set of condition-action rules, by measuring the performance of individual rules and then periodically using crossover and mutation to breed new rules from old ones. An early account can be found in [17], and a collaction of more recent work in [21,22,23,24,34,4].

In early learning classifier systems, rules occasionally did an action that earned external reward, and this contributed to the rule's fitness and to the fitness of those that enabled it to fire. Earned rewards were spread by the so-called 'bucket brigade algorithm' (effectively a trickle-down economy) or 'profit-sharing plan' (essentially a communal reward-sharing) or other such algorithm. However, in those early systems, a rule's fitness was a measure of the reward it might earn (when considering what rule to fire) and also a measure of the reward it had earned (when selecting rules for breeding). This caused various problems, notably that rules which fired very rarely but were crucial when they did, would tend to be squeezed out of the population by the evolutionary competition, long before they could demonstrate their true value. XCS [37] largely fixed this by instead valuing a rule for the accuracy rather than the size of its prediction of reward.

For this reason – because, in this application, there might be heuristics which were rarely used but crucial – it was chosen to use XCS rather than, say, Gold-berg's SCS.

3.3   The End Product

The idea behind XCS is that the end product of combining accuracy and a niche GA results in a complete and accurate mapping X x A => P from inputs and actions to payoff predictions. Further, as Wilson stated "XCS tends to evolve classifiers that are maximally general subject to an accuracy criterion, so that the mapping gains representational efficiency. In traditional classifier systems there is in theory no adaptive pressure toward accurate generalisation, and in fact accurate generalised classifiers have rarely been exhibited, except in studies using payoff regimes biased toward formally general classifiers (e.g. [36])".

The classic paper entitled "A Critical Review of Classifier Systems" [38] gives a good summary of the unsolved problems and new challenges that LCS faced in the late 80s. Since then, there have been great accomplishments in theoretical aspects (mapping performance and generalisation), of XCS solving a variety of single-step environments such as the boolean multiplexer and sequential environments (multi-step) like the woods-type of problems.

4   One-Dimensional Bin-Packing Problems

In the one-dimensional Bin Packing problem (1DBPP), there is an unlimited supply of bins, each with capacity c (a positive number). A set of n items is to be packed into the bins, the size of item i is $s_i > 0$, and items must not over-fill any bin:

$$\sum_{i \in \text{bin}(k)} s_i \leq c$$

The task is to minimise the total number of bins used. Despite its simplicity, this is an NP-hard problem. If M is the minimal number of bins needed, then clearly:

$$M \geq \lceil (\sum_{i=1}^{n} s_i)/c \rceil$$

and for any algorithm that does not start new bins unnecessarily, $M \leq$ bins used $< 2M$ (because if it used $2M$ or more bins there would be two bins whose combined contents were no more than c, and they could be combined into one).

Many results are known about specific algorithms. For example, a commonly-used algorithm is Largest-Fit-Decreasing (LFD): items are taken in order of size, largest first, and put in the first bin where they will fit (a new bin is opened if necessary, and effectively all bins stay open). It is known [19] that this uses no more than $11M/9 + 4$ bins. A good survey of such results can be found in [9]. A good introduction to bin-packing algorithms can be found in [25], which also introduced a widely-used heuristic algorithm, the Martello-Toth Reduction Procedure (MRTP). This simply tries to repeatedly reduce the problem to a simpler one, by finding a combination of 1-3 items that provably does better than anything else (not just any combination of 1-3 items) at filling a bin, and if so packing them. This may eventually halt with some items still unpacked; the remainder are packed using a 'largest first, best fit' algorithm.

Various authors have applied EAs to bin-packing, notably Falkenauer's grouping GA [12,14,13]; see also [20] for a different approach. Falkenauer also produced two of several sets of benchmark problems. In one of these, the so called *triplet problems*, every bin contains three items; they were generated by first constructing a solution which filled every bin exactly, and then randomly shrinking items a little so that the total shrinkage was less than the bin capacity (thus the same number of bins is necessary). Note that these algorithms solve each problem in independent runs for each one.

As ever, specific knowledge about problems can help greatly. Suppose you know in advance that each bin contains exactly three items. Take items in order, largest first, and for each item search for two others that come very close to filling the bin. A backtracking algorithm that considers such 'filler pairs', taking pairs in which the two members at most nearly equal in size first and permitting only limited backtracking, solves many of the Falkenauer triplet problems very quickly. See [16] for some questions about whether these problems are hard or not.

The reader may wonder if the simple strategy of searching for a combination of items which come as close as possible to filling a bin, thereby reducing the problem to a simpler one in which there seems to be more available slack, is a good one. But consider a problem in which bins have capacity 20 and there are six items: 12, 11, 11, 7, 7, 6. One bin can be completely filled $(7 + 7 + 6)$ but then three more bins are needed since the three largest items are each larger than half a bin. If bins are under-filled, then a three-bin solution is possible, for example $12 + 7$, $11 + 7$, $11 + 6$. This should help to convince the reader that even one-dimensional bin-packing problems have their interest. And they are worth studying, because bin-packing is a constituent task of many other optimisation problems; exam timetabling is just one such example.

4.1   Bin-Packing Benchmark Problems

The problems used in this work come from two sources. The first collection is available from Beasley's OR-Library [1], and contains problems of two kinds that were generated and largely studied by Falkenauer [13]. The first kind, 80 problems named uN_M, involve bins of capacity 150. N items are generated with sizes chosen randomly from the interval 20-100. For N in the set $(120, 250, 500, 1000)$ there are twenty problems, thus M ranges from 00 to 19. The second kind, 80 problems named tN_M, are the triplet problems mentioned earlier. The bins have capacity 1000. The number of items N is one of 60, 120, 249, 501 (all divisible by three), and as before there are twenty problems per value of N. Item sizes range from 250 to 499 but are not random; the problem generation process was described earlier.

The second class of problems studied in this paper comes from the Operational Research Library [2] at the *Technische Universität Darmstadt*. Their 'bpp1-1' set and their very hard 'bpp1-3' set were used in this paper. In the bpp1-1 set problems are named NxCyWz_a where x is 1 (50 items), 2 (100 items), 3 (200 items) or 4 (500 items); y is 1 (capacity 100), 2 (capacity 120) or 3 (capacity 150); z is 1 (sizes in 1...100), 2 (sizes in 20...100) or 4 (sizes in 30...100); and a is a letter in A...T indexing the twenty problems per parameter set. (Martello and Toth [25] also used a set with sizes drawn from 50...100, but these are far too easy to solve so they have been excluded from this work.) Of these 720 problems, the optimal solution is known in 704 cases and in the other sixteen, the optimal solution is known to lie in some interval of size 2 or 3. In the hard bpp1-3 set there are just ten problems, each with 200 items and bin capacity 100,000; item sizes are drawn from the range 20,000...35,000. The optimal solution is known in only three cases, in the other seven the optimal solution lies in an interval of size 2 or 3. These results were obtained with an exact procedure called BISON [32] that employs a combination of tabu search and modified branch-and-bound.

In all, therefore, 890 benchmark problems are used.

5   Combining Heuristics with XCS

This section is divided into three parts. The first subsection describes the heuristics used, addressing why they were selected. The next subsection describes the representation used within XCS. Finally, how XCS is used to discover a good set of rules is explained.

5.1   The Set of Heuristics

First, a variety of heuristics were developed and their performances were evaluated on the benchmark collection. Of the fourteen that were implemented and tested, some were taken directly from the literature, others were variants created by the authors. Some of these algorithms were always dominated by others; among those that sometimes obtained the best of the fourteen results on a problem, some were always first-equal, rather than being *uniquely* the best of the set. There is no space here to describe the full set, but four, whose performance seemed collectively to be representative of the best, were selected as follows:

- Largest Fit Decreasing (LFD), described in Section 4 above. This was the best of the fourteen heuristics in over 81% (compared with other heuristics tested and not with the results reported in literature) of the bpp1-1 problems, but was never the winner in the bpp1-3 problems.
- Next-Fit-Decreasing (NFD): an item is placed in the current bin if possible, or else a new bin is opened, becoming the current bin where the item is placed. This is usually very poor.
- Djang and Finch's algorithm (DJD), see [11]. This puts items into a bin, taking items largest-first, until that bin is at least one third full. It then tries to find one, or two, or three items that completely fill the bin. If there is no such combination it tries again, but looking instead for a combination that fills the bin to within 1 of its capacity. If that fails, it tires to find such a

combination that fills the bin to within 2 of its capacity; and so on. This of course gets excellent results on, for example, Falkenauer's problems; it was the best performer on just over 79% of those problems but was never the winner on the hard bpp1-3 problems.

- DJT (Djang and Finch, more tuples): devised by the authors, this corresponds to a modified form of DJD, considering combinations of up to five items rather than three items. In the Falkenauer problems, DJT performs exactly like DJD, as one would expect; in the bpp1-1 problems it is a little better than DJD.

In addition to these four, a 'filler' process was also used (coupled with each algorithm), which tried to find items to pack in any open bins, rather than moving on to a new bin. This might, for example, make a difference in DJD if a bin could be better filled by using more than three items once the bin was one-third full. Thus, in all, eight heuristics were used. The action of the filler process is described later.

5.2   Representing Problem State for XCS

As explained above, the idea is to find a good set of rules each of which associates a heuristic with some description of the current state of the problem. To execute the rules, the initial state is used to select a heuristic and that heuristic is used to pack a bin. The rules are then consulted again to find a heuristic appropriate to the altered problem state, and the process repeats until all items have been packed.

The problem state is reduced to the following simple description. The number of items remaining to be packed are examined, and the percentage R of items in each of four range is calculated. These ranges are shown in table 1.

Table 1. Item size ranges

Huge:	items over 1/2 of bin capacity
Large:	items from 1/3 up to 1/2 of bin capacity
Medium:	items from 1/4 up to 1/3 of bin capacity
Small:	items up to 1/4 of bin capacity

These are, in a sense, natural choices, since at most one huge item will fit in a bin, at most two large items will fit a bin, and so on. The percentage of items that lie within any one of these ranges is encoded using two bits as shown in table 2.

Thus, there are two bits for each of the four ranges. Finally, it seemed important to also represent how far the process had got in packing items. For example, if there are very few items left to pack, there will probably be no huge items

Table 2. Representing the proportion of items in a given range

Bits	Proportion of items
0 0	0 – 10%
0 1	10 – 20%
1 0	20 – 50%
1 1	50 –100%

Table 3. Percentage of Items Left

Bits	% left to pack	Bits	% left to pack
0 0 0	0 – 12.5	1 0 0	50 – 62.5
0 0 1	12.5 – 25	1 0 1	62.5 – 75
0 1 0	25 – 37.5	1 1 0	75 – 87.5
0 1 1	37.5 – 50	1 1 1	87.5 – 100

left. Thus, three bits are used to encode the percentage of the original number of items that still remain to be packed; table 3 illustrates this.

The action is an integer indicating the decision of which strategy to use at the current environmental condition, as shown in table 4. As mentioned earlier, the second four actions use a filler process too, which tries to fill any open bins as much as possible. If the filling action successfully inserts at least one item, the filling step finishes. If no insertion was possible, then the associated heuristic (for example, LFD in 'Filler+LFD') is used. This forces a change in the problem state. It is important to remember that the trained XCS chooses deterministically, so that it is important for the problem state (if not the state description) to change each time, to prevent endless looping.

Table 4. The action representation

Action	Meaning, Use	Action	Meaning, Use
000	LFD	100	Filler + LFD
001	NFD	101	Filler + NFD
010	DJD	110	Filler + DJD
011	DJT	111	Filler + DJT

The alert reader might wonder whether the above problem state description in some way made heuristic selection an easy task. However, when each of our 14 original heuristics were evaluated, it was found that many cases where two problems had the same initial state description, but different algorithms, were the winners of the 14-way contest. For each of the 14 algorithms it was tried using a perceptron to see whether it was possible to classify problems into those on which a given algorithm was a winner and those on which it was not a winner. In every case, it was not possible, and therefore the learning task faced by XCS was not a trivial one.

6   The Experiments

The core of the application is Martin Butz' version of XCS [6,7,8], freely available over the web from the IlliGAL site. The reward scheme, including *when* and *what* to reward, lies in a level above the general XCS implementation. Some modifications were made for multi-step environment, which will be explained below, but the core remained the same, in an effort of providing more evidence of the usefulness and generality capabilities of XCS. The XCS parameters used were exactly as in [37], unless otherwise stated.

6.1   Set-Up

Training and Testing Sets: Each set of bin-packing problems is divided into training and test sets. In each case, the training set contained 75% of the problems; every fourth problem was placed in the test set. Since the problems come in groups of twenty for each type, the different sorts of problem were well represented in both training and test sets. All types of problems were combined into one large set of 890 problems that divided in training and test sets in the same manner. The reports below focus only on results by this combined collection, in which the training set has 667 problems and the test set has 223 problems. Other results obtained with the different types of problems have been omitted due to space limitations. The combined set provides a good test of whether the system can learn from a very varied collection of problems.

Computational Effort: The experiments proceeded as follows. A limit of L for training cycles for XCS was set, where the values tried were $L = 10000, 15000, 20000, 25000, 30000$. During the learning phase, XCS first randomly chooses a problem to work on from the training set. One step, whether explore or exploit, usually corresponds to filling one bin (see below more about steps). In an explore step the action is chosen randomly, in an exploit step it is chosen according to the maximum prediction appropriate to the current problem state description. This is repeated until all the items in the current problem have been packed. A new random problem is then chosen. A cycle will always be considered as packing a single bin (some 'steps' can pack several bins). Clearly, a large problem such as one of the u1000_M will consume a great number of cycles.

Recording the Results: The best result obtained on each problem during this training phase is recorded. Remember, however, that training continues, so the rule set may change after such best result was found. In particular, the final rule set at the end of the training phase might not be able to reproduce the best result ever recorded on every problem. Note as well that, during training, there is some exploration that uses random choices. Nevertheless, it is reasonable to record the best result ever found during (rather than at the end of) training on each problem also, because these are still reproducible results, by re-running the training with the same seed, and easily so. That best result obtained during training will be termed as TRB. It represents a potential achievable result for the hyper-heuristics.

At the end of training, the final rule set is used on every problem in the same training set to assess how well this rule set works. This is done using exploitation all the time so the result may be worse than the aforementioned best result recorded during the training process. This reproducible training result will be referred to as TRR.

Of course, the classifier obtained is also applied to every problem in the test set to measure the generalisation capabilities of the process. These test results will be shown under the label TST.

In summary, there would be two values for the training problems set: best ever explored (TRB) and the best recalled (TRR). Likewise, there is one value for the testing problems set (TST). All results shown are averages over ten runs with different seeds, unless stated otherwise.

In previous work [30,27,33] comparisons where made not only against the best results reported in literature, but also with the best result obtained by any of the constituent heuristics used. This makes sense, as a fair measure of the synergistic effect of hyper-heuristics is to compare it with the best of all of its individual components for each problem. However, since results of these comparisons were excellent –always above 98% and for the extra 2%, only one extra bin was used–, demonstrating the convenience of using hyper-heuristics over single heuristics has been set aside in this work. So in what follows, only comparisons against *best results reported in literature* will be provided.

Rewards: The reward earned is proportional to how efficiently a bin was packed. For example, if a bin is packed to 94% of its capacity, then the reward earned is 0.94. Remember that 'packing' here means continuing to the point where the heuristic would switch bins, rather than optimally packing. A reward of 1.0 is paid for packing the final bin. Otherwise, an algorithm which, say, placed the final item of size 1 in a final bin in order to complete the packing, would earn only 0.01. The filler is rewarded in a slightly different way; it is rewarded in proportion to how much it reduces the empty space in the open bins.

Random Algorithm for Comparison: The problems where also solved using an algorithm that chooses randomly a heuristic to be applied. This is done to double check that the algorithm is learning indeed. The results are also averaged over 10 seeds as well. However, the random choice of the heuristic solved optimally only 50.4% and 54.9% of the problems on the training and test sets respectively, and the HH reaches over 80%. Since the value is so low, it is not included in the following figures. But note that the "random" word that appears in the results shown later, refers to random size of multi-step chains, not to the random algorithm; this will be explained.

6.2   Alternatives Explored

In the experiments performed, the focus was directed towards several aspects considered interesting to study. These are as follows:

Step by State/Action: In [30] one step is defined as packing one bin (LFD was modified to pack only one bin and stop). From deeper study of the sequences of actions applied and the status of the problems, it was clear that in many cases the filling of a single bin does not change the status itself, given the binary representation proposed. It is easy to see that, as what is represented is ranges, and not absolute values, slight changes may not be reflected immediately in a change of status. As an alternative, one step can be considered to be taking a 'single action' (the repetition of the same single bin packing action) until there is a change of state.

The reward is the average of the different rewards obtained in each action of packing a single bin, and is only applied once at the change of state. For example, if four actions of type 2 were applied to perform a single step by state $-2222-$ and each action has rewards of, say: 990, 560, 230, and 870, the reward given to the action set (the rules that suggested the action) would be 990+560+230+870 = 2650/4 = 662.5, and it will be rewarded only once.

Single-Step and Multiple-Step Environment: As mentioned, in single step environments, rewards are available at every step (being this step delimited by either a change of state or action). A multi-step environment allows to consider sequences of steps and be rewarded together as a chain. This means that the reward can be paid after performing a number of complete states, or individual actions (depending if step by state is used or not, respectively). In these cases is likely that several and different actions are applied in one multi-step. Several multi-step chain sizes have been tried, namely chains of size 2, 5 and 10 actions. In addition, a random length from 2 to 10, but biased towards small numbers (in particular the smallest of 3 equally probable random numbers between 2 and 10), was also tested. The results on the following tables refer to this random size of chain length, and should not be confused with the aforementioned result of a Random choice of heuristics.

Combining Single-Step and Multi-Step, with Step by Action or by State: Combining single or multiple step with step by action or state can be confusing. To clarify the procedure, let's consider the following example:

Given a hypothetical sequence of actions (dash represent a change in state) as follows:

$$-\underbrace{222}_{S_1}-\underbrace{11}_{S_2}-\underbrace{3333}_{S_3}-\underbrace{2}_{S_4}-\underbrace{111}_{S_5}-\underbrace{1}_{S_6}-$$

Here, there are 6 different states and 14 different actions of 3 types. Each of the 14 actions produces a potential reward r_i based, as mentioned, in how efficient was the packing of a single bin.

In the sequence below, underlines will represent where the reward pointed by the upwards arrow is applied. Overbraces will represent how the rewards to be used are obtained (by averaging). If Single Step and Step by Action is used, then each single one is rewarded with its own action reward value r_i, the reward routine is therefore called 14 times, as shown below:

$$\overbrace{}^{r_1}\ \overbrace{}^{r_2}\ \overbrace{}^{r_3}\ \overbrace{}^{r_4}\ \overbrace{}^{r_5}\ \overbrace{}^{r_6}\ \overbrace{}^{r_7}\ \overbrace{}^{r_8}\ \overbrace{}^{r_9}\ \overbrace{}^{r_{10}}\ \overbrace{}^{r_{11}}\ \overbrace{}^{r_{12}}\ \overbrace{}^{r_{13}}\ \overbrace{}^{r_{14}}$$

r_1	r_2	r_3	r_4	r_5	r_6	r_7	r_8	r_9	r_{10}	r_{11}	r_{12}	r_{13}	r_{14}
2	2	2	1	1	3	3	3	3	2	1	1	1	1
↑	↑	↑	↑	↑	↑	↑	↑	↑	↑	↑	↑	↑	↑
r_1	r_2	r_3	r_4	r_5	r_6	r_7	r_8	r_9	r_{10}	r_{11}	r_{12}	r_{13}	r_{14}

If Single Step and Step by State is used, then the rewards are averaged for each state, and rewards applied only once per state, therefore rewards were given 6 times, as follows:

R_1	R_2	R_3	R_4	R_5	R_6
(222)	(11)	(3333)	(2)	(111)	(1)
↑	↑	↑	↑	↑	↑
R_1	R_2	R_3	R_4	R_5	R_6

$$R_1 = \frac{r_1+r_2+r_3}{3} \qquad R_2 = \frac{r_4+r_5}{2}$$
$$R_3 = \frac{r_6+r_7+r_8+r_9}{4} \qquad R_4 = r_{10}$$
$$R_5 = \frac{r_{11}+r_{12}+r_{13}}{3} \qquad R_6 = r_{14}$$

If a Multi-Step (of, for example, chain size 2) and Step by Action is used, then each action is rewarded with the average of its multiple step. In this example it means that the two rewards of the multi-step are averaged, and then the result is awarded to each component individually. So there are 7 different reward values, but these are given 14 times (because there are 14 actions), as represented here:

R_1		R_2		R_3		R_4		R_5		R_6		R_7	
(2	2)	(2	1)	(1	3)	(3	3)	(3	2)	(1	1)	(1	1)
↑	↑	↑	↑	↑	↑	↑	↑	↑	↑	↑	↑	↑	↑
R_1	R_1	R_2	R_2	R_3	R_3	R_4	R_4	R_5	R_5	R_6	R_6	R_7	R_7

$$R_1 = \frac{r_1+r_2}{2} \quad R_2 = \frac{r_3+r_4}{2} \quad R_3 = \frac{r_5+r_6}{2} \quad R_4 = \frac{r_7+r_8}{2}$$
$$R_5 = \frac{r_9+r_{10}}{2} \quad R_6 = \frac{r_{11}+r_{12}}{2} \quad R_7 = \frac{r_{13}+r_{14}}{2}$$

Finally, if Multi-Step (size 2) and Step by State is used, then rewards are calculated as the averages of the rewards of the states (already averaged over its actions as explained in Step by State), and awarded only once per state. Therefore there are 3 different reward values, given 6 times:

	R_1			R_2			R_3	
M_1		M_2	M_3		M_4	M_5		M_6
(222)		(11)	(3333)		(2)	(111)		(1)
↑		↑	↑		↑	↑		↑
R_1		R_1	R_2		R_4	R_3		R_3

$$M_1 = \frac{r_1+r_2+r_3}{3} \quad M_2 = \frac{r_4+r_5}{2} \qquad M_3 = \frac{r_6+r_7+r_8+r_9}{4}$$
$$M_4 = r_{10} \qquad M_5 = \frac{r_{11}+r_{12}+r_{13}}{3} \quad M_6 = r_{14}$$
$$R_1 = \frac{M_1+M_2}{2} \qquad R_2 = \frac{M_3+M_4}{2} \qquad R_3 = \frac{M_5+M_6}{2}$$

A keen reader may realise that if the step is by action, it is not necessarily true, that the sequences of actions in the same state are of the same type, at least during training. This is so because in an exploration cycle, a random action is chosen regardless of what the classifier may be suggesting. Nevertheless, this does not invalidate the example, as it is easy to see, that when step by action is used, each action executed is finally rewarded individually. On the other hand, when using step by state the classifier is only consulted once for each time the state changes, no matter how many actions are required. Therefore, in step by state, all the actions for that state are of the same type. Likewise the classifier is rewarded only once (with the averages shown above) for the whole group of actions.

Exploration vs Exploitation Scheme: The results reported in [30] used a 50/50 explore/exploit ratio. In the present work it was considered interesting to try an alternative schema. It is simply lineally decreasing from 100/0 explore/exploit ratio at the beginning of the search, to a 0/100 explore/exploit at the end. This ratio is the epsilon parameter of the XCS implementation of Martin Butz.

6.3   Results

Exploration vs Exploitation Scheme: Figure 1 shows average runs with a multi-step of size 2 and with step by state for the two schemes for the epsilon value, namely decreasing or 50/50, drawn against the number of training steps. There is a rather consistent but slight advantage, when less training steps are allowed, of the 50/50 scheme over the decreasing one. This seems to point to the fact that very early in the training, some general rules are learnt, and its exploitation during these first stages has, in general, a positive effect. The bin packing problem is sensible to inappropriate early allocations, no matter how efficiently the last bins are packed. Excess of randomness usually produces poor results. This can have dramatic effects when trying to learn long chains. Nevertheless, as more training is allowed, the differences become less significant within the same chain length. Interestingly, as it will be seen later, the 50/50 scheme, when allowed enough time, seems to have a synergistic effect with step by state.

Computational Effort: Figure 2 shows average runs with exploration/ exploitation in decreasing mode and step by state for steps of size 1 and 2 drawn against the number of training cycles. As it can be expected, the more cycles allowed to the classifier, the better the results. Nevertheless, there are differences between the best ever obtained while training (TRB) and what can be later recalled (TRR) on the same training problems (see section 6.1). While the former is obtaining better and better results, the latter seems to level near the 80%. This points to the fact that it seems that although the heuristics can be combined to achieve the results given by TRB values, the classifier, in exploiting mode (TRR), is unable to reproduce them later. It seems that the representation of the state is too poor to provide enough information about situations where other

Chain Size 2, Step by State

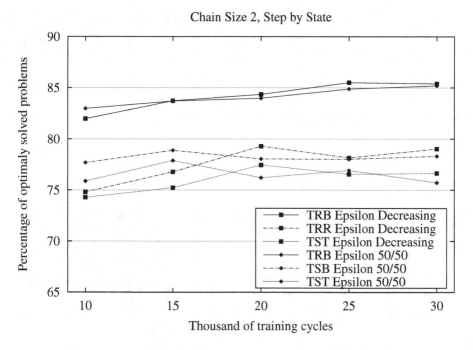

Fig. 1. Exploration vs Exploitation

actions are more fruitful. Remember that in exploiting mode the same action is performed until the state representation changes, as expected of a deterministic algorithm. There is sufficient evidence of the learning capabilities of XCS, even within the limitations of its binary representation, so research was conducted to evaluate the effects of using more information for the state representation [29,28] and with alternative learning methods using real numbers [27]. There was a substantial increase in performance, reaching values close to the best of training. This seems to support the idea that the state representation used can be improved and that XCS will then be provided with enough information to properly discriminate. Experiments with extended status representations using XCS are proposed as further research.

Single step seems to have a ceiling of performance at around 78% on TRR values (on the averaged results). This ceiling is reached with just 10000 cycles, except with Epsilon Decreasing and Step by Action, which needs 25000 cycles. More cycles do not improve the results significantly. Figure 3 shows, for a 50/50 Epsilon schema and step by state, the effect on performance for several different training efforts with regards to the length of the chains used in single and multi-step environments. There, the small impact of increasing the number of cycles on single step environments can be seen. It seems to learn fast what it can learn. But that is below what multi-step environments can achieve, albeit with more computational effort, as will be seen next.

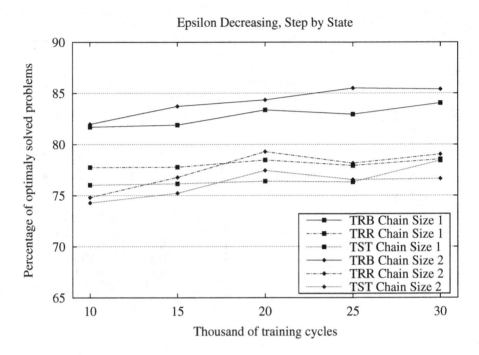

Fig. 2. Computational Effort

Step by Action vs Step by State: Figure 4 shows average runs with exploration/exploitation in decreasing mode and 25000 training cycles, using either step by action or step by state, drawn against the size of the multiple step used. Figure 5 shows the same, but using an exploration/exploitation scheme of 50/50. The most striking difference when comparing the type of step used is that TRB values are much better when step by state is used. This provides more evidence of the harmful effect of too much randomness. Remember that step by state means the same action during a particular state while by action it means that there can be several random exploration attempts for that same state. Consistent use of an action under the same conditions (same state) seems to be useful indeed. XCS is very robust to take advantage of less exploration if the new rules are accurate enough.

On the other hand, state by action performs worse than step by state on TRR and TST as well. Remember that these are the results provided by the classifier once the learning has finished. This is to be expected, as it makes easier the reinforcement learning by providing more consistent and smooth rewards for the same position, making predictions more accurate. State by action would focus on same set of rules several times, until there is a change in state. As reward variability can be high due to the limited information provided to XCS, it seems to have disruptive effects on accuracy of predictions, and it appears to slow down the learning.

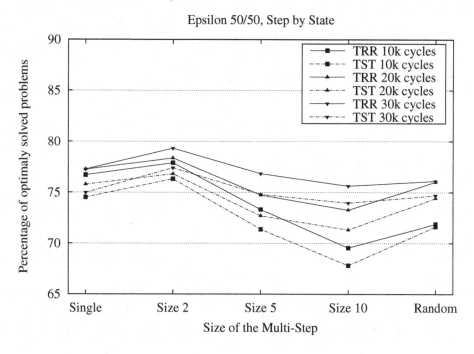

Fig. 3. Single vs Multiple step approach

Single vs Multiple Step Approach: Again, figures 4 and 5 show the effect on multiple step learning of both epsilon schemes and also the use of step by action and by state. In [33] early experiments with multiple step (size 2) and step by state provided promising results on its impact in hyper-heuristic bin packing using XCS. When, in this work, larger chains were tested, a decrease in performance was clearly revealed. While size 2 and step by state with 50/50 epsilon schema outperformed single step, it failed to generalise with longer chains. It needs many more training cycles to achieve closer results. Figure 3 shows how the longer the chain, the more training needed. Nevertheless, remember that in multiple step environments the rewards are usually given much less frequently than in single environment (or using action as the step). Training cycles are counted as bins packed, and not as rewards applied, and rightly so, as the bin packing procedure is much more costly than rewarding the classifier. It would be unfair to count rewards applied instead of bins packed, because it would hide the computational effort needed to achieve similar results.

On the other hand, as mentioned above, when the size of the chain to be learnt becomes larger, a more stable reward schema produces much better results and needs less training. Decreasing epsilon, with its many random choices, at the beginning and possibly insufficient exploration at the end, makes more difficult the learning of long chains. Martin Butz's 50/50 ratio seems a better schema, especially when combined with step by state. That effect is clear in Fig 5.

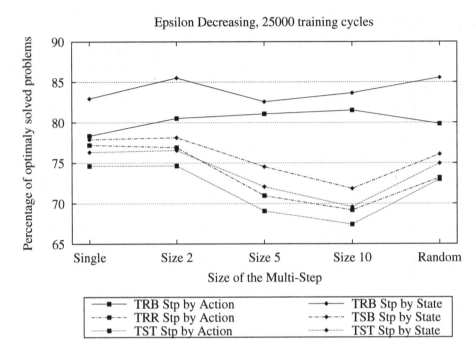

Fig. 4. Step by Action vs Step by State, Epsilon Decreasing

This result of a performance peak on size 2 chains while it decreases on longer chains is very interesting. It suggests that there must be pairs of actions that seem to be useful enough, one after the other, as to easily emerge after the reinforcement learning. This is so even though, as mentioned, the XCS receives less updates via rewards. Unfortunately, longer chains seem not to be as useful for the bin packing problem as to easily overcome this decrease in the application of rewards that the proposed multi-step environment require.

Of course, it may also be that 10000, the minimum number of cycles chosen, is enough for the size 2 chains to overcome the reduction in rewards applied, and that single environments have reached their ceiling some cycles before. But this then highlights the potential of multi-step environments, as the ceilings seem to be above single step ones. The issue then will be at which computational cost to achieve a gain. Here 10000 cycles is enough, under Epsilon 50/50 and Step by State (see fig. 3), for multi-step environments to start to show its potential.

Best Runs: While the results provided are averages, on several runs of the XCS training, to show the general behaviour expected, it is true that a single run is usually easily affordable (with respect to computer use) in applications that are not time-critical. This means that several runs with different seeds can be computed, and the best chosen as the final product. Figure 6 shows the best results (here only TRR and TST are shown) out of the 10 runs for Step by State, Epsilon 50/50 for the chain sizes 1, 2 and 10, drawn against the number

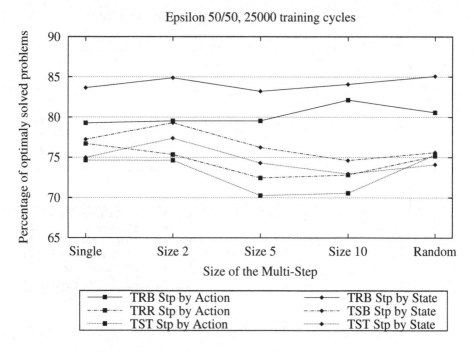

Fig. 5. Step by Action vs Step by State, Epsilon 50/50

of cycles allowed for training. Here it is more clear that longer chain schemes may catch up with shorter ones, but at the cost of much more training. This graph also hints that with the configuration used, the status chosen and the heuristics available, the best recallable XCS is at around the 82% mark, with 80% on the testing set. The best attainable ever found (that is, the best TRB result) reached an exceptional 86.1% of optimality.

6.4 Generalisation Properties

Although not previously mentioned, it seems clear from the figures that the results on the testing set where similar to those obtained in the training set, a good indication that the system is able to generalise well. Results of the exploit steps during training are very close to results using a trained classifier on new test cases. This means that particular details learnt (a structure of some kind) during the adaptive phase (when the classifier rules are being modified according to experience, etc.) can be reproduced with completely new data (unseen problems taken from the test sets). As an example, as already reported in [30], for one of Falkenauer's problems, DJD (and our DJT) produced a new best, and optimal result (this had already been reported in [11] where DJD was described). Even if this problem is excluded from the training set, the learnt rule set can still solve it optimally.

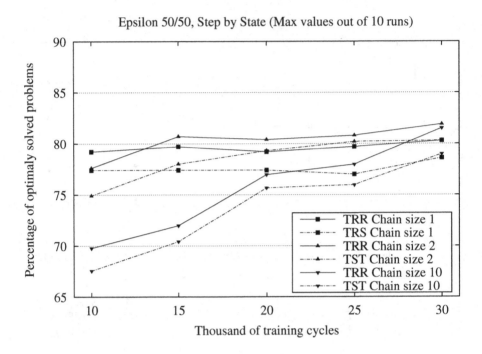

Fig. 6. Best results of some experiments

7   Conclusions and Further Work

This work has extended the experiments and applied some of the suggested enhancements of previous work [30,27,33] on applying hyper-heuristics to the bin packing problem using learning classifiers. Several reward schemes have been tested, in combination with alternate exploration/exploitation ratios, and several sizes and flavours of multi-step environments.

It will be appreciated that the scope of this work is broad. Attempts have been made to address several important issues, including narrowing the gap between the use of heuristic algorithms and evolutionary algorithms, and that of academic reports, and real-world applications. In summary, XCS was able to create a solution process that performed competitively well on a large collection of NP-hard, non-trivial problems found in literature. The system performed better than any of its constituent individual heuristics, and always performed better than the worst of the algorithms involved.

In particular, when using the heuristics independently, the best one of all (our own DJD version), achieved optimality in around 73% of the problems. However, the evolved XCS rule-sets were able to produce optimal solutions for over 78% of the problems (on unseen testing problems) in single-step environments, and over 81% in multi-step environments (also on unseen testing problems), while in the rest it produced a solution very close to optimal. The best attainable found was

86.1%, which indeed represents an exceptionally good result given the ingredients used, and a considerable improvement from previous work reported.

Experiments showed that multistep environments of small size, under the right conditions of step by state and a simple 50/50 exploration/exploitation ratio outperform single step results. Longer chain schemes need much more training, although it must be stressed that they receive less updating in the learning. This research also hints that schemes of even exploration/exploitation rates seem to help the XCS to learn long chains faster.

In addition, results seem to point out that multi-step environments can obtain better results than the single step environments could ever achieve. The ceiling, after which an increase of training cycles does not provide better results, appears to be higher in multi-step environments.

Throughout the paper some suggestions for further research have been outlined, including the following:

- The number of heuristics could be increased to give more options to the system. This can make learning harder, but could improve results a bit further.
- The number of available heuristics could be changed as conditions change. For example, some heuristics can be good at the first choices, while others are good finishers. Allowing different subgroups of heuristics in different general conditions can increase the effective options, while not burdening the system with many of them at the same time, as just a few would be active at a given moment.
- Adding more (or different) information could help XCS to discriminate more situations. This, again could result in more complex learning process, since each additional bit multiplies by 3 times the number of possible antecedents of the rules. Later work on related problems [29,28] have hinted that, for example, the percentage of items left, may not be as informative as expected, and that these bits could be used to provide other kind of information to the system.
- The hyper-heuristic explained here is constructive, meaning that it builds a solution step by step, until a complete solution is obtained. There is no tuning of that final solution. Hyper-heuristics, in principle, can also be designed to be used in a tuning process to be pipelined after the constructive one. Using the proposed approach of: "given a state, apply this heuristic", one must remember that statuses must eventually change, or endless looping would occur. As tuning is usually hill-climbing style, and decrease in performance is usually unaccepted, the suggested actions may not be applied and, if status is based just on the solution, no change is reflected. A simple way to avoid this would be to include a kind of *timer* in the status designed, which would force changes in the state, even if the solution is not being modified.
- One of the disadvantages mentioned of multi-step environments is that it may take much longer to achieve results similar to those achieved by shorter chain environments. A way to speed up the learning could be to create multi-action classifiers. It can be interesting to test learning classifiers with actions that would encode fixed sequences of actions instead of expecting them to emerge

from the reward schemes proposed for long chains of individual actions. They would move from point A to C without having to visit B, and making B to learn to go to C. Although this is, in fact, similar to revert to single environments, but giving them a higher variety of options. Besides, this is not as straightforward as it sounds, as it only can be applied to step by action. Remember that step by state can not foresee the number of actions to be taken needed to change the state. And of course, the idea behind step by state is, precisely, that each step taken changes the state.

Acknowledgements

Early developments of this work were funded by UK EPSRC research grant number GR/N36660 ending July 2004. The authors would like to thank Peter Ross and Emma Hart for their contributions during that period, and Stewart Wilson for useful comments and discussions.

Recent developments have been carried out by the authors sponsored by the Ramón y Cajal Research Program in the Intelligent Systems Research Group at the University of Murcia, Spain, and by Scottish Enterprise's Royal Society of Edinburgh Enterprise Fellowship, at the Centre of Intelligent Systems and their Applications, University of Edinburgh, Scotland.

References

1. http://www.ms.ic.ac.uk/info.html.
2. http://www.bwl.tu-darmstadt.de/bwl3/forsch/projekte/binpp/.
3. Lashon B. Booker. *Intelligent Behaviour as an Adaptation to the Task Environment.* PhD thesis, University of Michigan, 1982.
4. Larry Bull and Tim Kovacs, editors. *Foundations of Learning Classifier Systems*, volume 183 of *The Studies in Fuzziness and Soft Computing Series*. Springer-Verlag, Berlin, 2005.
5. Edmund Burke, Emma Hart, Graham Kendall, Jim Newall, Peter Ross, and Sonia Schulenburg. Hyper-heuristics: An Emerging Direction in Modern Search Technology. In Fred Glover and Gary Kochenberger, editors, *Handbook of Meta-heuristics*, chapter 16, pages 457–474. Kluwer, 2002.
6. Martin V. Butz. An Implementation of the XCS classifier system in C. Technical Report 99021, The Illinois Genetic Algorithms Laboratory, 1999.
7. Martin V. Butz. XCSJava 1.0: An Implementation of the XCS classifier system in Java. Technical Report 2000027, Illinois Genetic Algorithms Laboratory, 2000.
8. Martin V. Butz and Stewart W. Wilson. An Algorithmic Description of XCS. Technical Report 2000017, Illinois Genetic Algorithms Laboratory, 2000.
9. E.G Coffman, M.R. Garey, and D.S. Johnson. Approximation algorithms for bin packing: a survey. In D. Hochbaum, editor, *Approximation algorithms for NP-hard problems*, pages 46–93. PWS Publishing, Boston, 1996.
10. Kenneth A. DeJong. Learning with Genetic Algorithm: An Overview. *Machine Learning*, 3:121–138, 1988.
11. Philipp A. Djang and Paul R. Finch. Solving One Dimensional Bin Packing Problems. *Journal of Heuristics*, 1998.

12. Emanuel Falkenauer. A new representation and operators for genetic algorithms applied to grouping problems. *Evolutionary Computation*, 2(2):123–144, 1994.
13. Emanuel Falkenauer. A hybrid grouping genetic algorithm for bin packing. *Journal of Heuristics*, 2:5–30, 1996. http://citeseer.nj.nec.com/falkenauer96hybrid.html.
14. Emanuele Falkenauer. A Hybrid Grouping Genetic Algorithm for Bin Packing. Working Paper IDSIA-06-99, CRIF Industrial Management and Automation, CP 106 - P4, 50 av. F.D. Roosevelt, B-1050 Brussels, Belgium, 1994.
15. Michael R. Garey and David S. Johnson. *Computers and Intractability: a Guide to the Theory of NP-Completeness*. Freeman, 1979.
16. I. P. Gent. Heuristic Solution of Open Bin Packing Problems. *Journal of Heuristics*, 3(4):299–304, 1998.
17. David E. Goldberg. *Genetic Algorithms in Search, Optimization and Machine Learning*. Addison-Wesley, Reading, MA., 1989.
18. John H. Holland and J. S. Reitman. Cognitive Systems Based on Adaptive Algorithms. In D. A. Waterman and F. Hayes-Roth, editors, *Pattern-Directed Inference Systems*. Academic Press, New York, 1978. Reprinted in: Evolutionary Computation. The Fossil Record. David B. Fogel (Ed.) IEEE Press, 1998. ISBN: 0-7803-3481-7.
19. D.S. Johnson. *Near-optimal bin-packing algorithms*. PhD thesis, MIT Department of Mathematics, 1973.
20. Sami Khuri, Martin Schutz, and Jörg Heitkötter. Evolutionary heuristics for the bin packing problem. In D. W. Pearson, N. C. Steele, , and R. F. Albrecht, editors, *Artificial Neural Nets and Genetic Algorithms: Proceedings of the International Conference in Ales, France, 1995*, 1995.
21. Pier Luca Lanzi, Wolfgang Stolzmann, and Stewart W. Wilson, editors. *Learning Classifier Systems: From Foundations to Applications*, volume 1813 of *Lecture Notes in Artificial Intelligence*. Springer-Verlag, Berlin, 2000.
22. Pier Luca Lanzi, Wolfgang Stolzmann, and Stewart W. Wilson, editors. *Advances in Learning Classifier Systems*, volume 1996 of *Lecture Notes in Artificial Intelligence*. Springer-Verlag, Berlin, 2001.
23. Pier Luca Lanzi, Wolfgang Stolzmann, and Stewart W. Wilson, editors. *Advances in Learning Classifier Systems*, volume 2321 of *Lecture Notes in Artificial Intelligence*. Springer-Verlag, Berlin, 2002.
24. Pier Luca Lanzi, Wolfgang Stolzmann, and Stewart W. Wilson, editors. *Advances in Learning Classifier Systems*, volume 2661 of *Lecture Notes in Artificial Intelligence*. Springer-Verlag, Berlin, 2003.
25. Silvano Martello and Paolo Toth. *Knapsack Problems. Algorithms and Computer Implementations*. John Wiley & Sons, 1990.
26. Zbigniew Michalewicz. *Genetic Algorithms + Data Structures = Evolution Programs*. Springer-Verlag, Berlin Heidelberg New York, third edition, 1996.
27. Peter Ross, Javier Marín-Blázquez, Sonia Schulenburg, and Emma Hart. Learning a procedure that can solve hard bin-packing problems: A new ga-based approach to hyper-heuristics. In E. Cantú-Paz, J. A. Foster, K. Deb, D. Davis, R. Roy, U.-M. O'Reilly, H.-G. Beyer, R. Standish, G. Kendall, S. Wilson, M. Harman, J. Wegener, D. Dasgupta, M. A. Potter, A. C. Schultz, K. Dowsland, N. Jonoska, and J. Miller, editors, *Genetic and Evolutionary Computation Conference - GECCO 2003*, volume 2 of *Lecture Notes in Computer Sscience*, pages 1295–1306, Chicago, IL, USA, 2003. Springer.

28. Peter Ross and Javier G. Marín-Blázquez. Constructive hyper-heuristics in class timetabling. In *CEC 2005*, pages 1493–1500, Edinburgh, Scotland, 2-5 September 2005. IEEE Press.

29. Peter Ross, Javier G. Marín-Blázquez, and Emma Hart. Hyper-heuristics applied to class and exam timetabling problems. In *CEC 2004*, pages 1691–1698, Portland, Oregon, 20-23 June 2004. IEEE Press.

30. Peter Ross, Sonia Schulenburg, Javier G. Marín-Blázquez, and Emma Hart. Hyper-heuristics: learning to combine simple heuristics in bin packing problems. In *Genetic and Evolutionary Computation Conference - GECCO 2002*, pages 942–948, New York, NY, USA, 2002. Morgan Kauffmann Publishers.

31. Stuart J. Russell and Peter Norvig. *Artificial Intelligence: a modern approach*, chapter 4: Informed Search and Exploration, pages 116–120. Prentice Hall, Upper Saddle River, N.J., 2nd international edition edition, 2003.

32. Armin Scholl and Robert Klein. Bison: A fast hybrid procedure for exactly solving the one-dimensional bin packing problem. *Computers and Operations Research*, 1997.

33. Sonia Schulenburg. A Hyper-Heuristic Approach to Single and Multiple Step Environments in Bin-Packing Problems. In Pier Luca Lanzi, Wolfgang Stolzmann, and Stewart W. Wilson (workshop organisers), editors, *5th International Workshop on Learning Classifier Systems (IWLCS) 2002*. Unpublished contribution, presented, Granada, Spain, September 7-8, 2002.

34. Wolfgang Stolzmann, Pier Luca Lanzi, and Stewart W. Wilson (guest editors). Special Issue on Learning Classifier Systems. In *Journal of Evolutionary Computing*, volume 3. 2003.

35. Hugo Terashima-Marín, Peter Ross, and Manuel Valenzuela-Rendón. Evolution of Constraint Satisfaction Strategies in Examination Timetabling. In *GECCO-99: Proceedings of the Genetic and Evolutionary Computation Conference*, pages 635–642. Morgan Kaufmann, 1999. Early hyper-heuristics.

36. Stewart W. Wilson. Classifier Systems and the Animat Problem. *Machine Learning*, 2:199–228, 1987.

37. Stewart W. Wilson. Classifier Systems Based on Accuracy. *Evolutionary Computation*, 3(2):149–175, 1995.

38. Stewart W. Wilson and David E. Goldberg. A Critical Review of Classifier Systems. In J. D. Schaffer, editor, *Proceedings of the Third International Conference on Genetic Algorithms*, pages 244–255. Morgan Kauffmann, 1989.

Adaptive Value Function Approximations in Classifier Systems

Lashon B. Booker

The MITRE Corporation
7515 Colshire Drive
McLean, VA 22102-7508, USA
booker@mitre.org

Abstract. Previous work [1] introduced a new approach to value function approximation in classifier systems called *hyperplane coding*. Hyperplane coding is a closely related variation of tile coding [13] in which classifier rule conditions fill the role of tiles, and there are few restrictions on the way those "tiles" are organized. Experiments with hyperplane coding have shown that, given a relatively small population of random classifiers, it computes much better approximations than more conventional classifier system methods in which individual rules compute approximations independently. The obvious next step in this line of research is to use the approximation resources available in a random population as a starting point for a more refined approach to approximation that re-allocates resources adaptively to gain greater precision in those regions of the input space where it is needed. This paper shows how to compute such an adaptive function approximation.

1 Introduction

Considerable attention has been paid to the issue of value function approximation in the reinforcement learning literature [13]. One of the fundamental assumptions underlying algorithms for solving reinforcement learning problems is that states and state-action pairs have well-defined values that can be approximated and used to help determine an optimal policy. The quality of those approximations is a critical factor in determining the success of many algorithms in solving reinforcement learning problems.

One approach to improving approximation quality in classifier systems is to increase the computational abilities of individual rules so that they become more capable function approximators (e.g., [16,3]). Another idea is to look back to the original concepts underlying the classifier system framework and seek to take advantage of the properties of distributed representations in classifier systems [2]. This paper follows in the spirit of the latter approach, looking for ways to tap the distributed representational power present in a collection of rules to improve the quality of value function approximations.

Previous work [1] introduced a new approach to value function approximation in classifier systems called *hyperplane coding*. Hyperplane coding is a closely

X. Llorà et al. (Eds.): IWLCS 2003-2005, LNAI 4399, pp. 219–238, 2007.

related variation of tile coding [13] in which classifier rule conditions fill the role of tiles, and there are few restrictions on the way those "tiles" are organized. The basic idea is to treat rules as features that collectively specify a linear gradient-descent function approximator. The hypothesis behind this idea is that classifier rules can be more effective as function approximators if they work together to implement a distributed, coarse-coded representation of the value function.

Experiments with hyperplane coding have shown that it computes much better approximations than more conventional classifier system methods in which individual rules compute approximations independently. The results to date also demonstrate that hyperplane coding can achieve levels of performance comparable to those achieved by more well-known approaches to function approximation such as tile coding. High quality value function approximations that provide both data recovery and generalization are a critically important component of most approaches to solving reinforcement learning problems. Because hyperplane coding substantially improves the quality of the approximations that can be computed by a classifier system using relatively small populations of classifiers, it may provide the foundation for significant improvements in classifier system performance.

One open question remaining about hyperplane coding is how the quality of the approximation is affected by the set of classifiers in the population. A random population of classifiers is sufficient to obtain good results. Would a more carefully chosen population do even better? The obvious next step in this research is to use the approximation resources available in a random population as a starting point for a more refined approach to approximation that re-allocates resources adaptively to gain greater precision in those regions of the input space where it is needed. This paper shows how to compute such an adaptive function approximation. The goal is learn a population of classifiers that *reflects the structure of the input space* (Dean & Wellman, 1991). This means more rules (i.e. more tiles) should be used to approximate regions which are sampled often and in which the function values vary a great deal. Fewer rules should be used in regions which are rarely sampled and in which the function is nearly constant. We discuss how to adaptively manage the space in the population in a way that achieves this goal.

2   Value Function Approximations

Given a decision policy π, most approaches to solving reinforcement learning problems compute and store some explicit representation of the value function V_π. For very simple problems, a lookup table is an adequate way to represent the value function. In most cases of interest, however, the input space is too large to represent V_π exhaustively in tabular form so the function must be represented more compactly. Efficient storage is not the only important issue though. In a large state space the learning agent will only directly experience a relatively small number of inputs. The agent nevertheless needs to leverage that experience to determine how to behave when it encounters inputs that have not been seen before. This implies that generalization is a key issue for reinforcement learning problems with large state spaces. The most common approach to addressing these issues is to use function approximation techniques to compute a compact representation of V_π that generalizes well.

The approach to approximating V_π used in learning classifier systems belongs to a class of techniques known as soft state aggregation [11]. In the simplest forms of state aggregation, the states are partitioned into a set of disjoint groups or clusters. A reinforcement learning problem can be solved at the cluster level to compute a value function for the clusters. The value of a cluster is then used as the value for each of the states in that cluster. Soft state aggregation techniques allow a single state to belong to more than one cluster, providing for cluster overlap. This is accomplished by defining cluster probabilities $P(x|i)$ that specify the degree to which state i is associated with cluster x. The value for a state is given by a weighted average of the values of the clusters the state is associated with; that is,

$$V_\pi(i) = \sum_x P(x|i)V_\pi(x)$$

Rule input conditions designate the clusters of states used by learning classifier systems. Each condition represents a set of states whose value is summarized in various ways by the rule's utility measure. In XCS, for example, a cluster's value is represented by the prediction parameter of the corresponding rule. The cluster probabilities are given by the rule's fitness divided by the sum of the fitnesses of all the rules matching state i.

While state aggregation approaches to function approximation can be useful in some settings, they are known to have serious shortcomings [13]. First, they tend to scale poorly as the number of dimensions of the state space increases. Second, large numbers of clusters may be needed to represent smooth functions accurately. The most widely used approaches to function approximation for reinforcement learning avoid these problems by relying on linear gradient-descent methods.

The remainder of this section takes a brief look at linear gradient-descent methods and one important special case that uses binary features.

2.1 Linear Approximations and Coarse Coding

Linear gradient-descent methods for value function approximation begin with a linearly parameterized representation of the value function given by

$$V(x_t) = \sum_i w_i(t)\phi_i(x_t)$$

where the ϕ_i are features defined on the state space and the w_i are real-valued adjustable weight parameters. The weights are adjusted to try to reduce the error on the observed sample points x, and to generalize from that data to provide good approximations for other points that have not yet been seen.

Gradient-descent methods try to minimize error by adjusting the weights on each step in the direction that reduces error the most. In the linear case, the gradient descent update for adjusting the weights is given by

$$w_i(t+1) = w_i(t) + \alpha[v(t) - V(x_t)]\nabla_{w_i}V(x_t)$$

where $\nabla_{w_i}V(x_t) = \phi_i(x_t)$ is the gradient of the linear function with respect to weight parameter w_i and $v(t)$ is the true function value for x_t.

Linear gradient-descent methods are simple and they are particularly well-suited to reinforcement learning [13]. A key aspect determining how well these methods work in practice, though, is the quality of the features they use. The features must represent whatever task-relevant qualities of the state may be needed to discriminate one state from another, as well as any associated feature interactions that may be important.

2.2 Tile Coding

Coarse coding [8] is a general approach to defining a set of adequate features. In this form of representation, each feature corresponds to some subset of the state space (the feature's "receptive field"). For a given state, a feature is said to be activated if the state belongs to that receptive field. The representation of state is coarse coded in the sense that the receptive fields overlap to produce a distributed representation whose acuity is proportional to the number of features activated in a given state. One general-purpose way to define receptive fields suitable for efficient on-line learning is called tile coding [13].

Tile coding is a particular form of coarse coding in which the receptive fields for all features are organized into exhaustive partitions of the input space. The features are assumed to be binary, the receptive fields are called *tiles*, and each partition is called a *tiling*. The tilings are offset from each other in order to achieve the overlap needed for local generalizations. For a single input dimension, the offsets typically used in tile coding are given by $i(w/n)$ where i is the index of the tiling, w is the tile width, and n is the number of tilings ($0 \leq i < n$). This concept is illustrated in Figure 1.

There are several advantages to organizing the receptive fields in this way. Every point in the input space activates the same number of tiles, so there is strict control over the density of tiles and the resulting precision of the approximation. It is also easy to set the learning rate for a linear gradient-descent function approximator based on tile coding. Since the number of features active for each point is equal to the number of tilings m, the learning rate can be expressed intuitively as a fraction of the rate $1/m$ which gives exact one-trial learning. The weight update for activated features is given by

$$w_i(t+1) = w_i(t) + \frac{\alpha}{m}[v(t) - V(x_t)]$$

where α is the desired fraction.

Tile coding has been been used extensively for reinforcement learning, and the overall coarse coding approach is known to be capable of computing high quality

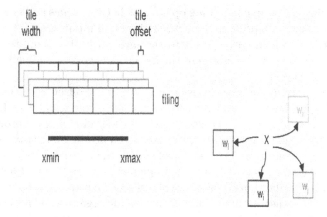

Fig. 1. Tile coding for a 1-dimensional input space. Every point x activates the same number of tiles.

approximations [10]. Recent work [1] has shown that there are aspects of the tile coding approach that can be leveraged to improve the value function approximations computed in classifier systems. That work introduced a new variation of tile coding called hyperplane coding. Initial experiments with hyperplane coding have shown that by carefully using the resources available in a random population of classifiers, continuous value functions can be approximated with a high degree of accuracy. The next section provides a brief overview of hyperplane coding, along with empirical results showing how its performance compares to other approaches.

3   Hyperplane Coding

Hyperplane coding is a closely related variation of tile coding in which classifier rule conditions fill the role of tiles, and there are few restrictions on the way those "tiles" are organized. The hypothesis behind this idea is that classifier rules can be more effective as function approximators if they collectively implement a distributed representation of the value function. The distributed representation is realized by treating individual rules as features rather than as independent function approximators whose estimates are pooled to compute an overall result.

3.1   Overview

The coarse coding idea requires the ability to represent patterns of contiguous inputs (the tiles arranged in a tiling) that can be offset from each other by arbitrary amounts. This requirement is trivial to fulfill in tile coding. The tiles are fixed sized intervals in each dimension, and the interval endpoints can be adjusted as needed. The standard syntax for the input condition of a classifier rule does not provide this kind of flexibility. It is not clear how to adjust that syntax to represent hyperplanes offset by arbitrary amounts in input space, while preserving the simple matching operations between rules and messages. For example, it is easy to represent the lower half of the input range $[0, 1]$ with the condition 0#...#, which corresponds to the interval $[0, 0.5]$ using the standard binary encoding. How do we represent the hyperplane corresponding to the offset interval $[0 + \epsilon, 0.5 + \epsilon]$?

One obvious way to manage this issue is to apply the offset to the input space, then define hyperplanes on that transformed space in the usual way. Looking at the offset interval $[0 + \epsilon, 0.5 + \epsilon]$ again, we can determine if some input value x belongs to that interval by checking if a message encoding the translated value $x - \epsilon$ matches the condition 0#...#. This leads to the following ideas for the way a population of classifiers is organized to implement coarse coding. Each classifier is assigned to a specific tiling[1], just like each tile belongs to a specific tiling under tile coding. In this case, though, there is no specific organization imposed on the tiling. Continuing with the analogy, we do associate a fixed offset with each tiling. The classifier system operating principles are also adjusted somewhat. Instead of having a single message matched against all rules on each cycle, we generate a separate message for each tiling. Each message is computed from the raw input by applying the offset associated with the tiling in question.

The only remaining details needing attention have to do with tile width and offsets. Since hyperplanes in general do not correspond to simple contiguous regions of the input space, some thought must be given to the issue of how to define tile width. There are several possibilities and we choose one of the simplest. The width of the smallest possible contiguous region defined by a hyperplane is given by the resolution size used to discretize the raw input. The width of every contiguous region matched completely by some hyperplane is a multiple of this resolution size. The resolution size is the simplest tile width that makes sense for classifier conditions with a specific bit at the lowest order bit position, and those classifiers occupy a large fraction of a random population. Larger tile widths are possible for the remaining classifiers, however, and the increased overlap can improve the acuity of the overall approximation. Consequently, classifiers are organized into two types of features, coarse and fine, which are stored in two separate groups of tilings. The fine features are those classifiers with a specific bit at the lowest order bit position. These classifiers use the resolution size as the tile width. The remaining classifiers are all treated as coarse features, which use a tile width equal to twice the resolution size. This organization results in a system that, on each cycle, generates up to $4n$ potentially distinct messages where n is the number of tilings used in a comparable tile coding scheme.

As noted previously, the offsets typically used in tile coding are given by $i(w/n)$ where i is the index of the tiling, w is the tile width, and n is the number of tilings $(0 \leq i < n)$. This translation scheme uses only positive offsets that translate tiles to the right. A point gets grouped with its neighboring points on the left when the adjacent tile on the left (that does not originally contain all

---

[1] We will call each major grouping of classifiers a tiling, even though the set does not partition the input space (i.e., the elements are not disjoint, and they may not span the entire space).

the points) gets translated to cover those points. This scheme does not work well in the classifier system setting, however. An unmodified input message matches classifiers representing the base tiles (i.e., tiling $i = 0$) covering a point x. If we adhere to the usual concept of a match set, the only way that x will be grouped into a tile with any other point is if the match set contains a classifier that matches both points. Offsets can change the groupings by excluding some points, but there is no way to include points that are not covered by the base match set. This makes it important to group the matched points in as many ways as possible. Accordingly, we use a more symmetric set of offsets given by $i(w/n)$ with $-n/2 \leq i < n/2$ so that points get grouped in both directions.

3.2   Basic Implementation

A basic implementation of linear function approximation based on hyperplane coding uses the concepts and principles described above in a simple, skeletal classifier system. This skeletal system has traditional ternary rules with no actions and no rule discovery mechanisms. On every step the system is presented with a data point x, and the reward received is the function value $f(x)$. The system forms a match set and proceeds to update the basic parameters for individual classifiers. This approach of using function values as rewards is commonly used to assess classifier system methods for function approximation [16].

Two other implementation details need to be mentioned. The first detail concerns the way inputs should be encoded. One of the important properties of approximation techniques like tile coding is that the generalizations they compute are localized. Points that are sufficiently close in input space will produce output values that are close. Moreover, values in widely spaced regions can be learned with relatively little interference. This property can be compromised somewhat with hyperplane coding since hyperplanes are not restricted to contain localized collections of points.

The Gray code is known to be a representation for bit strings that provides more localized collections of points [7]. In order to see why this is true, consider the classifier condition ##10. The bit strings matching that condition are 0010, 0110, 1010, and 1110. None of these points are contiguous under a binary coding. A binary reflected Gray code, however, groups these points into two clusters of consecutive points: (0010, 0110) and (1110, 1010). This example is illustrative of a more general phenomenon. A Gray code will never group bit strings matching some condition into more clusters of consecutive points than a binary code does. Furthermore, for some conditions, the Gray code will organize the points into fifty percent fewer clusters than the binary code (as in our simple example). See Faloutsos [7] for more details. This analysis suggests that inputs for the hyperplane function approximator should always be encoded using a binary reflected Gray code.

The second implementation detail concerns how to handle the issue of feature salience. Under tile coding, every point belongs to exactly one tile in every tiling. As noted previously, a point belongs to many tiles in each tiling under hyperplane coding. Because the hyperplanes in a tiling are so diverse, they may not all be equally useful for approximating the function. It might matter that some are more specific than others, some may correspond more closely to key regularities in the function, and so on. This presents the approximation algorithm with a feature selection problem that does not occur with tile coding. The problem is important because irrelevant features are a source of noise that can slow down learning of relevant features.

One way to address this feature selection problem is to use the frequency that features are activated to identify which rule-based features are most relevant in a given context. The idea is that features that are active across a large number inputs may over-generalize across a broad range of function values. The smaller the set of activating inputs, the less likely this will occur (generally speaking) and the more likely it is that the feature will be relevant to the approximation task. Classifier systems traditionally use specificity to measure the relevance of a rule condition in a given context. When inputs are uniformly distributed, specificity is directly correlated with frequency of activation. However, when the input distribution is not uniform, the activation frequency must be estimated from experience. Two parameters are used for this purpose. The experience parameter ε_i counts the number of times the feature ξ_i has been activated in a match set. The age parameter a_i records the time step when ξ_i was created. The activation frequency for ξ_i is therefore given by $\nu_i = \varepsilon_i/(t - a_i)$ where t is the current time step. The relevance of a feature is then estimated by the inverse function $1/\nu_i$. This relevance measure is used to change the weight update for activated features by replacing the factor α/m, which divides the step size α equally among all m features, with the term

$$\frac{\alpha}{\nu_i \sum_{j=0}^{m-1} \frac{1}{\nu_j}}$$

which biases the allocation of step size based on presumed relevance. This bias helps make learning and generalization more efficient[2].

3.3   Initial Empirical Results

Experiments have shown that hyperplane coding can provide high quality function approximations [1]. The test function suite for those experiments was taken from a set of functions proposed by Donoho and Johnstone [6] that has been widely used in the literature on statistical estimation and reconstruction of signals from data. Four one-dimensional functions were used — Blocks, Bumps, Doppler, and HeaviSine — that provide a good variety of spatial variability and smoothness (see definitions in the Appendix). The training data for each function was drawn from a set of 2048 equally spaced sample points. A separate distinct set of 2000 equally spaced sample points was set aside to use as a test set. Providing a separate set of data for testing is critically important. The

---

[2] The original implementation [1] used Sutton's [12] Incremental Delta-Bar-Delta (IDBD) method to adjust the learning rates for individual features. That algorithm works well, but the new approach based on activation frequency is just as effective in this setting and is much simpler to implement.

quality of a function approximator is not just determined by the size of the approximation error on the training data. Though that error is a good indication of how the function approximator performs data recovery, it tells us nothing about how well the approximator can generalize to new data. It also tells us very little about the quantity of approximation resources (e.g., the number of features) that was needed to achieve that level of performance. A good function approximator performs well on both data recovery and generalization, while minimizing the quantity of resources required.

The quality of an approximation was measured in terms of the average squared error at the sample points. More specifically, the performance measure was

$$R = n^{-1} \sum_{i=0}^{n-1} (\hat{f}(x_i) - f(x_i))^2$$

where $\hat{f}$ is the approximation and f is the true function. Performance was measured separately on the test data and the training data. In all of the experiments, learning proceeded over 100 trials with 10,000 steps per trial, and with a random data point selected from the training set on each step. This gave each function approximator ample time to converge to its most accurate output. Results were averaged over 10 replications, and statistical significance was assessed using a Student's t-test with significance level 0.05.

The experiments provided an empirical comparison of hyperplane coding with tile coding and with the widely used classifier system mechanisms in XCS [4] for predicting expected payoff[3]. The goal of the comparison was to assess how well each approach makes use of a fixed allocation of resources to approximate a value function. For tile coding this means that the number of tiles and the way they are organized was fixed. Each test function was approximated using 2048 grid-like tiles each having width 1/256. The tiles were organized into 8 tilings offset as described previously. The learning rate was specified by the assignment $\alpha = 0.2$. For the XCS prediction mechanism and for hyperplane coding, the population of classifiers was fixed at 2048 rules generated randomly using a probability of 1/3 for placing the # symbol at any given position in a rule condition. Each classifier condition was 8 bits, giving every classifier the same input resolution as one of the grid-like tiles.

The XCS prediction mechanism was implemented in one of the skeletal classifier systems described above. On every step the system forms a match set and proceeds to update the basic XCS parameters: experience, prediction, prediction error, and fitness. The system prediction is calculated in the usual way and that prediction becomes the system's estimate for the value of x. The parameter

---

[3] The XCS mechanism was chosen because it is the typical approach used to approximate value functions in classifier systems. Moreover, because each rule uses one adjustable parameter for the approximation, this mechanism can be directly compared to algorithms like tile coding and hyperplane coding that use the same amount of approximation resources. The issue of comparable approximation resources becomes problematic with alternative approaches like XCSF [16], which use more parameters and more computation per rule to compute approximations.

Table 1. Average square errors for XCS prediction, tile coding and hyperplane coding

Algorithm	Approximation Error							
	Blocks		Bumps		Doppler		HeaviSine	
	Train	Test	Train	Test	Train	Test	Train	Test
XCS prediction	3.0979	2.9054	23.841	24.793	2.0295	2.0252	0.10821	0.11192
Tiles	0.06988	1.7535	0.16809	0.93068	0.03579	0.08922	0.02327	0.06458
Hyperplanes	0.12150	0.29966	0.33694	0.99382	0.04145	0.07626	0.01473	0.02500

settings were consistent with those used for XCS in the literature [16]: learning rate 0.2, error threshold 0.2, fitness power 5.0, and fitness scale (i.e., α) 0.1. See Butz and Wilson [4] for details about these parameters and computations.

The linear function approximator based on hyperplane coding organized the population of 2048 random classifiers into 16 tilings of 128 classifiers each. Each classifier has a weight parameter w that is adjusted by gradient descent just as in tile coding. The learning parameter α for gradient decent was set to 0.2, again in agreement with the tile coding experiments. These choices gave the linear approximator based on hyperplane coding roughly the same amount of approximation resources to work with as the tile coding version had.

The overall results of those experiments are summarized in Table 1. All of the differences in performance between the XCS prediction and the other two approximation techniques are statistically significant. Tile coding and hyperplane coding are substantially more effective than the XCS prediction on these functions. Both tile coding and hyperplane coding show an impressive ability to reconstruct functions with respect to the training data. Their performance on the four test functions compares favorably with results on the same data achieved by more sophisticated approximation techniques like a discrete wavelet transform [6]. The reconstructions shown in Figure 2 show that these two methods have enough precision to pinpoint the location of abrupt changes in function values. Moreover, both techniques also have local generalization properties that are adequate enough to make the approximations fairly smooth.

The XCS prediction, on the other hand, does poorly from the standpoint of both precision and smoothness. There is a sense in which this is not surprising, since the mechanisms were intended to be used in combination with rule discovery to compute a good approximation of the value function. There is a dilemma with that arrangement, however. Rule discovery depends on guidance from the prediction computations in order to know what type of rules to generate. If that guidance is poor, then rule discovery will have to thrash around somewhat randomly until it discovers something that improves the approximation. Overall, the performance of the XCS mechanisms suggest that classifier systems typically do a poor job of approximating value functions as they solve reinforcement learning problems.

Tile coding has a statistically significant performance advantage over hyperplane coding on the training data for all of the test functions except HeaviSine, where hyperplane coding is far superior. Hyperplane coding has a statistically significant performance advantage on the test data for all of the test functions

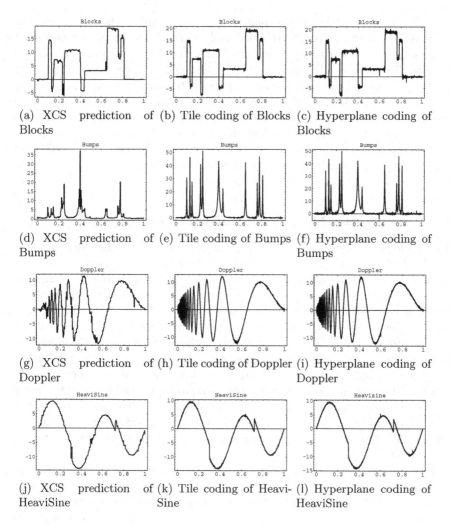

Fig. 2. Reconstructions computed by XCS prediction, tile coding, and hyperplane coding

except Bumps, where tile coding has a small but significant advantage. It appears that hyperplane coding offers an alternative for linear approximations that is comparable in performance to what can be achieved with a more conventional approach like tile coding.

4   Adaptive Hyperplane Coding

The hyperplane coding approach seems to offer more opportunities than tile coding for increasing precision without incurring significantly greater computational costs. The density of tiles in hyperplane coding is naturally higher than the density in tile coding. This contributes to more resolution in the final approximation. The precision of the approximation can also be increased by increasing the length of the classifier input conditions instead of by adding more tiles. Most importantly, the hyperplane coding scheme makes it possible to adapt the collection of tiles to achieve more precision. The obvious next step in this research is to use the approximation resources available in a random population as a starting point for a more refined approach to approximation that re-allocates resources adaptively to gain greater precision in those regions of the input space where it is needed. The remainder of this paper discusses how to adaptively manage the space in the classifier population in order to accomplish this objective.

4.1 Unsupervised Methods for Online Learning

The random population of classifiers used in hyperplane coding can be viewed as a kind of random representation that avoids the difficult questions about how to choose the "right" set of features by simply generating a sufficiently large variety of features. One important advantage of this approach to representation is that it is well-suited to online learning tasks. Sutton and Whitehead [14], for example, have shown that many approaches to online learning share a common two-layered structure. The first layer maps the input into a high-dimensional feature space. The second layer maps that feature space into the final answer using some simple supervised learning procedure like gradient decent. When a random representation is used in the first layer, unsupervised learning techniques can be very effective in adapting that representation online and improving the effectiveness of the overall system. Clearly, the combination of coarse-coded features with linear-gradient decent methods fits this characterization of a random representation.

This suggests that the most straightforward way to implement adaptive hyperplane coding is to use unsupervised methods to change various statistical properties of the random population. Sutton and Whitehead focused on reducing the size of a random representation by adapting the number of features active at one time and the frequency that any particular feature is activated. There are two related goals for adaptive hyperplane coding: to make sure minimal amounts of approximation resources are available throughout the input space; and, to provide more resources as needed in regions where the approximation error is high. One key statistical property relevant to both of these goals is the size of a match set.

A simple statistical argument provides a useful criterion for managing the size of a match set. Given an input message s, we can view a random population of N classifiers as if it were generated by a sequence of Bernoulli trials, each trial having probability p_m of producing a classifier matching s. This allows us to treat the size of the match set as a binomial random variable $\mathcal{M}$ with mean Np_m and variance $Np_m(1 - p_m)$. Assuming that specificity in the random classifiers is binomially distributed, then p_m is given by

$$p_m = \left(\frac{p_\# + 1}{2}\right)^l$$

where $p_\#$ is the probability of the # symbol appearing at any position in the condition, and l is the length of the condition. A reasonable range for the value of $\mathcal{M}$ in a random population should rule out values that are too large or too small. One computationally efficient way to specifying such a range is to simply use the interval containing values within one standard deviation of the expected value of $\mathcal{M}$.

Given this criterion for a desirable match set size as a starting point, adaptive hyperplane coding uses two mechanisms for implementing unsupervised learning: triggered rule creation and stochastic deletion. Learning triggers add a new random rule to a match set whenever some condition related to the input distribution becomes true. Various triggers are defined to manage the match set size as described below. The deletion mechanism is structured to minimize the impact that removing a feature will have on the approximation error. A feature has minimal impact on approximation error when its weight is small. Since the gradient descent procedure distributes responsibility for reducing error among all activated features, the features in large match sets will also tend to have a relatively small impact on the approximation error. We will consider a weight to be "small" if it is consistently smaller than the median weight in the match sets it belongs to. We will consider a match set "large" if it is more than one standard deviation larger than the expected match set size.

The deletion mechanism is accordingly defined as follows. Two additional parameters are defined for each feature. The match set size ς estimates the average size of the match sets a feature belongs to. The impact-below-average parameter ι estimates how frequently the absolute value of the feature weight is smaller than the median in the match sets the feature belongs to. These parameters are updated just like other typical classifier system parameters, using the standard Widrow-Hoff delta rule with the MAM technique [15] and learning rate $\beta = 0.05$. Deletion candidates are selected from those rules whose parameter ς is more than one standard deviation bigger than Np_m. Features are deleted stochastically, using a probability distribution derived from the quantity $\iota_i \varsigma_i$ for each candidate feature ξ_i. The next section describes experiments that empirically evaluate the performance of these mechanisms in an implementation of adaptive hyperplane coding.

5   Experiments

We evaluate the adaptive hyperplane coding approach described above by modifying the skeletal classifier system used in the previous experiments to include mechanisms for inserting and deleting rules online. This section presents a series of learning triggers that improve the performance of the function approximator. The experiments use the same parameter settings as the previous experiments

Table 2. Average square errors for variations of adaptive hyperplane coding on the Bumps function

| Algorithm | Approximation Error | |
Learning Trigger	Train	Test
Baseline	0.33694	0.99382
Minimal [M] size	0.33763	0.97809
[M] size for large errors	0.31676	0.95900
[M] size for outlier errors	0.28222	0.93383
Unbalanced tile offsets	0.26607	0.92007

with hyperplane coding, except that the following results are averaged over 20 replications. We initially focus our attention on the Bumps function, since tile coding out-performed hyperplane coding on both the test and training data for that function.

5.1 Minimal Match Set Size

The first learning trigger is designed to make sure that every match set contains some minimum amount of approximation resources. Whenever the size of a match set is less than one standard deviation below the expected value, a new random classifier matching the current input is generated. The new classifier is assigned to the tiling in this match set that contains the fewest number of existing classifiers. This learning trigger, like all subsequent triggers, is not operative until all classifiers in the match set have been activated at least $1/\beta$ times. By enforcing this restriction, each match set has an opportunity to absorb the effects of a new member before another new rule is inserted.

Maintaining a minimum level of resources in each match set should provide some increase in generalization performance on the test data. Table 2 shows that there is indeed some performance improvement on the test data, but it is not statistically significant. This first result is important, though, because it shows that the insertion and deletion mechanisms can operate online without adversely impacting the quality of the approximation.

5.2 Match Set Size for Large Errors

Once the function approximator has a minimal amount of resources to use for each input, it is reasonable to ask if the resources available at a particular point are adequate given the current quality of the approximation. An important goal for an adaptive function approximator is to use more resources in those input regions where approximation error is high. This section describes a simple error-driven trigger that adds more rules to a match set when the error is larger than expected.

The first step to defining the error-driven trigger is to identify which errors warrant the extra resources. The histogram in Figure 3 shows the distribution of squared errors at the end of a typical run of hyperplane coding on the training data for the Bumps function. The approximation error is small for the large

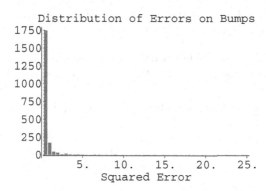

Fig. 3. Typical distribution of squared errors

majority of training points (out of 2048 points, 1746 points have squared error less than 0.5 and fall into the first bin.) It is also interesting to note that the error distribution has a long tail (the largest squared error is 24.75) These characteristics suggest that the points with the worst errors can be easily identified by using techniques for finding outliers in statistical data.

Since the exact properties of the error distribution are unknown, and the assumption of a normal distribution is clearly not justified, the most appropriate techniques to use here are those based on non-parametric statistics. The boxplot technique [9] is one widely used, robust approach to labeling outliers. This technique uses two measures computed from the data: the value of the 25th percentile (the lower quartile q_1), and the value of the 75th percentile (the upper quartile q_3). A box around the median is then specified by computing the interquartile range $q_3 - q_1$ and defining cutoff points (or *fences*) at $1.5(q_3 - q_1)$ above q_3 and the same distance below q_1. Any value that falls outside either fence is considered more extreme than "expected". This approach is effective in part because the interquartile range is a measure of variability that focuses on the spread of the data values near the median and ignores the variability in the tails. When applied to the data shown in Figure 3, the upper fence lies at the value 0.66. The upper tail defined by this value includes roughly 11% of the data points and accounts for about 68% of the total error. This result seems to capture the intuition about which errors are larger than normal in this case. The boxplot approach appears to be a good choice for identifying which errors need additional approximation resources.

The next question concerns how to compute the quartiles. Several approaches are available for computing incremental estimates of percentiles for a stream of data [5]. Since the percentiles used here are not extreme ones (i.e., not close to 0 or 1) requiring careful management of sample biases, a simple moving average should be sufficient for our purposes. Consequently, we maintain a moving window of data values. In each window, sample estimates of q_1 and q_3 are computed from the data and then used to update an exponentially weighted moving average as follows:

1. Initialize a window of $\mathcal{N}$ data values. Set $t = 0$.
2. Sort the values in the current window and compute sample estimates of $q_1(t)$ and $q_3(t)$.
3. Update the incremental estimates $Q_1(t)$ and $Q_3(t)$:

$$Q_1(t) \leftarrow \begin{cases} q_1(t) & \text{if } t == 0 \\ (1-\omega)Q_1(t-1) + \omega q_1(t) & \text{otherwise} \end{cases}$$

$$Q_3(t) \leftarrow \begin{cases} q_3(t) & \text{if } t == 0 \\ (1-\omega)Q_3(t-1) + \omega q_3(t) & \text{otherwise} \end{cases}$$

4. Remove the oldest data value from the window and add the next new value.
5. Set $t \leftarrow t + 1$ and go to Step 2.

The experiments described here used a window size $\mathcal{N} = 100$ and a step size $\omega = 0.001$.

The boxplot technique and the incremental quartile estimates were used to specify a learning trigger as follows. Whenever the approximation error in a match set is larger than expected (as determined by the boxplot method), a check is made to see if the match set size is also larger than expected (assuming a randomly generated population). If the match set size is smaller than expected, learning is triggered to generate one new rule for this match set. The results in Table 2 show that this simple trigger had the desired effect. Increasing the resources allocated to regions with large errors, in combination with the trigger maintaining a floor on match set sizes, gives a significant improvement in performance on the test data. Apparently, this re-allocation of resources leads directly to improved generalization for the function approximator. The increase in performance on the training data was not statistically significant.

5.3   Match Set Size for Error Outliers

It is important to recognize that the points outside the standard upper fence in a boxplot should not all be labeled as outliers. They are usually just called "outside values" that merit closer scrutiny. A conservative and reliable test for outliers sets additional fences at a distance $3.0(q_3 - q_1)$ beyond the lower and upper quartiles. Points beyond these values are clearly in a different category and can be confidently labeled as outliers. Adaptive hyperplane coding might perform even better if it allocated additional resources to the match sets corresponding to these error outliers.

In order to test this hypothesis, another trigger was defined using the additional upper fence for outlier identification. Whenever the approximation error in a match set is an outlier, a check is made to see if the match set size is at least half a standard deviation bigger than the expected size. If the match set size is smaller than this, learning is triggered to generate one new rule for this match set. This trigger is designed to make sure that the match set size for error outliers is in the upper tail of the distribution of sizes in the current population. Table 2 shows that this new trigger worked as expected. Significant performance improvements were obtained on both the training data and the test data.

5.4   Unbalanced Tile Offsets

The final learning trigger addresses the way approximation resources are organized in the input space. The classifiers in every match set are organized into tilings that are intended to be symmetrically offset from the unmodified input message. The offsets group together nearby inputs in both directions in many ways. The overlap of these tilings is a key factor underlying the precision and generalization that makes coarse coding effective. In a random population where the tilings are assigned randomly, the set of feature offsets in any particular match set might be far from symmetric. This could seriously limit the ability of the approximator to achieve the desired precision or generalization with respect to the affected inputs.

The final learning trigger was designed to address this issue. Whenever the match set size is adequate, which means that none of the other triggers apply, the number of features shifted to the left and to the right are counted. The expected number of features shifted in one direction can be viewed as a binomial random variable with an easily computed mean and variance. If the actual number of features shifted in one direction is more than 1.5 standard deviations from the expected number[4], the offsets are considered imbalanced. The learning trigger selects the classifier with the lowest impact (i.e., highest value of ι) from those offset in the overloaded direction, then changes the tiling of that classifier to be the least populated tiling shifted in the opposite direction. The results in Table 2 shows that this new trigger is effective. Performance improved on both the test data and the training data, though only the improvements on the training data were statistically significant.

6   Conclusions

The overall performance of this version of adaptive hyperplane coding on the test function suite is summarized in Table 3. Adaptive hyperplane coding is significantly more competitive with tile coding than plain hyperplane coding was. While tile coding retains a statistically significant performance advantage on the training data for Blocks and Bumps, that advantage is smaller than the one shown in Table 1 over plain hyperplane coding. Moreover, adaptive hyperplane coding performs significantly better than tile coding on the training data for Doppler while it increases the performance advantage on Heavisine. On the test data, adaptive hyperplane coding has a statistically significant performance advantage on all of the test functions except Bumps, where the small advantage

---

[4] A tighter criterion that keeps the number within 1 standard deviation is too restrictive and leads to so many trigger activations that it becomes counterproductive.

Table 3. Average square errors for tile coding and adaptive hyperplane coding

Algorithm	Approximation Error							
	Blocks		Bumps		Doppler		HeaviSine	
	Train	Test	Train	Test	Train	Test	Train	Test
Tiles	0.06988	1.7535	0.16809	0.93068	0.03579	0.08922	0.02327	0.06458
Hyperplanes	0.09157	0.26932	0.26607	0.92007	0.02833	0.06379	0.01161	0.02167

shown is not significant. This suggests that adaptive hyperplane coding supports generalizations better than tile coding.

This paper has shown that adaptive hyperplane coding offers a promising alternative for approximating value functions. Its performance is comparable to – and in many cases better than – a more conventional approach like tile coding. Adaptive hyperplane coding computes much better approximations than standard classifier system methods in which individual rules compute value function approximations independently. High quality value function approximations that provide both data recovery and generalization are a critically important component of most reinforcement learning algorithms. Future research will investigate how to increase the quality of value function approximations in classifier systems by integrating adaptive hyperplane coding techniques into a complete classifier system architecture. This is expected to provide significant improvements in classifier system performance on reinforcement learning problems. Future work will also compare hyperplane coding with stand-alone classifier system function approximation methods like XCSF [16].

Acknowledgments

This work is based on research originally funded by the MITRE Sponsored Research program. That support is gratefully acknowledged. The author's affiliation with The MITRE Corporation is provided for identification purposes only, and is not intended to convey or imply MITRE's concurrence with, or support for, the positions, opinions or viewpoints expressed by the author.

References

1. Lashon B. Booker. Approximating value functions in classifier systems. In Larry Bull and Tim Kovacs, editors, *Foundations of Learning Classifier Systems*. Springer, 2005.
2. Lashon B. Booker, David E. Goldberg, and John H. Holland. Classifier Systems and Genetic Algorithms. *Artificial Intelligence*, 40:235–282, 1989.
3. Larry Bull and Toby O'Hara. Accuracy-based neuro and neuro-fuzzy classifier systems. In W. B. Langdon, E. Cantú-Paz, K. Mathias, R. Roy, D. Davis, R. Poli, K. Balakrishnan, V. Honavar, G. Rudolph, J. Wegener, L. Bull, M. A. Potter, A. C. Schultz, J. F. Miller, E. Burke, and N. Jonoska, editors, *GECCO 2002: Proceedings of the Genetic and Evolutionary Computation Conference*, pages 905–911. Morgan Kaufmann Publishers, 9-13 July 2002.

4. Martin V. Butz and Stewart W. Wilson. An Algorithmic Description of XCS. In Pier Luca Lanzi, Wolfgang Stolzmann, and Stewart W. Wilson, editors, *Advances in Learning Classifier Systems*, volume 1996 of *LNAI*, pages 253–272. Springer-Verlag, Berlin, 2001.
5. Fei Chen, Diane Lambert, and Jose C. Pinheiro. Incremental quantile estimation for massive tracking. In *Proceedings of ACM SIGKDD International Conference on Knowledge Discovery and Data Mining*, pages 516–522, New York, 2000. ACM Press.
6. David L. Donoho and Iain M. Johnstone. Ideal spatial adaptation by wavelet shrinkage. *Biometrika*, 81:425–455, 1994.
7. Christos Faloutsos. Gray codes for partial match and range queries. *IEEE Transactions on Software Engineering*, 14(10):1381–1393, October 1988.
8. Geoffrey E. Hinton, James L. McClelland, and David E. Rumelhart. Distributed representations. In David E. Rumelhart, James L. McClelland, and CORPORATE PDP Research Group, editors, *Parallel distributed processing: explorations in the microstructure of cognition, vol. 1: foundations*, pages 77–109. MIT Press, 1986.
9. Boris Iglewicz and David C. Hoaglin. *How to Detect and Handle Outliers*. American Society for Quality Control Basic References in Quality Control: Statistical Techniques (Volume 16). ASQC Quality Press, Milwaukee WI, 1993.
10. W. Thomas Miller, Filson H. Glanz, and L. Gordon Kraft. CMAC: An associative neural network alternative to backpropagation. *Proceedings of the IEEE*, 78(10):1561–1567, October 1990.
11. Satinder P. Singh, Tommi Jaakkola, and Michael I. Jordan. Reinforcement learning with soft state aggregation. In G. Tesauro, D. Touretzky, and T. Leen, editors, *Advances in Neural Information Processing Systems*, volume 7, pages 361–368. The MIT Press, 1995.
12. Richard S. Sutton. Adapting bias by gradient descent: An incremental version of delta-bar-delta. In *Proceedings of the Tenth National Conference on Artificial Intelligence*, pages 171–176, 1992.
13. Richard S. Sutton and Andrew G. Barto. *Introduction to Reinforcement Learning*. MIT Press, Cambridge,MA, 1998.
14. Richard S. Sutton and Steven D. Whitehead. Online Learning with Random Representations. In *Machine Learning: Proceedings of the Tenth International Conference*, pages 314–321, San Mateo, CA, 1993. Morgan Kaufmann.
15. Gilles Venturini. *Apprentissage Adaptatif et Apprentissage Supervisé par Algorithme Génétique*. PhD thesis, Université de Paris-Sud., 1994.
16. Stewart W. Wilson. Classifiers that approximate functions. *Natural Computing*, 1(2-3):211–234, 2002.

Appendix - Test Functions

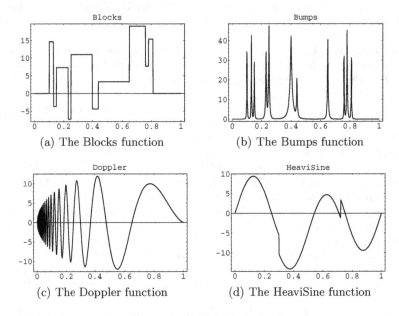

(a) The Blocks function   (b) The Bumps function

(c) The Doppler function   (d) The HeaviSine function

Fig. 4. The Four Donoho Test Functions

Blocks

$$f(t) = 3.65948 * \sum h_j K(t - t_j) \text{ where } K(t) = (1 + \text{sgn}(t))/2$$

$(t_j) = (0.1, 0.13, 0.15, 0.23, 0.25, 0.4, 0.44, 0.65, 0.76, 0.78, 0.81)$
$(h_j) = (4, -5, 3, -4, 5, -4.2, 2.1, 4.3, -3.1, 2.1, -4.2)$

Bumps

$$f(t) = 10.5174 * \sum h_j K((t - t_j)/w_j) \text{ ,where } K(t) = (1 + |t|)^{-4}$$

$(t_j) = (0.1, 0.13, 0.15, 0.23, 0.25, 0.4, 0.44, 0.65, 0.76, 0.78, 0.81)$
$(h_j) = (4, 5, 3, 4, 5, 4.2, 2.1, 4.3, 3.1, 5.1, 4.2)$
$(w_j) = (0.005, 0.005, 0.006, 0.01, 0.01, 0.03, 0.01, 0.01, 0.005, 0.008, 0.005)$

Doppler

$$f(t) = 24.2158 * \sin(2\pi(1 + \epsilon)/(t + \epsilon))\sqrt{t(1 - t)} \text{ , where } \epsilon = 0.05$$

HeaviSine

$$f(t) = 2.3564 * [4\sin(4\pi t) - \text{sgn}(t - 0.3) - \text{sgn}(0.72 - t)]$$

Three Architectures for Continuous Action

Stewart W. Wilson

Prediction Dynamics, Concord MA 01742 USA
Department of General Engineering
The University of Illinois at Urbana-Champaign IL 61801 USA
wilson@prediction-dynamics.com

Abstract. Three classifier system architectures are introduced that permit the systems to have continuous (non-discrete) actions. One is based on interpolation, the second on an actor-critic paradigm, and the third on treating the action as a continuous variable homogeneous with the input. While the last architecture appears most interesting and promising, all three offer potential directions toward continuous action, a goal that classifier systems have hardly addressed.

1 Introduction

Typically, classifiers in learning classifier systems have fixed, discrete actions, e.g., "turn left", "turn right", "yes", "no". Classifiers have not been introduced in which the action is *continuous* in the sense that it depends on and is a continuous function of the input. Continuous actions, however, are desirable in domains where fine reactions and control are important, such as robotics. This paper describes three distinct classifier system architectures that permit continuous actions.

The first architecture, termed "interpolating action learner" (IAL) [12], has one classifier system observing a second classifier system's optimal (discrete) actions and learning them as a smooth function of the input. IAL is perhaps the simplest approach to continuous action. The second approach, "continuous actor-critic" (CAC) is a continuous-action, classifier-system version of the well-known actor-critic architecture [9]. CAC is quite intricate but is a plausible approach to direct learning of optimal continuous actions. The third architecture, a "generalized classifier system" (GCS) [11], breaks new ground by aggregating a continuous action with the input, and learning payoff as a function of both together. In GCS, the optimal action is chosen as the action that maximizes this function. All three architectures make use of a recently introduced function-approximating classifier system called XCSF.

XCSF [14,15] extends XCS [13,3], a classifier system with fitness based on prediction accuracy, to real-valued inputs and continuous payoff landscapes. XCSF learns an approximation to $P(x, a_k)$, the environmental payoff function of input x and discrete actions a_k. Denoting the approximation $\hat{P}(x, a_k)$, and given a particular input x, XCSF maximizes payoff by picking the action a_k^* that maximizes $\hat{P}(x, a_k)$. However, XCSF may be used to approximate functions other than payoff, a flexibility that is exploited here.

X. Llorà et al. (Eds.): IWLCS 2003-2005, LNAI 4399, pp. 239–257, 2007.

The three architectures are illustrated using a simple one-dimensional but non-linear testbed problem introduced in [15]. In this so-called "frog" problem, a system (frog) senses an object (fly) via a signal that monotonically decreases with the distance between them. The frog has available a continuous range of jump lengths and is supposed to learn to jump exactly the distance to the fly. Following the jump, which may undershoot or overshoot the fly, payoff is given by the resulting sensory signal.

The paper begins with a more detailed description of the frog problem, providing a concrete context for the three architectures. Then XCSF is briefly reviewed. Further sections describe and test the architectures themselves, followed by conclusions as to their merits and suggestions for further work.

2   Frog and Fly: A 1-D Continuous-Action Problem

As noted in Section 1, the frog should learn to pounce on the fly in one jump. Let d be the frog's distance from the fly, with $0.0 \le d \le 1.0$. For simplicity, we assume x, the frog's sensory input, falls linearly with distance: $x(d) = 1 - d$. The frog "sees the fly better" the closer it is.

The payoff should be any function of x and action a that is bigger the smaller the distance d' that remains after jumping. That is, $P(x, a)$ should monotonically increase with smaller d'. A natural choice is to let the payoff equal the sensory input *following* the jump, as though the frog is rewarding itself based on what it "sees". Then, with the sensory function above, $P = 1 - d'$.

To write the payoff in terms of x and a, we need to make one assumption. Suppose the frog's jump overshoots, i.e., the frog lands *beyond* the target fly. In this case we assume that d' equals the amount of the overshoot (taken as a positive number). Thus $d' = d - a$ for $a \le d$ and $d' = a - d$ for $a \ge d$. Substituting for d' in $P = 1 - d'$, then using $d = 1 - x$ and rearranging, we get

$$P(x, a) = \begin{cases} x + a & : & x + a \le 1 \\ 2 - (x + a) & : & x + a \ge 1 \end{cases} \tag{1}$$

Figure 1 illustrates $P(x, a)$. For a given sensory input x, the payoff in general first rises, reaches a peak, then falls as a increases. The "ridge" of the function's tent shape in fact corresponds to the maximum payoff and thus optimal action over x's range. Note that the payoff function is highly non-linear—albeit composed of two linear planes.

To solve the frog problem, a system must learn to choose, given x, the value of a corresponding to maximum payoff. In the context of classifier systems (and reinforcement learning (RL) [9] in general) this is a non-trivial problem. The reason is that the only feedback allowed the system is in terms of payoff; the system gets no "error signal" with respect to its action choice. From the RL point of view, this restriction models the actual learning situation of many natural and artificial systems. Classifier systems (as well as other RL systems) deal with the restriction by, in effect, observing and estimating the payoffs associated with different action choices, then, given an input, choosing the action that appears

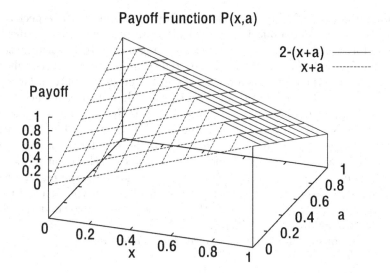

Fig. 1. Frog problem payoff function $P(x, a)$

to pay the best. This process is quite well understood within the framework of discrete actions. We build on this understanding in the three architectures for continuous action.

3 Brief Review of XCSF

XCSF is similar to XCS. This section assumes familiarity with XCS and discusses only differences between XCSF and the earlier system.

The essential innovation in XCSF with respect to XCS is that classifier predictions in XCSF are *calculated* instead of being fixed scalars. Each classifier has a *weight vector* w which is combined with the input vector x to form a prediction that is a linear function of the input. As a result, continuous, real-valued payoff landscapes can be approximated more efficiently [15] than would be the case with fixed scalar predictions, which allow only piecewise-constant approximations. Polynomials of order higher than linear may also be used in XCSF [7].

The classifier condition in XCSF is a truth function, denoted $t(x)$, which defines a subspace of the real-valued (or integer-valued) input space such that a classifier matches if an input is in that subspace. In the simplest case, $t(x)$ is a concatenation of "interval predicates", $int_i = (l_i, u_i)$, where l_i ("lower") and u_i ("upper") are reals. A classifier matches an input x with components x_i if and only if $l_i \leq x_i \leq u_i$ for all x_i. However, $t(x)$ can be any evolvable function that defines a subspace of the input space.

With these definitions, a classifier in XCSF can be conveniently notated

$$t(x) : w : a_k \Rightarrow p(x, a_k),\qquad(2)$$

where a_k is the kth discrete action. The expression $p(x, a_k)$ is the classifier's prediction given x and a_k.

In operation, XCSF differs significantly from XCS only with respect to prediction calculation and updating. To calculate its prediction, a classifier with action a_k forms $p(x, a_k) = w \cdot x'$, where x' is x augmented by a constant x_0, i.e., $x' = (x_0, x_1, ..., x_n)$. Just as in XCS, the prediction is only produced when the classifier matches the input. As a result, $p(x, a_k)$ in effect computes a *hyperplane approximation* to the payoff function $P(x, a_k)$ over the subspace defined by $t(x)$. The value of x_0 is chosen to be of the same order as the component values of x.

As in XCS, XCSF forms a *system prediction* $\hat{P}(x, a_k)$ for each a_k as a fitness-weighted average of the predictions $p(x, a_k)$ of the classifiers that match the input. Then, if XCSF is in *explore* mode, one of the a_k is chosen at random and sent to the environment. If in *exploit* mode, the action with the highest system prediction, a_k^* is chosen. In explore mode, the payoff actually received, $P(x, a_k)$, is used to update the predictions of the classifiers that advocated the chosen action (i.e., the *action set* classifiers).

The predictions are updated using a *modified delta rule*

$$\Delta w_i = (\eta / |x'|^2)(t - o)x_i \tag{3}$$

in which t ("target") is the current payoff $P(x, a_k)$ and o ("output") is the current calculated value of $p(x, a_k)$. The factor η is a rate constant; the normalization $|x'|^2$ makes the change in output strictly proportional to $(t - o)$ and controllable by η. More powerful techniques of updating have been investigated [6].

Other differences from XCS adjust the genetic algorithm (GA) and the covering of unmatched inputs to accord with the structure of $t(x)$. Further details on this and the above material may be found in [15]. To be noted at this point is the fact that XCSF can be employed either as a discrete-action classifier system or, by reducing the a_k to a single, dummy, action, as a function approximator. In the latter case, the prediction becomes just $p(x)$, with $P(x)$ representing a function to be learned by approximation. XCSF is used both ways in the three architectures.

4   Interpolating Action Learner (IAL)

4.1   IAL Concept

In IAL, a second classifier system learns the action choices of a first classifier system. The overall system, shown schematically in Fig. 2, consists of two classifier systems S1 and S2 working in tandem.

The lower system S1 is in contact with an environment *Env* that provides real state vectors x to S1, receives discrete actions a_k from S1, and provides real payoffs $P(x, a_k)$ to S1 in response. The system S1 is based on XCSF. For each possible x, S1 learns to approximate the payoff $P(x, a_k)$ to be expected for each a_k it might take.

While S1 is learning, the system S2 *observes* S1's *prediction array* (the array containing S1's system predictions). Thus, given x, S2 can see what payoff S1 expects for each a_k. In particular, S2 notes which a_k has the *maximum* expected

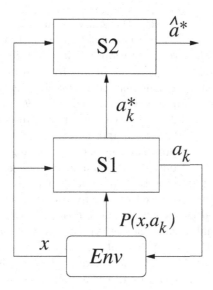

Fig. 2. Interpolating Action Learner. Both S1 and S2 are XCSF systems, but S2 observes S1's optimal actions and learns a continuous approximation to $a^*(x)$, the optimal policy for environment *Env*.

payoff, and uses this action value, a_k^*, as an input. The other input to S2 is x. S2's objective is to learn to predict, given x, the correct value of a_k^*.

The combined system S1-S2 has two innovations. First, once S2 has learned to predict correctly, it represents a *direct* mapping from x to the optimal action, thus embodying the optimal policy $a^*(x)$. Ordinarily, RL systems do not directly map from x to $a^*(x)$. Instead, as with S1, they learn expected payoffs for each a_k, then, given an x, choose the a_k with the maximum expectation and send it to the environment. In contrast, S2 sees x and predicts $a^*(x)$ directly.

Second, the architecture permits the output of S2 to be *continuous-valued*, despite the fact that S2 is only capable of discrete actions. Like S1, S2 is a classifier system based on XCSF, but it is configured for just one, dummy, action. In this configuration a_k^* is treated formally like a *payoff to be learned*, with $\hat{a}^*$ the predicted value. Thus S2 in effect acts as a function approximator approximating $a^*(x)$. If S2's error threshold ϵ_0 is small (strict) then S2's approximation $\hat{a}^*$ should be similar to $a_k^*(x)$ which—for the frog problem employed here—is a staircase-like function of x. However, if ϵ_0 is not small, S2's approximation should be less discontinuous, becoming smoother with larger ϵ_0. Consequently, as x changes continuously, $\hat{a}^*$ should also change continuously so that S2 will approximate $a^*(x)$, the optimal *continuous* policy for *Env*.

4.2   IAL Experiments

The IAL system was tested on a version of the frog problem in which the frog had five discrete jump lengths: 0.0, 0.25, 0.50, 0.75, and 1.0. This is the same problem learned by XCSF in [15].

Parameter settings for S1 were the same as in [15]: population size $N = 500$, learning rate $\beta = 0.2$, error threshold $\epsilon_0 = 0.01$, fitness power $\nu = 5$, GA threshold $\theta_{GA} = 48$, crossover probability $\chi = 0.8$, mutation probability $\mu = 0.04$, deletion threshold $\theta_{del} = 50$, fitness fraction for accelerated deletion $\delta = 0.1$. Also, mutation increment $m_0 = 0.1$, covering interval $r_0 = 0.1$, $\eta = 0.2$ and $x_0 = 1.0$. GA subsumption was enabled, with $\theta_{GAsub} = 100$. The same parameter set was used in S2, except that the error threshold ϵ_0 was different in each of three experiments: 0.01, 0.05, and 0.10 [1].

In an experiment, each problem consisted of placing the fly at a random distance $(0.0 \leq d \leq 1.0)$ from the frog and having S1 go through a standard explore cycle in which it matched the input, chose an action at random, got payoff according to the resulting distance from the fly, updated the action set classifiers, and possibly performed the GA. At the same time, S2 watched S1's prediction array, noted the optimal action (the one with the largest predicted payoff), and learned that action, a_k^*, by approximating it via the XCSF mechanism, as a function of x. That is, given x, S2's classifiers predicted S1's action (instead of a payoff) and were updated according to the actual value of a_k^*. (Those classifiers thus had the format, $t(x) : w \Rightarrow a(x)$, where $a(x)$ took the place of $p(x, a)$ and the classifiers themselves had no action variable per se.)

Each experiment consisted of 400,000 problems to be sure the system had stabilized. At the end of each experiment, the input range was *scanned* with a resolution of 0.01 and S2's predicted action $\hat{a}^*$ was calculated for each point and plotted. Figure 3 shows the results for each value of ϵ_0.

For the smallest ϵ_0, the plot of $\hat{a}^*$ is an almost perfect staircase corresponding to the best discrete a_k vs. x. With the middle value of ϵ_0, the plot shows a tendency to flatten, and with the largest value, 0.10, it is a straight line corresponding to the optimum *continuous* action (except that the line is displaced slightly upward!).

Examination of the final populations for the three cases showed 14 classifiers for $\epsilon_0 = 0.01$, 39 for 0.05, and exactly one classifier for $\epsilon_0 = 0.10$. That classifier's weight vector is directly indicated by the slope and intercept of the straight-line plot in Figure 3. The population for $\epsilon_0 = 0.10$ reached less than a dozen classifiers by 50,000 problems and fell to one classifier around 200,000 problems. Maximum accuracy in approximating $a^*(x)$ was reached much quicker, in a few thousand problems.

4.3   IAL Discussion

The experiments suggest that the IAL system can do what was intended, i.e., approximate $a^*(x)$ from the payoffs to a set of discrete actions. However, the function to be approximated—a straight line over a 1-D input domain—is certainly about the simplest imaginable. Because the IAL method achieves smoothness of approximation through a large error threshold, approximation accuracy in functions with higher bandwidth may be limited. A second drawback is that while

---

[1] The values given are the actual error threshold divided by the maximum possible payoff, in this case 1000.

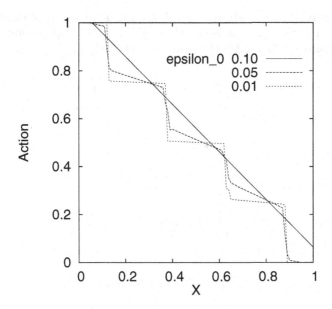

Fig. 3. Frog problem results for IAL. Predicted optimal action $\hat{a}^*$ (jump length) vs. input x (sensory input) as learned by S2 for three values of error threshold ϵ_0. Larger values permit closer approximation of optimal continuous action.

the optimal continuous action is learned, nothing is learned about the value of non-optimal actions. That is, the rest of the continuous payoff landscape remains unknown to the system. Depending on the goals for the system, this may not matter, however.

The fall of the population size to a single classifier in the straight-line case (and to small numbers in the others) is encouraging because it confirms the ability of XCSF to evolve classifiers that approximate functions with high efficiency. This property of XCS-like systems has been investigated theoretically [1]. On the other hand, the time—tens of thousands and more of problem instances—to reach this efficiency seems very long. That may be a consequence of working in the real domain—instead of the binary, as with XCS—and calls for further investigation.

Overall, IAL seems interesting but not outstanding as an architecture for achieving continuous action. Better would seem an architecture that more directly developed an approximation to $a^*(x)$, instead of indirectly as a secondary process. A step in this direction is taken in the next section.

5   Continuous Actor-Critic (CAC)

5.1   CAC Concept

In the actor-critic approach to reinforcement learning problems, one system, termed "actor", produces actions in response to an environmental input, x, and a second system, "critic", adjusts the actor with the intention of improving the actor's response. On what basis can the critic do this? The general scheme is shown in Figure 4, where S3 is the actor and S4 the critic.

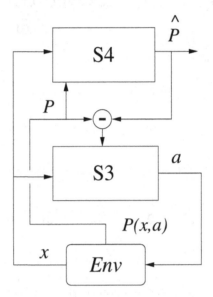

Fig. 4. Continuous Actor-Critic. S4 ("critic") is an XCSF system, learning $P(x)$, but S3 ("actor") is a special system with classifiers of the form $t(x) : w \Rightarrow a(x)$. S4's prediction error modifies probabilities of activation of S3's classifiers.

While S3 interacts with *Env*, S4, an XCSF system, learns to predict *Env*'s payoffs. Specifically, S4 learns an approximation to $P(x)$. On every learning step, there will in general be a difference between S4's prediction and the actual payoff received. If, for instance, the actual payoff is less than the predicted one, it is likely that S3 could have chosen a better action than it did. This information, though possibly erroneous, is used to *reduce the probability* of that action being chosen in the future. Conversely, if the payoff is larger than predicted, the probability of the current action is increased.

If S3 were simply choosing actions randomly, these corrective signals from S4 would be meaningless. However, if the probability of S3's actions is affected as above by S4's prediction errors, there should be a convergence in which S4's predictions become more and more accurate and S3's actions produce payoffs that are increasingly likely to be maximal. This is the essence of the actor-critic idea [9]. As a method in RL, it has been somewhat displaced by the introduction of Q-learning [10]. However, actor-critic retains computational advantages where the action is computed as a continuous function of the input, as it is here (in the language of RL, "the policy is explicitly stored" [9]).

CAC is an actor-critic architecture implemented with classifier systems. As noted, S4 is XCSF used as a function approximator to predict $P(x)$. S3, however, is not an XCSF system. It contains classifiers with the format $t(x) : w \Rightarrow a(x)$ each of which has an associated probability of activation π_i (i indexes the classifier). Given x, classifiers form a match set as usual, but then one of the matching classifiers is chosen based on their relative probabilities (the π_i are first normalized). This classifier calculates an action using x and its weight vector w, then the action is sent to the environment. The environment returns payoff, but this is unusable directly by S3 since S3 emitted an action, not a payoff prediction. Instead, the payoff is sent to S4, where it is used to adjust the payoff prediction of S4. Finally, the difference between S4's prediction and the actual payoff is used to adjust the probability π of the activated classifier.

A straightforward method would adjust the probability as follows:

$$\pi \leftarrow \begin{cases} \pi + g(\Delta P / P)\pi & : \quad \Delta P \leq 0 \\ \pi + g(\Delta P / P)(1.0 - \pi) & : \quad \Delta P > 0 \end{cases} \tag{4}$$

Here g ($0 < g \leq 1$) is a gain factor and ΔP is the error in S3's prediction, i.e., ($\hat{P} - P$). The adjustment is intended to have the effect that the best matching classifier's probability tends toward 1.0 and the probabilities of other classifiers tend toward 0.0.

S3's classifiers are generated by a genetic algorithm acting on the match sets, using the probabilities π_i as fitnesses. Thus classifiers whose actions result, on the average, in above-average payoffs should tend to be selected and reproduced. Furthermore, there should be a pressure to preferentially reproduce higher-probability, more-*general* (having large $t(x)$ domains) classifiers because such classifiers will occur in more match sets than more-specific classifiers [1]. As a result, S3's population should tend toward efficient coverage of the input space.

An important difference between S3 and XCSF-like systems is that the weight vector w must be *evolved* along with $t(x)$; it cannot be adjusted based on environmental feedback since the classifiers compute actions, and the feedback is in terms of payoff.

5.2 CAC Discussion

Unfortunately, significant experiments on CAC have not been done. As a preliminary test, an XCSF system was investigated in which the weight vector was evolved instead of adjusted. The adaptation was slow and irregular, suggesting that adjustment might be superior to evolution in this case. However, recent work by Hamzeh and Rahmani [4] evolved weight vectors successfully. Also, evolutionary techniques other than the GA may prove better for this kind of smooth maximization problem.

Assuming the CAC concept can work, it is an advance over the first technique, IAL, because it should result in classifiers in S3 that closely approximate $a^*(x)$, instead of achieving continous action through large approximation error. However, like IAL, CAC still fails to learn anything about the payoff landscape associated with non-optimal actions. The third architecture, presented in the next section, remedies this and though it too has problems, it may turn out to be the best-founded direction for continuous action.

6   Generalized Classifier System (GCS)

6.1   GCS Concept

As a discrete-action classifier system, XCSF learns $P(x, a_k)$. It uses classifiers of the form, $t(x) : w : a_k \Rightarrow p(x, a_k)$, in which the variable a_k is not only discrete, but treated separately from the variable x. However, the underlying payoff landscape is simply $P(x, a)$; not only is a continuous but it is functionally homogeneous with x. Why not learn $P(x, a)$ directly? Why not have classifiers of the form $t(x, a) : w \Rightarrow p(x, a)$, in which the condition depends on, and the weight vector refers to, both x and a? We term this sort of classifier system "generalized" in recognition of the homogeneous treatment of input and action, with both permitted to be real variables [2].

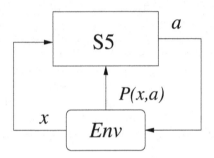

Fig. 5. Generalized Classifier System. S5 is a new system partly based on XCSF with classifiers of the form $t(x, a) : w \Rightarrow P(x, a)$.

Learning, or exploration, in GCS is straightforward and similar to learning in XCSF. Given an input x, the system picks an action a according to its current exploration regime, e.g., it might try actions randomly. Then a match set [M] is formed of classifiers satisfying $t(x, a)$. The predictions $p(x, a)$ of each member of [M] are calculated, using the member's w, just as in XCSF. The action is performed in the environment and payoff P received. P is then used to adjust the error and fitness of each classifier in [M], and to update the w's of [M], again just as in XCSF. New classifiers are formed by the GA, and classifiers created by covering if needed, just as done in XCSF, except of course that now the condition and weight vector include a as well as x.

It is in the "exploit" side of GCS that substantially new steps are required. The object is to determine, given an input x, what action a will produce the

---

[2] As defined here, GCS is not, of course, fully general since operation in multi-step tasks is not discussed, and the system has no provision for internal state variables, as in [8]. Neither extension appears problematic.

greatest payoff. The simplest way conceptually would appear to be to scan all possible actions a, form an [M] for each, and choose a^* as the one for which the system prediction $\hat{P}(x, a)$ was highest. But there are difficulties. In the first place, scanning the range of a inevitably means discretizing the range, which in turn implies that the system's action is not truly continuous. Secondly, the higher the scan's resolution the greater the number of match sets [M] that must be formed, so that higher approximate continuity in a trades off against efficiency. For GCS, a new approach to picking a^* seems desirable.

Consider the exploit situation. An input x comes in. The action a has not been determined, so how does the system determine which classifiers belong in [M]? The answer is: a classifier belongs in [M] if there exists an a such that that classifier's $t(x, a)$ is satisfied. For example, if $t(x, a)$ employs interval predicates, x need only satisfy the predicates devoted to x's components (a can be any value satisfying its predicate). For other forms of $t(x, a)$ the answer to the question and the computational expense of answering it depend on the form. An important form of $t(x, a)$ for which matching is quick will be mentioned shortly.

Once the match set [M] is built, the system moves on to determine the highest-paying action a^*. Consider first a classifier in [M] and suppose for concreteness that interval predicates are employed in $t(x, a)$. Let (a_l, a_u) be the action predicate. Since, $p(x, a)$ is computed *linearly*, i.e., using a weight vector w, it is clear that *either* a_l or a_u will yield a higher prediction than any other value of a that satisfies the interval. Thus, no scanning is necessary to determine this classifier's best action, a_{best}; it is only necessary to compare $p(x, a_l)$ with $p(x, a_u)$. Finally, for a^*, the system picks the highest-paying a_{best} of the individual classifiers of [M].

At this point it is necessary to note a disabling drawback of basing $t(x, a)$ on interval predicates. Consider a classifier in [M]. For this classifier, a_{best} is a constant; it equals either a_l or a_u, independent of x. Similarly, the values of a_{best} for the other members of [M] are also constant. The implication is that if x changes slightly, a^* will either not change, or it will change abruptly as a different a_{best}, from another classifier in [M], becomes a^*. Thus the use of interval predicates has the consequence that a^* is discontinuous with respect to x.

This effect will be especially evident if the payoff landscape, $P(x, a)$ is *oblique*, as in Figure 1. The "ridge" corresponding to the maximum of the function is not parallel with either axis. Consequently, approximating the function with conditions based on (axis-parallel) interval predicates would require that classifiers overlap in staircase fashion at the ridge, so that a^* could approximate continuity only by employing large numbers of such classifiers. On the other hand, if the classifier conditions could have non-axis-parallel contours then much or all of the staircasing discontinuity could be avoided.

Consider, as a sort of thought-experiment, the possibility of evolving conditions with arbitrary contours. Then imagine a classifier whose $t(x, a)$ exactly covered the portion of the x-a plane lying below the line $x + a = 1$. This classifier would match any x and, significantly, its $a_{best} = 1 - x$, which is both perfectly continuous and exactly the correct solution to the frog problem. To represent the complete landscape, just one more classifier would be needed; its condition would cover the portion of the plane lying above the $x + a = 1$ line. Thus it appears that GCS can achieve both continuous action and exact solutions—as well as economy of representation—if classifier conditions can be evolved that have arbitrary contours. While this thought-experiment is illuminating, evolution of arbitrary $t(x, a)$ is difficult (see [5] for an initial attempt). The essential requirement, however, is for conditions that can evolve so that their contours align with oblique features of the payoff landscape. The rest of this section outlines an example approach using orientable elliptical conditions that appears to offer considerable continuity, precision, and economy.

Using XCSF, Butz [2] investigated classifiers having general hyperellipsoidal conditions and showed they could approximate highly oblique functions more efficiently (smaller populations, lower system error) than classifiers having hyperrectangular interval conditions. The general hyperellipsoids had the property that they could rotate with respect to the coordinate axes, permitting them to orient to the contours of the function [3]. Butz's work applied to function approximation, but the idea appears promising for continuous action as well. The following describes GCS with orientable elliptical conditions and its application to the frog problem.

In the frog problem, $P(x, a)$ is 2-dimensional so that conditions can be represented with ordinary ellipses. We define such a condition using the parameters $< m_x, m_a, l_x, l_a, \theta >$. In general, the ellipse is not centered at the origin, and it is rotated with respect to the x-axis. To obtain the actual ellipse using the parameters, we imagine it "starts" at the origin and that its semi-axis in the x direction is aligned with that axis and has length l_x; similarly, the a semi-axis has length l_a. Now, rotate the ellipse (counterclockwise) with respect to the x-axis through angle θ and displace its center to the point (m_x, m_a) to obtain the actual ellipse used in the condition. The equation for that ellipse is

$$\left(\frac{(x - m_x) \cos \theta + (a - m_a) \sin \theta}{l_x} \right)^2 + \left(\frac{-(x - m_x) \sin \theta + (a - m_a) \cos \theta}{l_a} \right)^2 = 1 \tag{5}$$

This may be expressed compactly as

$$(\mathbf{EV})^2 = 1, \tag{6}$$

where $\mathbf{E}$ equals the ellipse matrix

$$\begin{bmatrix} \cos \theta / l_x & \sin \theta / l_x \\ -\sin \theta / l_a & \cos \theta / l_a \end{bmatrix}$$

and $\mathbf{V}$ is the column vector

$$\begin{bmatrix} (x - m_x) \\ (a - m_a) \end{bmatrix}$$

---

[3] For hyperellipsoids, rotation (which is not possible to visualize) is considered more generally as the introduction of cross-product terms, e.g., $x_4 x_7$, in the hyperellipsoid definition, in analogy to the definition of a rotated ellipse.

The condition $t(x, a)$ is represented by the ellipse of Eqn. 6; it is satisfied if

$$(\mathbf{EV})^2 \leq 1, \tag{7}$$

i.e., if the point (x, a) is *inside* the ellipse.

In explore mode, as noted earlier, GCS combines the input x with a randomly chosen a and forms the match set [M] with classifiers whose conditions $t(x, a)$ satisfy (7). In exploit mode, [M] consists of classifiers for which, given x, there exists an a such that (7) holds. This is accomplished by setting $\mathbf{V} = \begin{bmatrix} (x - m_x) \\ 0 \end{bmatrix}$, i.e., by choosing a equal to the center a-value of the condition (if the condition does not hold for this value, it will certainly not hold for any other value).

The next step, in exploit, is to determine the a_{best} of each classifier in [M]. Consider the condition of a matching classifier. The input x defines a straight line crossing the ellipse parallel to the a axis. The line intersects the ellipse twice, at two values of a. By the reasoning used earlier in this section in connection with interval predicates, one of these values must be a_{best}. The two values are found by solving Eqn. 5, a quadratic equation with two roots, for a. The root with the larger predicted payoff, as calculated using w, is a_{best}. Finally, the largest a_{best} of the classifiers in [M] becomes the system's chosen action, a^*.

It is important to observe that, for a given matching classifier the value of a_{best} will in general change—continuously—as x changes. The reason is that since $t(x, a)$ is elliptical, its contour varies continuously with x. Furthermore, since the ellipses can rotate and elongate, they may evolve to align with and along a considerable length of the oblique ridge in $P(x, a)$. Thus the transition between different a^*s as x changes may be substantially continuous and not staircased. Moreover, the ability of the conditions to rotate holds out the possibility, in the frog problem, of evolving very small populations, since in principle just two extremely elongated ridge-parallel classifiers could cover the whole payoff landscape.

This completes the description of the GCS concept. In summary, the landscape $P(x, a)$ is approximated using an XCSF-like method in which a is an "input" along with x. Optimal actions are chosen as the best of the best of recommendations of individual classifiers. The actions are largely continuous with respect to x because the condition contours are continuous with respect to x and the conditions themselves can orient to align with landscape features. In the next section we examine the results of an experiment with GCS.

6.2   GCS Experiments

Experimental investigations of GCS are work in progress. Reported here is one of the first experiments that showed some aspects of what the concept suggested should happen. In fact, it took a number of experiments to arrive at an implementation of the concept that seemed correct. This was due largely to the unaccustomed use of the action variable as a kind of input.

Parameters for the experiment were as follows: population size $N = 2000$, learning rate $\beta = 0.5$, error threshold $\epsilon_0 = 0.01$, fitness power $\nu = 5$, GA threshold $\theta_{GA} = 48$, crossover probability $\chi = 0.8$, mutation probability $\mu = 0.04$, deletion threshold $\theta_{del} = 50$, fitness fraction for accelerated deletion $\delta = 0.1$, $\eta = 0.2$ and $x_0 = 1.0$. Neither GA nor action-set subsumption was enabled.

Mutation was handled differently according to the ellipse parameter being mutated. For ellipse center coordinates m_x and m_a, the value was changed by a random quantity from [-0.1,0.1]. For ellipse axis lengths l_x and l_a, the value v was changed (following [2]) by a random quantity from $[-0.5v, 0.5v]$. For ellipse angle θ, the value was changed by a random quantity from [-0.5,0.5] (radians). In covering: m_x was given by the x input; m_a was the selected action if it was an explore problem, otherwise a random value from [0.0,1.0]; the axis lengths were random from [0.0,0.1]; and θ was random from [0.0,1.0].

In each problem of an experiment, the fly was placed at a random distance $(0.0 \leq d \leq 1.0)$ from the frog. A conventional design was used in which with probability 0.5 the problem was an explore problem, otherwise an exploit problem. Figure 6 shows the results of one run that ended at 100,000 explore problems. Each curve plots a moving average over the past 50 exploit problems.

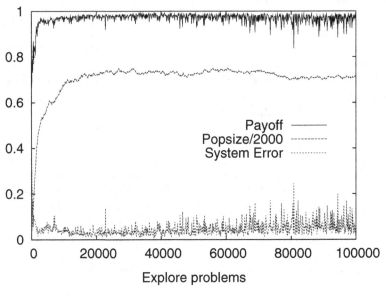

Fig. 6. Results for GCS in a frog problem experiment (curve order same as in legend)

The Payoff curve shows that by a few thousand problems, the system was regularly receiving payoff greater than 0.95. Thus, since $P = 1 - d'$ in this experiment, the frog was regularly jumping to a position less an 0.05 from the fly. The System Error curve approximately complements the Payoff curve, though not precisely since system error is the difference between actual and predicted payoffs. In both curves there appears to be a slow degradation toward greater volatility. The population size curve rises to about 70% of N and then declines very gradually.

Inspection of the population at 100,000 problems (as well as much earlier) showed that the w_1 and w_2 components of the weight vectors of most classifiers had converged either to (1.0,1.0) or to (-1.0,-1.0), depending on whether the classifier's condition applied below the $a = 1 - x$ diagonal or above it. Thus most classifiers were indeed approximating the payoff function (Eqn. 1). In fact, most classifiers had an error less than .0005, even though the system error was much higher.

The conditions of the population's ten highest-numerosity classifiers were plotted to get an idea of their size and orientation (Fig. 7). The figure indicates that the conditions have evolved so that their major axes are approximately aligned with the $a = 1 - x$ landscape ridge. The ellipse sizes appear reasonable for collectively approximating the regions above and below the ridge.

As a direct indication of the system's ability to choose the best action a^* and to gauge a^*'s continuity with respect to x, the input was scanned with increment 0.001 and the resulting a^* plotted (Fig. 8). The plot lies close to the diagonal and shows intervals of continuous change broken by abrupt small discontinuities suggesting the system is switching from one highest-predicting classifier to another.

6.3   GCS Discussion

The limited experimental results with GCS are mixed but quite promising. The condition ellipses indeed seem to be evolving to align with the diagonal (oblique) feature of the frog problem environment. Their sizes are reasonable: not so small as to require large numbers to approximate the diagonal nor so large—or fat— as to produce a poor approximation. At the same time, we did not observe the evolution of just *two* dominant ellipses, one covering each landscape region, each very long so that its side was practically a straight line. This would have been the ideal result, and it is not clear why the system's generalization pressure [1] did not cause larger- and particularly longer-condition accurate classifiers to win out until just two dominant ones remained.

In the binary domain, generalization involves substituting #'s (don't-cares) for specified alleles in the condition. Each additional # in fact doubles the subspace that the condition matches. If corresponding inputs occur, the classifier becomes twice as active and thus has twice the reproductive opportunity compared with classifiers that are just one # less general. In the real-valued domain, in contrast, mutation of a single allele normally results in a condition that is at most only slightly more general than the next-more-specific classifiers, so that the evolutionary pressure over them is considerably less than in the binary case. It may be that this weakness of generalization pressure explains the fact that

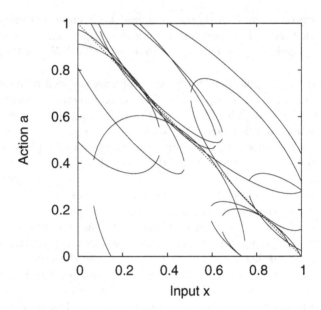

Fig. 7. Elliptical conditions of the 10 highest-numerosity classifiers in GCS frog problem experiment. Diagonal $(a = 1 - x)$ corresponds to "ridge" in payoff landscape. Breaks in ellipses are a plotting artifact.

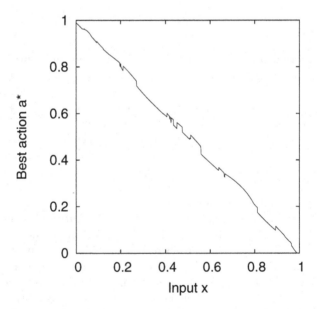

Fig. 8. Scan of best action a^* vs. input x for GCS frog problem experiment (x increment 0.001)

the population did not appreciably "condense" (shrink) in our experiment, and may explain other similar real-domain classifier system results. One of the most attractive features of accuracy-based classifier systems is their ability, under the right conditions, to generalize, i.e., condense down to a set of classifiers that quite directly expresses the regularities of the payoff environment. It seems important for future research to understand how this process can be as effective in the real domain as it is in the binary—clearly the present frog problem offers the possibility of very substantial generalization: to just two classifiers!

Apart from the important issue of generalization, the GCS concept appears, from the experiments so far, to be relatively sound. It is not understood why the performance became more volatile with time but more intimate tracking of populations should yield clues. Often deterioration of this sort is due to increasing overgeneralization—which in this instance would mean classifiers whose conditions fall across the landscape diagonal. Yet, as seen, the highest-numerosity classifiers at 100,000 problems were fine in this respect.

While volatile, the performance did rise quickly to near-optimal. Why it did not go all the way to optimal (payoff 1.0) appears to depend at least in part on the specific classifier that ends up providing its a_{best} for a^*. The edge of this classifier's condition lies close to the landscape diagonal: it may fall slightly short of the diagonal, or go over a bit. In some cases the classifier may be newly generated and inexperienced and thus not accurate; choosing its a_{best} could mean a large error. In earlier experiments it was found that system error improved if classifiers with prediction errors greater than a threshold were excluded from the a^* competition; this criterion was in force in the experiment reported.

Unlike the frog problem, most practical problems involve more than one input dimension. With a two-dimensional input, GCS would need ellipsoidal conditions, which require three (Euler) angles to specify their orientation. Above that, direct visualization is lost and (if keeping an ellipsoid-like basis) one would need to move to general hyperellipsoids of the kind already investigated by Butz [2]. It is interesting, however, that even for higher input dimensionality, the calculation of a_{best} still only requires solution of a quadratic equation.

7   General Conclusion

Three architectures for continuous action in classifier systems have been introduced and examined. None appears unreasonable, and the third, GCS, appears to be perhaps the most interesting direction since it treats the action homogeneously with the input—reflecting the nature of the payoff landscape—and it is the only one that learns the landscape and not just its highest-paying region. Yet all three architectures as presented are quite crude and unrefined and need much more work (including, probably, correction of errors!). Perhaps the most useful perspective on the offering here is as a smorgasbord from which with careful selection and combination new insights toward continuous action might be gleaned.

References

1. Martin Butz, Tim Kovacs, Pier Luca Lanzi, and Stewart W. Wilson. Toward a theory of generalization and learning in XCS. *IEEE Transactions on Evolutionary Computation*, 8:28–46, 2004.
2. Martin V. Butz. Kernel-based, ellipsoidal conditions in the real-valued XCS classifier system. In Hans-Georg Beyer, Una-May O'Reilly, Dirk V. Arnold, Wolfgang Banzhaf, Christian Blum, Eric W. Bonabeau, Erick Cantu-Paz, Dipankar Dasgupta, Kalyanmoy Deb, James A. Foster, Edwin D. de Jong, Hod Lipson, Xavier Llora, Spiros Mancoridis, Martin Pelikan, Guenther R. Raidl, Terence Soule, Andy M. Tyrrell, Jean-Paul Watson, and Eckart Zitzler, editors, *GECCO 2005: Proceedings of the 2005 conference on Genetic and evolutionary computation*, volume 2, pages 1835–1842, Washington DC, USA, 25-29 June 2005. ACM Press.
3. Martin V. Butz and Stewart W. Wilson. An Algorithmic Description of XCS. In Pier Luca Lanzi, Wolfgang Stolzmann, and Stewart W. Wilson, editors, *Advances in Learning Classifier Systems*, volume 1996 of *LNAI*, pages 253–272. Springer-Verlag, Berlin, 2001.
4. Ali Hamzeh and Adel Rahmani. An evolutionary function approximation approach to compute prediction in XCSF. In *Proceedings of the European Conference on Machine Learning*, pages 584–592, 2005.
5. Lanzi P. L. and S. W. Wilson. Classifier conditions based on convex hulls. Technical report, Illinois Genetic Algorithms Laboratory, University of Illinois at Urbana-Champaign, 2005.
6. P. L. Lanzi, D. Loiacono, S. W. Wilson, and D. E. Goldberg. Generalization in the XCS classifier system: analysis, improvement, and extension. Technical report, Illinois Genetic Algorithms Laboratory, University of Illinois at Urbana-Champaign, 2005.
7. Pier Luca Lanzi, Daniele Loiacono, Stewart W. Wilson, and David E. Goldberg. Extending XCSF beyond linear approximation. In Hans-Georg Beyer, Una-May O'Reilly, Dirk V. Arnold, Wolfgang Banzhaf, Christian Blum, Eric W. Bonabeau, Erick Cantu-Paz, Dipankar Dasgupta, Kalyanmoy Deb, James A. Foster, Edwin D. de Jong, Hod Lipson, Xavier Llora, Spiros Mancoridis, Martin Pelikan, Guenther R. Raidl, Terence Soule, Andy M. Tyrrell, Jean-Paul Watson, and Eckart Zitzler, editors, *GECCO 2005: Proceedings of the 2005 conference on Genetic and evolutionary computation*, volume 2, pages 1827–1834, Washington DC, USA, 25-29 June 2005. ACM Press.
8. Pier Luca Lanzi and Stewart W. Wilson. Toward optimal classifier system performance in non-Markov environments. *Evolutionary Computation*, 8(4):393–418, 2000.
9. Richard S. Sutton and Andrew G. Barto. *Reinforcement Learning: An Introduction*. The MIT Press/Bradford Books, Cambridge, MA, 1998.
10. C. J. C. H. Watkins. *Learning From Delayed Rewards*. PhD thesis, Cambridge University, 1989.
11. Stewart W. Wilson. Continuous action. Slides presented at the Eighth International Workshop on Learning Classifier Systems (IWLCS-2005). Available at prediction-dynamics.com.
12. Stewart W. Wilson. Optimal continuous policies: a classifier system approach. Extended Abstract, International Workshop on Learning Classifier Systems (IWLCS-2004). Available at prediction-dynamics.com.

13. Stewart W. Wilson. Classifier Fitness Based on Accuracy. *Evolutionary Computation*, 3(2):149–175, 1995.
14. Stewart W. Wilson. Function approximation with a classifier system. In Lee Spector, Erik D. Goodman, Annie Wu, W.B. Langdon, Hans-Michael Voigt, Mitsuo Gen, Sandip Sen, Marco Dorigo, Shahram Pezeshk, Max H. Garzon, and Edmund Burke, editors, *Proceedings of the Genetic and Evolutionary Computation Conference (GECCO-2001)*, pages 974–981, San Francisco, California, USA, 7-11 July 2001. Morgan Kaufmann.
15. Stewart W. Wilson. Classifier systems for continuous payoff environments. In Kalyanmoy Deb, Riccardo Poli, Wolfgang Banzhaf, Hans-Georg Beyer, Edmund Burke, Paul Darwen, Dipankar Dasgupta, Dario Floreano, James Foster, Mark Harman, Owen Holland, Pier Luca Lanzi, Lee Spector, Andrea Tettamanzi, Dirk Thierens, and Andy Tyrrell, editors, *Genetic and Evolutionary Computation – GECCO-2004, Part II*, volume 3103 of *Lecture Notes in Computer Science*, pages 824–835, Seattle, WA, USA, 26-30 June 2004. Springer-Verlag.

A Formal Relationship Between Ant Colony Optimizers and Classifier Systems

Lawrence Davis

President, VGO Associates
Senior Fellow, NuTech Solutions
ddavis@vgoassociates.com

Abstract. This paper demonstrates that, with minimal modifications, a classifier system can be made to operate just as an ant colony optimizer does for solving the TSP. The paper contains a formal proof of this result, and suggests that the modifications made could be useful in other ways. In effect, the paper suggests that there may be a new role for classifier systems in optimization, inspired by the way that ant colony optimizers have achieved their successes. The paper also suggests that there may be ways suggested by classifier systems to modify ant colony optimization practice.

1 Introduction

This paper centers on three topics. First, it shows how to transform any ant colony optimizer that solves the Traveling Salesman Problem, as specified in initial papers by Marco Dorigo and his collaborators, into a functionally equivalent classifier system. Second, it considers whether a reverse transformation, from any classifier system into an equivalent ant colony optimizer, can be produced in a similar way. Finally, it suggests that the formal results in the paper have some practical implications for the future direction of both ant colony optimizer and classifier system applications.

2 A Formal Specification of ACO-TSP

In the seminal papers Dorigo *et al* 1996 and Dorigo *et al* 1997, Marco Dorigo and his collaborators went a long way toward creating the current field of ant systems and ant colony optimization. In those papers the authors described ant colony optimizers—ingenious optimization algorithms inspired by the way ants find short paths to and from food sources. The initial ant colony optimizer was used to solve the Traveling Salesperson Problem—also a problem having to do with minimizing path lengths—and it had some features that were new to the world of algorithms. These novel features included a new source of algorithmic metaphor—chemical deposition and evaporation as carried out by foraging ants; new ways of using solution evaluations in the optimization process; and a specification of the current state of the Traveling Salesperson Problem (henceforth, "TSP") optimization process that was completely contained in two vectors, one of fixed and one of variable weights.

X. Llorà et al. (Eds.): IWLCS 2003-2005, LNAI 4399, pp. 258–269, 2007.

Let us call the ant colony optimizer that Dorigo described in his paper the ACO-TSP. The next two sections of this paper describe the TSP and the ACO-TSP in a formal way.

3   A Formal Description of the TSP

A central point of this paper is a proof that any ant colony optimizer of the type ACO-TSP can be transformed into a functionally equivalent classifier system. In order to state the proof clearly, we require some formalism. First, let us consider some notation that describes the TSP itself.

Let G be the graph for which short paths containing all nodes are to be found.
Let N be the set of nodes in G.
Let n_i be the ith member of N.
Let L be the set of links in G such that every pair of nodes in N has exactly one associated member of L, and let l_i be the ith member of L.
Let J be the number of nodes in n.
Let K be the number of links in L.
Let T, a solution to the TSP, be a permutation of N.

There are several functions used in the definition of the TSP.
Let $A(n_i, n_j)$ be the link in L associated with nodes n_i and n_j.
Let $D(n_i, n_j)$ be the distance between nodes n_i and n_j, such that $D(n_i, n_j) = D(n_j, n_i)$.
Let $E(T)$ be the evaluation of a solution T. For every solution T, where each n_i is a member of T, $E(T) = D(n_1, n_2) + D(n_2, n_3) + ... + D(n_{J-1}, n_J) + D(n_1, n_J)$.

Given these definitions, we can then say that the TSP is the task of finding an ordering T of the nodes of G such that $E(T)$ is minimized.

4   A Formal Description of ACO-TSP

Now let us describe the structure of ACO-TSP (please note in what follows that, in order to expedite exposition, this description is not identical to Dorigo and his co-authors' practice with regard to variable names and with regard to terminology.)
ACO-TSP has a number of parameter settings. These include:

a, an exponent used when the strength of a link is computed;
b, an exponent used when the strength of a link is computed;
d, a weight decay rate;
u, the value of each fixed weight;
m, the total number of solutions to build during optimization;
x, the number of solutions to build before updating weights; and
s, the seed for the random number generator used by ACO-TSP.

ACO-TSP uses two data structures to hold weights. Each is a vector of length K:

F, a list of fixed weights, one for each link in L, and
V, a list of variable weights, one for each link in L.

The values in F are set upon initialization and do not change during the optimization process. The values in V are given starting values on initialization and are modified during the optimization process during a weight update process.

ACO-TSP uses two data structures to hold solutions and their scores between weight updates. Each is a vector of length x:

X, a list of solutions, and
Y, a list of evaluations of these solutions.

ACO-TSP uses two data structures to hold the best solution found and its evaluation:

B holds the best solution found.
E is the evaluation of B.
Q is ACO-TSP's pseudorandom number generator, with a state initialized by s.
$O(l_i, n_j)$ is the other node on a link. That is, where l_i connects nodes n_1 and n_2,
$$O(l_i, n_1) = n_2 \text{ and } O(l_i, n_2) = n_1.$$

5 How ACO-TSP Works

This section contains the steps that ACO-TSP goes through in order to produce a solution to a TSP.

Step 1: Initialization
  1. For each field f_i in F, $f_i = u$
  2. For each field v_i in V, $vi = 1/D(n_i, n_j)$, where n_i and n_j are the two nodes connected by l_i.
  3. $B = $ null.
Step 2: Optimization
Repeat this process until m solutions have been produced:
  1. $i = 1$
  2. Create T, a solution to the TSP, executing the Solution Generation Procedure described below
  3. If B is null or if $E(T) < E(B)$, set B to T
  4. $X_i = T$ and $Y_i = E(T)$
  5. Increment i
  6. If i = updateInterval, execute the Weight Update Procedure described below and go to 1; otherwise, go to 2

When m solutions have been produced, B holds the best solution found, and E is its evaluation.

The Solution Generation Procedure, referred to above, proceeds as follows. It generates a solution T:

  1. $j = 1$
  2. Set each field of T to a null value
  3. Set T_1 to a node randomly chosen from the members of N
  4. Compute a probability vector P of length K, such that for each p_i in P, if l_i connects to n_p and if $O(l_i, n_p)$ is not a member of T, then $p_i = Fi^a + Vi^b$. If l_i does not connect to n_p or if $O(l_i, n_p)$ is a member of T, then $p_i = 0$.

5. Choose i from the distribution based on P, such that each i has a chance of being chosen proportional to P_i. (This is what evolutionary algorithm practitioners will recognize as a *roulette wheel* selection process).
6. Set $n_p = O(l_i, n_p)$.
7. Set $T_j = n_p$.
8. Increment j.
9. If $j = J$, stop; else, go to 5.

The output of this procedure is T, a set of J nodes that are a solution to the TSP. The Weight Update Procedure, referred to above, proceeds as follows.

1. For each weight v_i in V, set v_i to be $v_i * d$
2. For each solution x_i in the list of recently generated solutions X, and for each evaluation v_i in the list of evaluations for those solutions, use the Solution Weight Update procedure to update x_i's weights.
3. Clear X and V.

The Solution Weight Update procedure, referred to above, proceeds as follows. It updates the variable weights on the links that are determined by a solution T with evaluation $E(T)$:

1. For each pair of nodes n_j, n_k in T, including the pair consisting of the first and last nodes in T,
   a.   let $L_i = A(n_j, n_k)$.
   b.   Increment V_i by $1/E(T)$.

This concludes the description of ACO-TSP.

6   Can an Equivalent Classifier System Be Produced?

An examination of the preceding description of ACO-TSP suggests that an attempt to create a functionally equivalent classifier system is unlikely to succeed. Here are some differences between ant colony optimizers and classifier systems that would lead us to believe that such a project is infeasible:

- Ant colony optimizers have no functional units corresponding to classifiers.
- Classifiers have fitnesses, which are missing in ant colony optimizers.
- Classifier systems use an evolutionary algorithm to create new structures, and the ant colony optimizer does not.
- No new structures are added by the ant colony optimizer during optimization, but they are added during the run of a typical classifier system.
- Classifier systems do not optimize; instead, they form cooperating and competing groups of rules that generate responses to external stimuli.
- Ant colony optimizers do not typically generate responses to external stimuli; instead, they optimize.

In what follows, these differences between ant colony optimizers and classifier systems will be resolved in one of two ways. Either the classifier system will be reduced in functionality to eliminate capabilities that are not needed to carry out the transformation of ACO-TSP into an equivalent classifier system, or the classifier system will be extended with capabilities that it does not have in standard practice, but that are required to support the equivalence result. In a later section of this paper, we will consider the question whether the exercise has produced any result of practical value, given that there have been modifications made to the traditional classifier system in order to cause its functionality to match that of the ant colony optimizer.

7 A Formal Description of CS-TSP

Let us now consider CS-TSP, a classifier system that is functionally equivalent to ACO-TSP. First, some general remarks about CS-TSP and the intuitions underlying it.

First, following general classifier system practice, we describe two entities involved in the optimization. One is an external *environment* that presents stimuli to CS-TSP, receives CS-TSP's responses, and provides CS-TSP with feedback from time to time. The other is CS-TSP itself.

The environment will do two things. During the process of building solutions, it will present CS-TSP with binary input patterns and use CS-TSP's responses to build routes. It will also present CS-TSP with the evaluation of each route that is produced.

CS-TSP will store solutions and their evaluations, just as ACO-TSP did. Let us use the same terminology as in our description of ACO-TSP for the vector of solutions and the vector of solution evaluations used by the environment—they will be referred to as X and E. CS-TSP will also store its best solution and its best solution's evaluation. As we did for ACO-TSP, we will denote these as B and E.

Now let's look at CS-TSP in detail. CS-TSP contains these components:

C, a list of classifiers, two for each li in L;
X, a list of solutions,
Y, a list of evaluations of these solutions;
B, the best solution found;
E, the evaluation of B; and
Q, CS-TSP's pseudorandom number generator, shared with its environment, with a state initialized by s.

Each classifier in CS-TSP has these components: a match pattern, an output pattern, a fixed strength, and a variable strength.

A classifier's match pattern is a concatenation of two symbol patterns, each of length J. The first half of a match pattern contains $J-1$ zeros and one "1", which can occur at any position in this part of the list. For example, if $J=8$, then this is a possible first half of a match pattern for a CS-TSP classifier: 0 0 0 0 1 0 0 0. The symbol "1" represents the current node in the process of constructing a route. If this pattern forms the first part of a match pattern, then the classifier having that match pattern will match a case when the 5th node is the one just arrived at in the current construction of the route.

The second half of the match pattern is a pattern containing $J-1$ "don't care" symbols (represented as "#" here) and one "1", which can occur at any position in this part of the list, except the position occupied by the "1" in the first half of the match pattern. For example, if $J=8$, then this is a possible second half of a match pattern for a CS-TSP classifier: # 1 # # # # # #. The symbol "1" represents the next node to go to in the process of constructing a route. If this pattern and the one described above form the match pattern for a classifier, then the match pattern would be written as

Example match pattern: 0 0 0 0 1 0 0 0 # 1 # # # # # #,

and the interpretation of this match pattern would be that if node 5 is the current node, and if node 2 is a legal next node for the route at this point, then the next node in the route should be node 2.

The output pattern for a classifier is a bit string J bits long, like the second half of the match pattern, except that "0"s are substituted for "#"s. Thus, the output pattern for the classifier in the example just discussed would be

Example output pattern: 0 1 0 0 0 0 0 0.

The fixed strength of a classifier will represent the fixed strength of a link in ACO-TSP. The variable strength of a classifier will represent the variable strength of a link in ACO-TSP.

The fixed weight and variable weight of a classifier correspond to their values stored in vectors in ACO-TSP. These weights may be somewhat reminiscent of the way that fitness, strength, accuracy, and other measures of performance are associated with classifiers in standard classifier system practice.

A classifier containing the components listed above is associated with a link in L. The classifier's fixed and variable strengths will be equal to the fixed and variable strengths of that link during the corresponding path construction and weight update procedures in ACO-TSP.

Now let us define a function that will associate classifiers in CS-TSP with links in the TSP. Let c be a CS-TSP classifier. Then $L(c)$ is the link associated with the classifier. If i is the position of the "1" in the first half of the classifier's match pattern and j is the position of the "1" in the second half of the classifier's match pattern, and if n_i is the ith member of N and n_j is the jth member of N, then $L(c)$ is $A(n_1, n_2)$. That is, the link associated with a classifier is the link connecting the node it references in the first half of its match pattern to the node it references in the second half of its match pattern.

The two correlate classifiers for a link L are in a set that is the value of the function $C(L)$. Assume that L connects nodes n_i and n_j. Then C(L) is c_k and c_l, where the only two 1's in the match pattern of c_k are in the ith and $J+k$th positions and the only two 1's in the match pattern of c_l are in the kth and $J+i$th positions.

Finally, let us define the correlate of a classifier. $U(c)$ is the correlate of c. That is, if $L(c) = l_i$, then $U(c)$ is the classifier c_k not equal to c such that c_k is a member of $C(l_i)$.

8 How CS-TSP Works

Now let us consider how CS-TSP optimizes.

Step 1: Initialization.
1. Let S, the set of classifiers in CS-TSP, contain all classifiers c such that the first half of the match pattern of c is of the form described above, the second half of the match pattern of c is of the form described above, and the output pattern of c is derived from the second half of the match pattern of c as described above. S will contain no duplicate classifiers, and no classifiers other than those just described. S is a set of classifiers that represents all the links in G, with two classifiers for each link—one for each direction of traversal of the link.
2. Set the fixed strength of each classifier c in S to equal the fixed strength of $L(c)$ in ACO-TSP. Set the variable strength of each classifier c in S to equal the variable strength of $L(c)$ in ACO-TSP when ACO-TSP is initialized.
3. Order the classifiers in S so that the two classifiers $C(l_1)$ are first, then the two classifiers $C(l_2)$, and so on. Now the order of the classifiers (and their weights) matches the order of the weights in the ACO-TSP weight vectors.
4. Set B to null.

Step 2: Optimization

Repeat this process until m solutions have been produced:
1. $i = 1$
2. Create T, a solution to the TSP, executing the Solution Generation Procedure described below
3. If B is null or if $E(T) < E(B)$, set B to T and set E to $E(B)$
4. The environment passes T and $E(T)$ to CS-TSP, and CS-TSP sets $X_i = T$ and $Y_i = E(T)$
5. Increment i
6. If $i = x$, execute the Weight Update Procedure described below and go to 1; otherwise, go to 2

When m solutions have been produced, B holds the best solution found, and E is its evaluation.

The Solution Generation Procedure, referred to above, proceeds as follows. It generates a solution T:
1. $j = 1$
2. Initialize each field of T to be null
3. The environment sets n_1 to a node randomly chosen from the members of N, using one call to the random number generator to select the initial node.
4. The environment constructs a message to send to CS-TSP by concatenating these two bit strings:
    a. J bits all of which are 0 except the pth bit, where p is the index of the current node in the path, and
    b. J symbols all of which are "1" except for a "0" at the position of the index of each node already on the path
5. CS-TSP forms the set of classifiers matching the environment's message.

6. CS-TSP chooses the winning classifier from the match set by using a roulette wheel selection process, where the selection weight associated with each classifier matching the message equals $f^a + v^b$. This results in one call to the random number generator.
7. The output message of the winning classifier is sent to the environment.
8. Where i is the position of the "1" in the output message, the environment sets $T_j = n_i$.
9. Increment j.
10. If $j = J$, stop. Otherwise, go to 5.

The Weight Update Procedure, referred to above, proceeds as follows:

1. For each classifier c in S and variable weight v in c, $v = v*d$.
2. For each solution T_i in the list of recently generated solutions X, and for each evaluation v_i in the list of evaluations for those solutions, use the Solution Weight Update procedure to update classifier weights.
3. Clear X and V.

The Solution Weight Update procedure, referred to above, proceeds as follows. It updates the variable weights on the classifiers associated with a solution T_i with evaluation E_i:

1. Compute w, the new weight to be used in the update, equal to $1/E_i$.
2. For each pair of adjacent nodes n_1 and n_2 in T_i, including the pair consisting of the first and last nodes of T_i,
    a. Let L_j be the link $D(n_1, n_2)$
    b. Let c_j and c_k be the classifiers in the set $C(L_j)$
    c. Add w to the variable weight of c_j and c_k

(Note that since two classifiers are correlated with each link, in order to match the functioning of ACO-TSP, the weights on both classifiers must be incremented.)

Table 1 is an example of the way that CS-TSP would work for a three-node TSP, assuming $a=1$ and $b=1$.

Table 1. Table of classifiers

Associated link	Match pattern	Output pattern	Fixed strength	Variable str.
1-2	1 0 0 # 1 #	0 1 0	1.6	23.9
2-1	0 1 0 1 # #	1 0 0	1.6	14.7
1-3	1 0 0 # # 1	0 0 1	1.6	12.0
3-1	0 0 1 1 # #	1 0 0	1.6	4.2
2-3	0 1 0 # # 1	0 0 1	1.6	37.3
3-2	0 0 1 # 1 #	0 1 0	1.6	21.6

Let us suppose that the environment chose node 2 as the first node in the route. It then sends the message "0 1 0 1 0 1" to CS-TSP. The 2-1 and 2-3 classifiers are the only ones that match. The 2-1 classifier recommends node 1 with a strength of 16.3

and the 2-3 classifier recommends node 3 with a strength of 38.9. Let's suppose that the environment calls the random number generator for a number between 0 and 55.2 (= 16.3 and 38.9). The returned value is 24.6, so the 2-3 classifier wins and node 3 is the next node in the route. The environment updates the route so that it goes from 2 to 3. Now it wants to know what node to go to next. It sends the message "0 0 1 1 0 0" to CS-TSP. (Note that the last half of the message restricts the route from revisiting nodes 2 or 3.) The only classifier to match is classifier 3-1. Node 1 is added to the route and B and E are updated if this route is better than the current B. The route construction is complete. The route and its evaluation are stored in the vectors X and Y.

One important point to note in this simple example is that the message sent from the environment activates all and only those classifiers representing links that could be used at this point in the route construction for ACO-TSP. The weights on the classifiers determine which link is followed through a roulette wheel selection process, just like the one that occurs in ACO-TSP.

9   A Proof of the Functional Equivalence of ACO-TSP and CS-TSP

Given the notation introduced in the last two sections, we can now show that ACO-TSP and CS-TSP are functionally equivalent. In what follows, it may be helpful to assume that we are watching runs of computer programs executing both ACO-TSP and CS-TSP on the same problem, with both ACO-TSP and CS-TSP using pseudo-random number generators with the same structure. That is, if their random number generators are in the same state, then each algorithm will generate the same sequence of random numbers from that point forward. We assume also that CS-TSP and its environment share the same random number generator, so that a call to the random number generator by either CS-TSP or its environment changes the state of the same random number generator.

The proof proceeds as follows.

Definition: ACO-TSP and CS-TSP are functionally equivalent if and only if:

1. The pseudorandom number generators for the two algorithms are in the same state,
2. The routes under construction for the two algorithms contain the same nodes in the same order,
3. The X and Y vectors for the two algorithms contain identical entries,
4. The fixed weights for the classifiers in CS-TSP equal the entries in F, and
5. For each link L_i, V_i in ACO-TSP equals the variable weights for the two classifiers in $C(L_i)$.

Theorem 1: ACO-TSP and CS-TSP are functionally equivalent after initialization.

Proof: All five requirements of functional equivalence are met on initialization: the random number generators for the two algorithms are initialized by the same seed; there are no routes under construction; there are no X and Y entries; and all fixed and variables weights in CS-TSP were set equal to their correlates in ACO-TSP. Hence, the theorem is true.

Theorem 2: if ACO-TSP and CS-TSP are functionally equivalent before a selection of a node for a route, then they are functionally equivalent after the selection of a node for a route.

Proof: If the node being selected is an initial route node, then it is selected by a similar call to Q, which is in an identical state by hypothesis. If the node being selected is a later node in the route, then it is selected by a call to Q with a roulette wheel selection. By inspection, we see that each link considered by ACO-TSP has a corresponding classifier in the match set constructed by CS-TSP. The order of the classifiers matches the order of the links, and the weights of the classifiers in CS-TSP equal the weights of the links in ACO-TSP, so the roulette wheel selection process will make identical node choices for the two algorithms, with a single call. Parts 1 and 2 of the definition of functional equivalence are maintained by this process, and parts 3, 4, and 5 are unaffected, so the theorem is proved.

Theorem 3: if ACO-TSP and CS-TSP are functionally equivalent before constructing a route, then they are functionally equivalent after constructing a route.

Proof: The construction of a route involves iterating the selection of a node to add to the route. Theorem 2 proves that this process does not violate the state of functional equivalence. Hence, the theorem is proved.

Theorem 4: if ACO-TSP and CS-TSP are functionally equivalent before updating weights, then they are functionally equivalent after updating weights.

Proof: The weight update process involves decrementing all variable weights in both algorithms by the same constant. It then involves adding a quantity associated with each route in X to the variable weights (in ACO-TSP) and the variable weights of classifiers (in CS-TSP). By inspection, we see that for each update of a variable weight in ACO-TSP, there is a corresponding update of the variable weights of the two corresponding classifiers in CS-TSP. Parts 1-3 of the definition of functional equivalence are unaffected by this process. Parts 4 and 5 are maintained. Hence, the theorem is proved.

Theorem 5: ACO-TSP and CS-TSP are functionally equivalent throughout their runs on the same problem.

Proof: Theorem 1 tells us that the two algorithms begin their runs in a state of functional equivalence. The runs consist of only two processes: constructing routes and updating weights. Theorem 3 tells us that constructing routes preserves functional equivalence. Theorem 4 tells us that updating weights preserves functional equivalence. Hence, functional equivalence is preserved throughout the runs, and the theorem is proved.

10 Is the Reverse Mapping Possible?

What about the reverse process? Could we create an ant colony optimizer that mimics the operation of any classifier system? This is an interesting question. It would seem not, since there are a number of features of classifier systems that the basic ant colony optimizer does not include. These were mentioned earlier, and the evolutionary algorithm is one of the most significant. Importing the machinery of an evolutionary algorithm into an ant colony optimizer seems a larger-scale expansion than any of the modifications of the classifier system carried out above. Nonetheless, it is interesting to consider what such expansions would be like, for they would involve new types of optimization algorithms, with possible niches in which they could perform better than any algorithms we currently use. Consideration of what it would take to produce such a reverse mapping would a good subject for another paper.

11   Is There Any Practical Relevance of the Proof Presented Here?

There are two areas in which it seems the proof above may help us to extend or improve the practice of classifier systems and ant colony optimizers.

The first is the addition of a solution "memory" to classifier systems. The proof given above shows how to convert a basic ant colony optimizer into an equivalent classifier system. The conversion required us to reduce classifier system functionality (particularly with regard to using an evolutionary algorithm to find new classifiers) and to add a solution maintenance and reinforcement mechanism that has not been used in standard classifier system practice. This mechanism involved maintaining several complete solutions and their evaluations, and reinforcing classifiers from multiple solutions at a single time. Classifier systems typically reinforce their classifiers after a single complete solution has been produced—a single reinforcement event. Dorigo notes in his paper that this multiple-solution approach to reinforcement works much better than reinforcement after a single solution is produced, so a classifier system based on an ant colony optimizer and lacking the capability to store information on multiple solutions will perform worse than one that has this capability.

Once we accept the need to add this module in order to match ACO-TSP capabilities, could we consider adding such a module to classifier systems for other reasons? It seems so. In Davis, Wilson, and Orvosh 1993, the authors show that if a classifier system maintains a list of examples that were incorrectly classified together with their solutions, and if the classifier system reruns those examples "internally" at some low frequency in addition to the examples that are generated by the environment, this process increases the classifier system's learning speed. Similarly, it would appear that "remembering" routes and their evaluations could provide similar benefits for a self-reflective classifier system that is solving a TSP. And similar benefits could obtain for an animat if it were given a memory of prior examples and their outcomes. In each of these cases, it would appear that the classifier system can gain performance benefits if it has the extra memory available to "remember" prior cases.

The point made here is that the reinforcement mechanism added to CS-TSP is a special case of adding a memory for examples and their evaluations to the classifier, and this capability has already been shown in other contexts to provide benefits to classifier systems. One's hope is that the present result suggests other benefits to be gained, if one were to apply classifier systems seriously to combinatorial optimization problems, the area in which ant colony optimizers have most commonly operated.

A second practical consequence of the formal result above may be relevant to ant colony optimizers. As the size of a TSP increases, the number of links grows quickly. If the ant colony optimizer has memory limitations, then it may be necessary to reduce the number of links under consideration. If we suppose that links are maintained locally—that the links in memory are those between closest nodes in the problem—and if we suppose that the coding of the links reflects proximity, perhaps in a fashion akin to Gray coding, then it could be of use to an ant colony optimizer to use an evolutionary algorithm to generate new links that could be tried by the ant colony optimizer. This could also be a useful approach in problems of real-valued optimization rather than combinatorial optimization, where the number of possible positions is infinite. One promising approach might be to add an evolutionary algorithm module to the ant colony optimizer that provides it with new route points to try as it attempts to construct optimal answers.

12 Conclusions

This paper has shown a way to create a classifier system that functions in the same way as an ant colony optimizer for solving the TSP. The translation process includes modifying the capability of the classifier system approach, through capability deletion and addition. Nonetheless, it is interesting to see how little addition of capability is required in order to accomplish the transformation. The results above suggest an interesting approach to extending classifier system practice, and an interesting approach to extending ant colony optimizer practice. Finally, it may be interesting to recall that Marco Dorigo was a long-time, experienced classifier system practitioner before he began his seminal work on ant colony optimizers. The translation procedure described here might have a historical, as well as theoretical, grounding.

References

1. Ant Colony Optimizers: A web site maintained by Marco Dorigo that provides information about the field of ant colony optimizers can be found at http://www.aco-metaheuristic.org.
2. Classifier Systems: A web site maintained by Alwyn Barry that provides access to the field of learning classifier systems can be found at http://lcsweb.cs.bath.ac.uk.
3. Davis L., Wilson S., and Orvosh, D.: Temporary Memory for Examples Can Speed Learning in a Simple Adaptive System. Proceedings of the Second International Conference on the Simulation of Adaptive Behavior, Stewart Wilson editor. MIT Press, 1993.
4. Dorigo M. & Gambardella L.M. : Ant Colonies for the Traveling Salesman Problem. Bio-Systems, 43:73-81 (1997).
5. Dorigo M., Maniezzo V. and Colorni A.: The Ant System: Optimization by a Colony of Cooperating Agents. IEEE Transactions on Systems, Man, and Cybernetics-Part B, 26(1):29-41 (1996).

Detection of Sentinel Predictor-Class Associations with XCS: A Sensitivity Analysis

John H. Holmes

University of Pennsylvania School of Medicine
Philadelphia, Pennsylvania 19104 USA
jholmes@cceb.med.upenn.edu

Abstract. Knowledge discovery in databases has traditionally focused on classification, prediction, or in the case of unsupervised discovery, clusters and class definitions. Equally important, however, is the discovery of individual predictors along a continuum of some metric that indicates their association with a particular class. This paper reports on the use of an XCS learning classifier system for this purpose. Conducted over a range of odds ratios for a fixed variable in synthetic data, it was found that XCS discovers rules that contain metric information about specific predictors and their relationship to a given class.

1 Introduction

A number of study designs exist for the collection and analysis of epidemiologic surveillance data. These range from static, one-time observational (prevalence) studies of an existing population or sample, to ongoing observation of a given population or sample. The latter, commonly called a *cohort study*, is of particular interest, in that it provides the ability to investigate the incidence of outcomes, such as specific diseases, injuries, or other clinical events over time. Because of this added informational dimension, it is possible to create more robust models of causation that as would be the case in simple observational studies. Individuals are enrolled into a cohort based on their exposure to a putative agent or *risk factor*. Ingestion of a particular drug, cigarette smoking, or employment in a chemical factory are examples of risk factors or *exposures* that could be investigated in a cohort study. Often, more than one risk factor, and almost certainly many covariates, will be included in the set of predictor variables under investigation. Once enrolled, the members of the cohort are observed over time to determine if an outcome of interest occurs. Cohorts may be followed *prospectively* or *retrospectively*, where the investigation begins before or after, respectively any outcomes have occurred. A prospective cohort study may be used to evaluate risk factors associated with outcomes having short latency, or time-to-event after exposure. They tend to be expensive, as they require sophisticated follow-up methods to ensure the maintenance of the cohort, and even in outcomes with short latency they can require extended follow-up time. Retrospective cohort studies can be more cost-effective; however, as they rely on data that has already been collected, often for another purpose, they can suffer from poor data quality. Even

X. Llorà et al. (Eds.): IWLCS 2003-2005, LNAI 4399, pp. 270–281, 2007.

with these limitations, cohort studies are very common in epidemiologic surveillance and research.

One of the essential analytic parameters in the cohort study is the *relative risk*, sometimes called the risk ratio, or *RR*. The RR is simply the ratio of the incidence of an outcome in those with a given exposure to the incidence of the outcome in those without the exposure. The RR indicates the degree to which an individual is at risk of an outcome given an exposure, compared to one who has not experienced the exposure[6]. Restated, the RR indicates the degree to which a specific exposure influences the risk of developing a given outcome. For dichotomous exposure and outcome variables, the RR is calculated as shown in Figure 1.

Exposure	Class	
	Outcome	No outcome
Exposed	A	B
Not Exposed	C	D

Fig. 1. A 2x2 table showing outcome by exposure status

The relative risk is calculated as follows:

$$RR = \frac{A/(A+B)}{C/(C+D)} \tag{1}$$

A RR less than 1 indicates a *protective effect* of the exposure of interest on clinical outcome; while a RR exceeding than 1 indicates a positive association of the outcome with the exposure. If the RR is 1.0, the predictor has no association, positive or protective, with the outcome. The RR described in the above figure is commonly referred to as the *crude risk ratio*, because it has not been adjusted for other variables that may be important confounders or effect modifiers. While the crude RR is useful, the adjusted RR, typically derived by means of a Poisson regression or generalized linear model, is of more importance in ascertaining the true effect of a given variable on an outcome. However, these multivariate models are not without some degree of weakness: small sample sizes, missing data, and model complexity that may cause failure of the model to converge are some of the reasons why important variables might not be identified as statistically significant. This is especially the case in the early phases of epidemiologic surveillance, where the number of exposure-outcome associations may be small and apparently random. The discovery of features (variables as well as specific values of variables) that act as sentinels in alerting investigators to potential relationships is of great importance to epidemiologic surveillance and research.

This paper describes an investigation into the ability of an XCS-type learning classifier system to identify such exposure-outcome associations in a simulated cohort study under strict experimental conditions. Of particular interest is the detection of *sentinel features*, or those variables that may provide "early warning" as to the association between an exposure and outcome. To evaluate this ability, a sensitivity analysis was performed for this investigation, focusing on a single candidate sentinel feature nested in an array of covariate exposure variables. A series of 10 different cohorts were simulated over a range of RRs from 0.5 (highly protective) to 5.0 (high risk).   These cohorts were experimentally constrained to ensure that their size, outcome distribution, and covariate value distributions were maintained across the range of RRs. One variable was selected as the exposure of interest and therefore was modified in progressing across the range of RRs. Thus, this investigation simulated a series of retrospective cohort studies. Three research questions motivate this work:

1. Do the magnitude and direction of relative risk affect classification accuracy in XCS?
2. Do the magnitude and direction of relative risk affect rule and rule set complexity?
3. Is there a detection threshold at which XCS can detect predictive risk factors, defined here as sentinel features, at experimentally controlled relative risk levels?

2 Methods

2.1 Data

Baseline Data. A series of retrospective cohort studies were simulated by first creating a baseline dataset consisting of 10 dichotomous exposure variables and one dichotomous outcome variable. The value of each predictor was randomly generated using a binomial distribution, where 0="Absent" and 1="Present." The outcome variable was also randomly generated, using a binomial distribution, and constrained such that the dataset contained 1000 records, with equivalent class frequency to yield 500 outcome-positive and 500 outcome-negative records. The outcome variable was linguistically the same as the exposure variables. Then, a single exposure variable, referred to here as *V5*, was recoded to ensure that it yielded a RR of 1.0. The RR associated with V5 is designated as RR_{V5}. Figure 2 demonstrates the crosstabulation of *V5* with outcome in the baseline dataset.

V5	Outcome	
	1	0
1	250	250
0	250	250

Fig. 2. Crosstabulation of the exposure of interest, *V5*, with the outcome variable

The crude and adjusted RRs for each exposure variable were calculated and found to be approximately 1.0 and generally not statistically significant.   Variable 8 demonstrates mild association with the outcome with respect to the crude RR, but on adjustment this association weakens. The variable was left unaltered in the baseline dataset as a result.   The baseline dataset is described below in Table 1; results obtained from analyses using the STATA SE9.0 procedures cs and poisson [8].

Table 1. The baseline dataset used for this investigation. The putative exposure, *V5*, is highlighted in the table.

Variable	Crude Relative Risk [95%CI]	Adjusted Relative Risk [95%CI]
V1	1.03 [0.91, 1.17]	1.03 [0.86, 1.23]
V2	1.10 [0.96, 1.24]	1.09 [0.92, 1.30]
V3	1.04 [0.92, 1.18]	1.06 [0.89, 1.26]
V4	0.90 [0.79, 1.02]	0.89 [0.75, 1.07]
V5	1.00 [0.88, 1.13]	1.01 [0.85, 1.20]
V6	1.07 [0.95, 1.22]	1.08 [0.91, 1.29]
V7	1.02 [0.90, 1.15]	1.02 [0.86, 1.22]
V8	1.15 [1.02, 1.30]	1.16 [0.97, 1.38]
V9	1.04 [0.91, 1.17]	1.04 [0.87, 1.24]
V10	0.93 [0.82, 1.05]	0.93 [0.78, 1.12]

Incremental Data. The goal of this investigation was to evaluate the ability of XCS to discover a single sentinel feature over a range of relative risks. The baseline dataset was altered to decrease or increase the relative risk for a single variable, Attribute 5, labeled RR_{V5}. This was accomplished by sequentially selecting records from the baseline dataset and changing the value of *V5*, depending on the value of the outcome variable. All other variables, including the outcome, were not changed; thus RR_{V5} was rigorously controlled over the entire suite of simulation datasets. Figures 3 and 4 illustrate the crosstabulation of *V5* with the outcome at $RR_{V5}=0.5$ and $RR_{V5}=5.0$, respectively. As seen in these figures, the proportion of records where *V5*=1 and the outcome is 1 increases from low to high RRs.

Each iteration was successively written to a different dataset for evaluation. Thus, each dataset contained the alterations of the one created before it, in addition to the alteration at that iteration. A total of 10 such datasets were created, with RR_{V5} ranging from 0.5 to 5.0 in increments of 0.5.

V5	Outcome	
	1	0
1	166	334
0	334	166

Fig. 3. Crosstabulation of *V5* with the outcome variable at $RR_{V5}=0.5$

V5	Outcome	
	1	0
1	417	82
0	82	417

Fig. 4. Crosstabulation of *V5* with the outcome variable at $RR_{V5}=5.0$

The system. EpiXCS [5] was used as the learning classifier system for the experiments. Using the Lanzi XCSLib kernel [7], EpiXCS outputs condition-action rule sets in IF-THEN syntaxas well as graphical display for ease of reading and visualization. In addition, EpiXCS provides evaluation metrics commonly used in the decision sciences literature (described below). The system parameters were set as follows. The population size, number of iterations, and initial prediction error were evaluated on the baseline dataset as well as those at $RR_{V5}=0.5$ and $RR_{V5}=5.0$ in order to "frame" the parameters and evaluate their effects on classification accuracy. This was important given that the exposure variables were randomly generated and this could have an effect reflecting the value assigned to ε_0. No such effect was noted, and the final parameters were set as in [1], except that the optimal population size was determined to be 2500. Each experiment took place over 1000 iterations; even though each dataset contained 1000 records, extending the training phase beyond 1000 iterations did not substantially improve the classification accuracy or the quality of the rules returned by EpiXCS. As this investigation focused on the identification sentinels in the training period, no separate testing set was used; classification performance was determined using the training set.

Experimental procedure. This investigation focused on rule discovery occurring during the training phase in a supervised learning environment. As a result, no training-testing set pairs were created; however, 10-fold cross-validation was used to ensure random and complete exposure to each record in the datasets. Each dataset was trained in EpiXCS over 20 runs, wherein each run comprised 1000 iterations. The rule sets described here represent the conflation of these sets into a "meta-set" that was created by ranking the rules by their predictive value. The rule sets derived by EpiXCS were evaluated using the visualization tool provided in this software.

Evaluation metrics. Classification accuracy was ascertained by the area under the receiver operating characteristic curve (AUC) and the positive and negative predictive values (PPV and NPV, respectively). AUC provides a single measure of classification accuracy, wherein high values (approaching 1.0) indicate that the system discriminates between positive and negative exposure-outcome associations accurately. The AUC reflects the prior probability, and answers the question "will XCS accurately detect a given condition?" where in this investigation, the "condition" is the sentinel feature. An AUC, averaged over the 20 runs, is reported for each RR_{V5}. The predictive values provide posterior probabilities, and answer the question "given that XCS has detected an exposure as a sentinel feature, is this detection accurate?" The predictive values are calculated in two ways. First, they are reported for each RR_{V5}, as is AUC. Second, a predictive value is calculated for each rule: a PPV is calculated for an outcome-positive rule, and a NPV for an outcome-negative rule.

In addition to AUC and the predictive values, the rule sets returned by EpiXCS at each RR_{V5} level were evaluated on the following metrics. *Rule set size* is the number of rules in the set, stratified by the value of the outcome in the rules. *Rule size* represents the number of conjuncts in a rule, and is averaged separately for each outcome value. *Emergence* is the proportion of a rule set where *V5* is *predictive* (positive), and the outcome is either positive (thus *V5* emerges as a risk factor) or negative (*V5* emerges as a protective factor). Thus, there two emergence metrics: *risk factor emergence* and *protective factor emergence*. The emergence metrics are calculated separately for positive- and negative outcome-rules having a predictive value of 0.60 or greater:

$$\text{Risk Factor Emergence} = \frac{\sum_{j=1}^{n} \text{Rule}_j \mid V5 = 1 \text{ and Outcome} = 1}{\sum_{j=1}^{n} \text{Rule}_j} \quad (2)$$

$$\text{Protective Factor Emergence} = \frac{\sum_{j=1}^{n} \text{Rule}_j \mid V5 = 1 \text{ and Outcome} = 0}{\sum_{j=1}^{n} \text{Rule}_j} \quad (3)$$

3 Results

3.1 Classification Accuracy

Generally, AUC and the predictive values correlated linearly with RR_{V5}, as shown in Figures 5-7. Overall, these accuracy point estimates are not particularly high, probably due to the randomness of the exposure variables other than $V5$. Nine random variables, against one nonrandom, will place disproportionate pressure on these estimates. The wide variances at lower RR_{V5} would be expected, given the relative lack of pressure placed on the risk estimate by $V5$ at these levels. This is

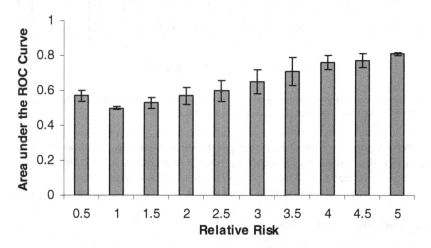

Fig. 5. Area under the receiver operating characteristic curve at each relative risk of $V5$ tested. Results based on average of 20 runs. Error bars represent plus/minus one standard deviation.

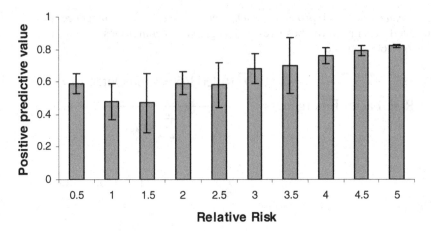

Fig. 6. Positive predictive value at each relative risk of *V5* tested. Results based on average of 20 runs. Error bars represent plus/minus one standard deviation.

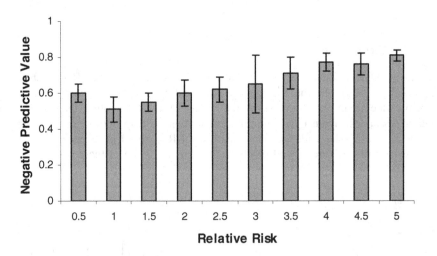

Fig. 7. Negative predictive value at each relative risk of *V5* tested. Results based on average of 20 runs. Error bars represent plus/minus one standard deviation.

further supported by the dips in the point estimates at $RR_{V5} = 1.0$, as the data were randomly distributed on all exposure variables at this level. Note, however, the decrease in variance in all three point estimates as RR_{V5} approaches 5.0, suggesting a threshold at which XCS begins to stabilize with respect to classification accuracy.

3.2 Rule Discovery by XCS

Rule set size. The number of rules, both those advocating positive as well as those advocating negative outcomes, was remarkably similar across the range of RR_{V5}, as shown in Figure 8. The slight differences at RR_{V5} of 2.5.and 3.0 are not significant with respect to the rule set sizes at the other RR_{V5} levels. It is remarkable that the number of rules in each set was lower than one may expect, given the size of the classifier population. However, this is probably due to the small number of variables, and the dichotomous coding scheme.

Rule size. The mean number of conjuncts per rule is shown in Figure 9, grouped by outcome. There are obvious differences in rule size across the range of RR_{V5}, and

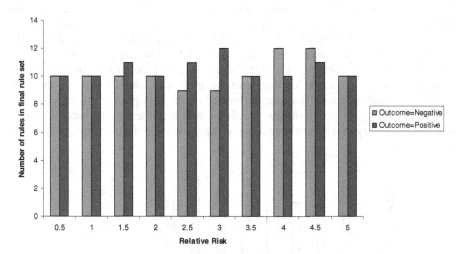

Fig. 8. Rule set size for positive and negative outcomes, grouped by RR_{V5}

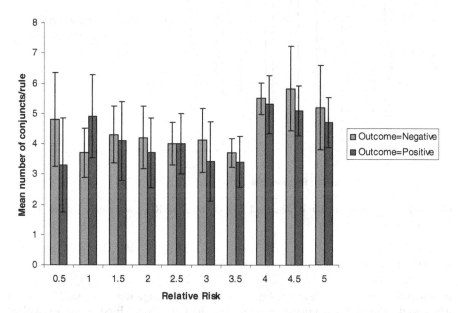

Fig. 9. Rule size, expressed as average number of conjuncts per rule, for positive and negative outcomes. Error bars represent plus/minus one standard deviation.

these are supported statistically, although the correlations are not strong. Spearman rho (the nonparametric equivalent of Pearson correlation, r, was 0.36 (p<0.001) for the negative outcomes, and 0.26 (p<0.01) for the positive outcomes. Kruskal-Wallis nonparametric one-way analysis of variance supported the observation that there are visually obvious differences in rule set size between the RR_{V5}s.

Sentinel feature emergence. Figure 10 illustrates the emergence score obtained from rules including *V5* as a conjunct, where the value of *V5* is 1. With the exception of the RR_{V5} =0.5 all rules in both classes contained *V5*. As can be seen in the figure, *V5* is a very powerful predictor of the outcome. The reason for the slight drop in the risk factor emergence score between RR_{V5}=1.5 and 2.0 is not clear, and may be due to such factors as the randomness of the initial classifier population or interactions between the exposure variables that were not previously identified in the bivariate analysis for quality control that was performed on the data prior to the experiments. Regardless, RR_{V5} is highly correlated with the risk factor emergence score (r=0.72, p<0.001) and negatively correlated with the protective factor emergence score (r=0.57, p<001).

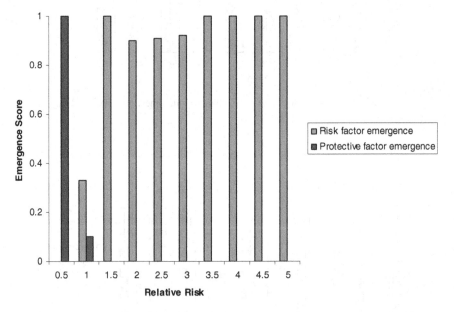

Fig. 10. Emergence scores for *V5* obtained over the range of RR_{V5}, separately for positive and negative outcomes

4 Discussion

In discussing the results and implications of this investigation, it is worthwhile to return to the three research questions that motivated this work. First, there was a clear relationship between the magnitude and direction of the relative risk with classification accuracy. Although this study was limited to ten simulation datasets of a small number of variables, the results suggest that the relationship is largely linear, with increasing classification accuracy associated with increasing relative risk. However, it appears that a bimodal distribution may underlie this relationship. The observed relationship is not completely unexpected: a relative risk of 0.5 represents a very strong protective effect, much more strong than a corresponding increase of the relative risk from 1.0 to 1.5, because of the exponential function underlying relative risk. Even so, this observation bears further investigation with an extended sensitivity analysis at more finely grained relative risks, especially below 1.0. More potentially troubling is the apparently random distribution of the classification accuracy metric variances across all values of RR_{V5}. The reason for this is not clear, and one could argue that the observed pattern of variances is artifact, resulting from aberrations in the data. The validity of this argument would be valid were it not for the fact that all of the exposure variables except for $V5$ were kept constant in each of the 10 datasets. Furthermore, the distribution of positive and negative values for $V5$ was carried over from one dataset to the next, the only difference between them being that the values were switched from positive to negative and vice-versa, depending on the value of the outcome and the RR_{V5} for that data increment. More likely is the possibility that interactions occur between $V5$ and the other exposure variables as the value of $V5$ is changed in each increment. For example, an interaction between $V5$ and another exposure variable (or combination thereof) could explain the increase in variance of the AUC in creating the $RR_{V5} =1.5$ dataset from the baseline data ($RR_{V5} =1.0$). The subsequent decrease in variance in creating the dataset at $RR_{V5} =2.0$ could be explained by a washout of the interaction at $RR_{V5} =1.5$. Again, a fine-grained sensitivity analysis, as well as the addition of rigorous interaction evaluation (not performed here) could solve this problem.

Second, it was found that the magnitude and direction of the relative risk of a given exposure variable, $V5$, did affect the size of the rule sets and the complexity of the rules, as determined by the average number of conjuncts per rule. However, this association was moderately weak, although statistically significant. It is important to note that no post-hoc rule reduction algorithms, such as that suggested by Wilson [9], were applied in this version of XCS. Rather, the rule sets discovered by the software represent the rules as they exist in the macroclassifier population at the end of training. Even so, the size of the rule sets remained fairly constant at approximately 10 for both positive and negative outcome rules. This is probably due to the type of feature coding (dichotomous) and the small number of variables. The results presented in Figure 9 are somewhat surprising, not so much that the number of rules in the rule sets is small, but that the number is so constant across the values of RR_{V5}. One explanation for this could be that even very a highly significant association such as that between the outcome and $V5$ at $RR_{V5} =5.0$ is suppressed by the pressure exerted by the other variables, expressed as conjuncts in the rules. In fact, the number of conjuncts fails to decrease appreciably as RR_{V5} increases. One would expect the rules to become smaller in size, focusing on the exposure that is most associated with the outcome, much like a default hierarchy. This was clearly not the case in this study, evidenced by Figure 9, where the observed pattern of rule size is the opposite of what might be expected.

Third, even in this highly controlled experimental study, a clear threshold at which XCS could identify the sentinel feature was not found. Again, this is probably due to the need for a more finely grained sensitivity analysis, particularly with relative risks between 0.1 and 2.0. In a sense, this is good news, because XCS was clearly able to identify the obvious relationship found at RR_{V5}=1.5 (see Figure 10), which represents at fairly low threshold of association in real life. It is probably reasonable to assume that XCS could identify a sentinel feature earlier on, but this ability may be data-dependent and this bears further investigation. Clearly, the lack of any concordant rules at RR_{V5} =0.5 troubling. In addition, another approach to the issue of emergence is to use some of the measures of surprisingness used in data mining and pattern detection. Work by Freitas and colleagues provides possible approaches to this problem [2, 3, 4].

5   Conclusion and Implications for Future Work

This paper reports on an ongoing investigation into the use of EpiXCS as a knowledge discovery tool in epidemiologic surveillance. The results indicate that EpiXCS is capable of discovering values of specific attributes that are associated with a particular class in a supervised learning problem. The importance of this finding is substantial. Heretofore, various univariate and multivariate methods, such as crosstabulation, logistic and Poisson regression and generalized linear models have been the analytic tools of choice for this problem. However, these analyses do not provide semantically useful rule ensembles that can be used easily for hypothesis generation. The single, mathematical model provided by regression analysis is highly useful, but the advantage to the knowledge discovery process that is provided by the multiple disjunctions in a rule set is lost. In addition, as noted earlier, these methods are not without their shortcomings, emanating from statistical assumptions that cannot always be met with the dataset at hand.

Even though the focus in this investigation is in epidemiologic surveillance, any problem domain to which data mining could be applied may benefit from the findings presented here.

This is very much a nascent study, and much needs to be done in the area of applying learning classifier systems, particularly XCS (but others as well) to the problem of discovering attribute-class associations. First, larger, more complex datasets need to be developed and evaluated. The small size of the datasets used here represents a real limitation to interpreting the results of this study. A related enterprise would be to simulate prospective studies using data overlays onto existing surveillance datasets. In this approach, the relative risks are adjusted by adding new cases to the datasets with specific associational patters that will cause the relative risk to increase (or decrease). Simulations using these types of datasets are becoming more popular in evaluating pattern recognition algorithms for detecting disease outbreaks and bioterrorism events.

Second, more work needs to be done with regard to parameterization. At lower relative risks, in the presence of more complex (read, conflicting) data, it is probably that population size, number of iterations, and perhaps genetic algorithm parameters will need to be adjusted. Finally, there is the intriguing possibility that the relative risk could be useful as a learning parameter. There could be at least two reasons why a relative risk would be low: the data don't support a higher value, or it represents a rare, but emerging association. In the latter situation, the relative risk could be considered as an interestingness metric that would help drive reinforcement.

Third, but not finally, is the extension of this work to prospective studies that capture data in real time. In these studies, time series analysis is often used to evaluate historical data on the cohort and to develop accurate forecasts of impending clinical events.

References

1. Butz, M.V. and Wilson S. W.: An algorithmic description of XCS. In Lanzi, P. L., Stolzmann, W., and S. W. Wilson (Eds.), Advances in Learning Classifier Systems. Third International Workshop (IWLCS-2000), Lecture Notes in Artificial Intelligence (LNAI-1996). Berlin: Springer-Verlag (2001).
2. Carvalho, D.R., Freitas, A.A., Ebecken N.F.F.: A critical review of rule surprisingness measures. Proc. Data Mining IV - Int. Conf. on Data Mining, pp.545-556, Rio de Janeiro, Brazil, Dec. 2003. WIT Press, 2003.
3. Fabris C.C. and Freitas A.A.: Discovering surprising instances of Simpson's paradox in hierarchical multidimensional data. To appear in Int. Journal of Data Warehousing and Mining, 2(1), pp. 26-48, Jan-Mar 2006.
4. Freitas, A.A.: On objective measures of rule surprisingness. Principles of Data Mining & Knowledge Discovery (Proc. 2nd European Symp., PKDD'98. Nantes, France, Sep. 1998). Lecture Notes in Artificial Intelligence 1510, 1-9. Springer-Verlag, 1998.
5. Holmes, J.H.: The architecture of EpiXCS: An XCS-based learning classifier system for epidemiologic research. Sixth International Workshop on Learning Classifier Systems, Chicago, IL July 2003.
6. Hulley S.B., et al: Designing Clinical Research. Philadelphia: Lippincott Williams and Wilkins, 2001, 122-123.
7. Lanzi P.L.: http://xcslib.sourceforge.net/
8. STATA Corporation: STATA/SE 9.0 for Windows. College Station, TX.
9. Wilson, S.W. Compact Rulesets from XCSI. In Advances in Learning Classifier Systems, 4th International Workshop, volume 2321 of Lecture Notes in Artificial Intelligence, pages 197–210. Springer, 2002.

Data Mining in Learning Classifier Systems: Comparing XCS with GAssist

Jaume Bacardit[1] and Martin V. Butz[2]

[1] ASAP, School of Computer Science and IT, University of Nottingham, Jubilee
Campus, Wollaton Road, Nottingham, NG8 1BB, UK
jqb@cs.nott.ac.uk
http://www.cs.nott.ac.uk/ jqb/
[2] Department of Cognitive Psychology, University of Würzburg, 97070 Würzburg,
Germany
butz@psychologie.uni-wuerzburg.de
http://www-illigal.ge.uiuc.edu/ butz/

Abstract. This paper compares performance of the Pittsburgh-style
system GAssist with the Michigan-style system XCS on several datamining problems. Our analysis shows that both systems are suitable for
datamining but have different advantages and disadvantages. The study
does not only reveal important differences between the two systems but
also suggests several structural properties of the underlying datasets.

1 Introduction

Successful data mining applications are important for modern-day learning classifier systems (LCSs). Additionally, the study and comparison of different types
of data miners on various data sets may enable the identification of strengths
and weaknesses of the respective data miners. Several types of problem difficulty
can be distinguished in data mining including data volume, search space size
and type, complexity of the concept, noise in the data, the handling of missing
values, or the problem of over-fitting.

Successful datamining applications of learning classifier systems have been
shown in the past [5] investigating and comparing performance of the accuracy-
based Michigan-style LCS XCS [11] and the Pittsburgh-style LCS GALE [10].
Both systems showed competent performance in comparison to six other machine
learning systems.

Recently, new systems have appeared in the LCS field, like the Pitt-style LCS
GAssist [2]. Also, there are improved versions of already established systems,
like the XCS with tournament selection [8]. The objectives of this paper are two-
fold: (1) We provide further performance results of GAssist and XCS on several
interesting datasets. (2) We compare and investigate performance of the two
systems revealing problem dependencies, suitability of the respective approaches,
as well as over-fitting or over-generalization tendencies.

X. Llorà et al. (Eds.): IWLCS 2003-2005, LNAI 4399, pp. 282–290, 2007.
© Springer-Verlag Berlin Heidelberg 2007

2   Framework

Before we start with the datamining analysis, this section provides a short introduction to the LCSs under investigation.

2.1   GAssist

GAssist [2] is a Pittsburgh genetic-based machine learning system descendant of *GABIL* [9]. The system applies a near-standard *GA* that evolves individuals that represent complete problem solutions. An individual consists of an ordered, variable-length rule set. Bloat control is achieved by a combination of a fitness function based on the minimum description length (*MDL*) principle and a rule deletion operator [3].

The knowledge representation used for real-valued attributes is called *adaptive discretization intervals* rule representation (*ADI*) [1]. This representation uses the semantics of the *GABIL* rules (conjunctive normal form predicates), but applies non-static intervals formed by joining several neighbor discretization intervals. These intervals can evolve through the learning process splitting or merging among them potentially using several discretizers at the same time.

The system also uses a windowing scheme called *ILAS* (incremental learning with alternating strata) [4]. This scheme stratifies the training set into s subsets of equal size and approximately uniform class distribution. Each *GA* iteration uses a different strata to perform its fitness computation, using a round-robin policy. This method showed to introduce an additional implicit generalization pressure to GAssist. [1]

Figure 1 presents the pseudocode of *ILAS*. This kind of scheme is reported to apply some extra generalization pressure to the system, which is an interesting feature for data mining domains.

2.2   XCS

The XCS classifier system [11,12,7] evolves online a set of condition-action rules, that is, a *population* of *classifiers*. In difference to GAssist, in XCS the population as a whole represents the problem solution. XCS differs in two fundamental ways to other Michigan-style LCSs: (1) Rule fitness is derived from rule accuracy

---

[1] GAssist's parameters were set as follows: Crossover probability 0.6; tournament selection; tournament size 3; population size 400; probability of mutating an individual 0.6; initial number of rules per individual 20; probability of "1" in initialization 0.75; Rule Deletion Operator: Iteration of activation: 5; minimum number of rules: number of classes of domain +3; MDL-based fitness function: Iteration of activation 25; initial theory length ratio: 0.075; weight relax factor: 0.9. ADI knowledge representation: split and merge probability: 0.05; reinitialize probability at initial iteration: 0.02; reinitialize probability at final iteration: 0; merge restriction probability: 0.5; maximum number of intervals: 5; set of uniform discretizers used: 4, 5, 6, 7, 8, 10, 15, 20 and 25 bins; iterations: maximum of 1500. Results are averaged over 150 experiments.

Procedure Incremental Learning with Alternating Strata
Input : $Examples, NumStrata, NumIterations$
Initialize GA
Reorder $Examples$ in $NumStrata$ parts of approximately
equal class distribution
$Iteration = 0$
$StrataSize = size(Examples)/NumStrata$
While $Iteration < NumIterations$
 If $Iteration = NumIterations - 1$ **Then**
  $TrainingSet = Examples$
 Else
  $CurrentStrata = Iteration$ mod $NumStrata$
  $TrainingSet=$ examples from
   $Examples[CurrentStrata \cdot StrataSize]$ to
   $Examples[(CurrentStrata + 1) \cdot StrataSize]$
 EndIf
 Run one iteration of the GA with $TrainingSet$
 $Iteration = Iteration + 1$
EndWhile
Output : Best individual (set of rules) from GA population

Fig. 1. Pseudocode of the incremental learning with alternating strata (ILAS) scheme

instead of rule reward prediction. (2) GA selection is applied in the subsets of currently active classifiers resulting in an implicit pressure towards more general rules.

Due to the variable properties of the investigated datasets including real values, nominals, and binary features, we use a hybrid XCS/XCSR approach that can handle any feature combination as done before in [5]. Additionally, we apply tournament selection, which proved to result in more robust fitness pressure toward accurate rules [8]. In the investigated problems, a reward of 1000 is provided if the classification is correct, and 0 otherwise. [2]

3   Experiments

3.1   Setup

In Table 1 we show the most important properties of the datasets we have selected from the University of California at Irvine (UCI) repository [6]. The selected datasets are:

---

[2] XCS's parameters are set as follows: $N = 6400$, $r_0 = 4(100)$, $P_\# = 0.6$, $\beta = 0.2$, $\chi = 1.0$ applying uniform crossover, $\mu = 0.04$, $m_0 = 0.2$, $\theta_{GA} = 48$, $\tau = 0.4$, $\varepsilon_0 = 1$, $\delta = 0.1$, $\theta_{del} = 50$, GA Subsumption is applied with $\theta_{sub} = 50$. Experiments are run applying either 100,000 learning steps (averaging over 150 experiments) or 500,000 learning steps (averaging over 20 experiments).

- Annealing Data (*ann*)
- 1985 Auto Imports Database (*aut*)
- Balance Scale Weight & Distance (*bal*)
- Contraceptive Method Choice (*cmc*)
- Horse Colic (*col*)
- German Credit (*cr-g*)
- Glass Identification (*gls*)
- Cleveland Heart Disease (*h-c*)
- Hungarian Heart Disease (*h-h*)
- Johns Hopkins University Ionosphere database (*ion*)
- Sonar, Mines vs. Rocks database (*son*)
- Wisconsin Breast Cancer database (*wbcd*)
- Wisconsin Diagnostic Breast Cancer (*wdbc*)

The selection of datasets gives a representative overview over the phenomena we were able to detect while comparing GAssist with XCS.

Table 1. The dataset properties indicate complexity, size, and data distributions in the respective datasets. #Inst. = Number of Instances, #Attr. = Number of attributes, #Real = Number of real-valued attributes, #Nom. = Number of nominal attributes, #Cla. = Number of classes, Dev.C = Deviation of class distribution, Maj.C. = Percentage of instances belonging to the majority class, Min.C. = Percentage of instances belonging to the minority class, MV I. = Percentage of instance with missing values, MV A. = Number of attributes with missing values, MV V. = Percentage of values (#*instances* · #*attr*) with missing values.

Dataset Properties										
Name	#Inst	#Attr	#Real	#Nom	#Cla	Dev.C	Maj.C	Min.C	MV I	MV A	MV V
ann	898	38	6	32	5	28.28	76.17	0.89	—	—	—
aut	205	25	15	10	6	10.25	32.68	1.46	22.44	7	1.11
bal	625	4	4	—	3	18.03	46.08	7.84	—	—	—
cmc	1473	9	2	7	3	8.26	42.70	22.61	—	—	—
col	368	22	7	15	2	13.04	63.04	36.96	98.10	21	22.77
cr-g	1000	20	8	12	2	20.00	70.00	30.00	—	—	—
gls	214	9	9	—	6	12.69	35.51	4.21	—	—	—
h-cl	303	13	6	7	2	4.46	54.46	45.54	2.31	2	0.17
h-h	294	13	6	7	2	13.95	63.95	36.05	99.66	9	19.00
ion	351	34	34	—	2	14.10	64.10	35.90	—	—	—
son	208	60	60	—	2	3.37	53.37	46.63	—	—	—
wbcd	699	9	9	—	2	15.52	65.52	34.48	2.29	1	0.23
wdbc	569	30	30	—	2	12.74	62.74	37.26	—	—	—

The test design for GAssist has two goals: Comparing the effect of using both different number of iterations and different degrees of generalization pressure. The latter goal is achieved by using the *ILAS* windowing scheme. However, our goal is not run-time reduction, but rather the maximization of the generalization pressure introduced by the *ILAS* scheme. Thus, we will increase the number of iterations when using windowing proportional to the number of strata used. This means having constant number of learning steps (using the Michigan-LCS meaning of the term). We will also test another stratified setup using a number of iterations that makes it equivalent in run-time compared to the non-windowed setting (1 strata).

3.2   Results

Results of GAssist and XCS are shown in Table 2. The comparison is not meant to determine which system is better in general but rather to show in which problem types which system appears to have advantages. Our comparison starts with a general data observation and then investigates separate datasets with respect to specific phenomena.

A look at the overall performance shows that XCS and GAssist show comparative performance results indicating the general difficulty of the respective datasets. XCS tends to learn the training data much more precisely which however is not necessarily advantageous for performance on the test data (using stratified ten-fold cross-validation). The solution representation differs (as expected) very significantly between GAssist and XCS: The number of rules in the best individual in GAssist is much smaller than the number of rules in XCS. However, it should be noted that GAssist maintains 400 individuals and thus the overall number of rules is actually similar to the number of rules in XCS. While we did not make explicit speed comparisons it appears that XCS runs take longer than GAssist's. Again, this is expectable since XCS is an online learner that learns from each problem instance separately and iteratively. Thus, the number of necessary learning iterations are higher.

Taking a closer look at the particular datasets we see that in the anneal (ann) dataset, performance of both systems reaches a similar level if XCS is run long enough. As also indicated by XCS's smaller population size in longer runs, generalization appears important and requires sufficient learning time. Generalization is even more important in the autos (aut) problem indicated by XCS's poor performance when starting specific and its improved test performance and smaller population size in longer runs as well as in GAssist's slight performance improvement and rule number decrease when using three strata. Additionally, the higher population size of XCS compared to the anneal problem indicates a general higher complexity of the problem. Balance-scale (bal) is a typical problem which can be over-fitted easily: XCS's performance is worse when starting more specific and when performing longer runs. Note that the population size of XCS actually increases when starting general and running more iterations—a clear indication of over-fitting. GAssist's performance points in the same direction in that generalization can slightly improve performance but longer runs are not helpful. The cmc problem appears to be a tough problem in general. XCS over-fits the data more than GAssist showing higher train performance but worse test performance. In the colic (col) as well as in the heart-h (h-h) problem, performance of XCS is significantly worse compared to GAssist. The major reason for this appears to be the missing value policy. While in GAssist a missing value is replaced by the majority value for the nominal case or by the average value in the real-valued case, XCS assumes a match in the missing value case. The

Table 2. Train and test performance results of GAssist and XCS using 10-folded cross-validation. Besides the performance results, we show the number of rules in the best individual of GAssist and the number of (macro-)classifiers in XCS (at the end of a run). The different GAssist runs distinguish a different application of strata (1 vs. 3 strata) as well as number of iterations (609, 1827, and 1447, respectively). In XCS, we compare long (500,000 learning iterations) and short learning runs (100,000 learning iterations) as well as a general ($r_0 = 100$) and specific ($r_0 = 4$) initialization of the population.

Data	Res.	GAssist			XCS (500,000)		XCS (100,000)	
		1 strata	3 s.(steps)	3 s.(time)	$r_0 = 100$	$r_0 = 4$	$r_0 = 100$	$r_0 = 4$
ann	Train	97.4±2.2	97.8±3.3	97.9±2.5	99.6±.46	100±.18	94.3±2.0	98.9±.61
	Test	97.0±2.6	97.4±3.5	97.5±2.8	98.4±1.6	98.6±1.5	91.2±2.7	91.7±2.9
	#rules	6.9±.9	6.3±.7	6.3±.5	2507±232	3211±146	4440±87	5426±51
aut	Train	85.5±2.9	84.7±3.2	82.8±3.7	99.8±.23	99.6±.39	99.3±.67	99.4±.56
	Test	67.5±9.8	68.8±9.7	67.5±9.5	71.5±9.5	68.8±12	64.7±9.6	13.4±6.9
	#rules	12.8±2.7	7.8±1.1	7.8±1.0	3403±98.5	4679±217	4281±87.3	5426±36.9
bal	Train	87.7±.49	86.0±.69	85.9±.73	98.4±.72	98.6±.64	90.6±2.2	97.9±.86
	Test	79.0±4.2	78.8±3.8	79.2±4.4	81.4±3.6	81.0±3.8	84.6±3.3	82.0±3.5
	#rules	13.1±2.0	9.6±1.6	9.8±1.6	2061±73.2	2014±59.8	1611±169	2465±65.9
cmc	Train	59.8±.96	59.6±1.1	59.8±1.1	70.5±1.9	77.6±2.0	57.0±1.8	71.5±2.2
	Test	54.8±4.2	54.6±4.0	54.9±4.1	53.6±4.0	52.9±4.7	50.1±4.7	53.6±3.6
	#rules	7.7±1.4	9.3±3.0	9.1±2.9	3261±88.1	3210±84.3	3958±91.4	3929±64.7
col	Train	99.7±.34	99.6±.48	99.5±.50	94.6±1.2	95.5±1.3	91.7±1.6	95.0±1.1
	Test	93.0±4.7	93.8±4.6	94.1±4.3	84.4±5.0	83.7±5.8	84.5±5.8	84.8±5.6
	#rules	7.4±1.6	7.0±1.4	7.0±1.4	3102±156	3685±84.2	3612±169	4100±96.3
cr-g	Train	82.0±.76	83.7±.94	84.3±.83	98.2±1.2	99.6±.34	89.7±3.2	94.4±1.4
	Test	72.3±3.6	72.0±4.2	72.2±3.8	70.2±3.6	72.3±4.2	71.4±3.9	72.5±3.1
	#rules	6.8±1.5	11.3±3.0	13.1±2.1	2016±69.2	2623±75.4	3217±106	4401±104
gls	Train	82.1±1.8	80.4±1.9	79.9±1.8	98.8±.64	99.6±.67	89.7±2.8	96.6±1.4
	Test	68.2±9.3	69.4±9.2	68.4±9.9	74.7±7.7	71.2±8.7	70.7±8.2	70.7±8.4
	#rules	8.8±1.4	6.6±0.8	6.6±0.8	1808±86.5	2143±78.4	3093±134	3137±92.7
h-cl	Train	93.4±.82	91.4±.98	92.6±.86	99.9±.25	100±0.0	99.5±.46	100±0.0
	Test	80.2±7.0	80.0±6.8	80.28±6.5	76.4±6.7	79.6±6.5	77.7±6.8	68.9±8.6
	#rules	9.3±1.5	6.9±1.1	7.4±1.2	2043±69.2	2808±89.6	2854±99.6	2907±68.2
h-h	Train	99.7±.32	99.0±.48	99.0±.50	99.7±.44	100±0.0	95.4±2.2	100±0.0
	Test	95.5±4.4	95.7±4.4	95.8±3.3	78.7±9.0	76.6±6.9	79.4±7.7	70.8±6.9
	#rules	6.1±0.7	6.3±0.5	6.0±0.2	2072±103	2686±71.9	3091±136	2861±67.5
ion	Train	98.2±.46	96.8±.63	96.8±.59	99.9±.19	99.7±.41	99.7±.32	99.8±.34
	Test	92.5±4.9	92.7±4.7	93.0±4.8	89.3±4.8	57.4±6.4	90.7±5.3	57.1±6.8
	#rules	3.9±0.8	2.2±0.7	2.2±0.8	2935±93.5	5613±28.9	3479±97.6	5685±31.4
son	Train	97.0±1.0	96.6±1.2	96.3±1.2	100±0.0	100±0.0	99.9±.30	100±0.0
	Test	74.4±8.9	76.8±9.0	77.5±9.2	78.4±7.4	82.6±8.3	77.3±8.1	81.6±7.9
	#rules	8.3±1.4	6.8±1.1	6.9±1.1	4959±120	4168±142	5148±107	4473±89.8
wbcd	Train	99.1±.27	97.8±.50	97.9±.47	99.8±.24	100±0.0	97.7±.89	99.9±.13
	Test	95.2±2.9	96.1±2.6	96.0±2.4	96.1±2.8	96.2±2.2	96.2±2.2	96.5±1.9
	#rules	5.0±1.0	2.4±0.6	2.4±0.6	1562±96.8	2131±52.9	1108±144	3137±81.8
wdbc	Train	98.6±.5	97.6±.68	97.6±.78	100±.09	100±0.0	99.8±.22	99.9±.24
	Test	94.1±3.0	94.2±2.9	94.1±2.8	96.1±2.5	96.7±2.2	95.9±2.6	92.9±3.3
	#rules	6.0±1.3	3.8±0.7	3.9±0.9	4104±112	5051±50.9	4485±85.5	5551±87.7

latter strategy appears mediocre in the investigated data mining experiments explaining XCS's poor performance in these settings.

Performance in the credit-g problem (cr-g) indicates that over-fitting is unlikely but in order to reach higher performance more specific initialization is helpful. Again, XCS reaches a much higher train performance but test performance is hardly influenced.

XCS's behavior in the glass problem (gls) is similar to that of credit-g. However, generalization is more important as also indicated by the performance improvement in GAssist when using three strata. Similar to the autos problem, XCS outperforms GAssist in the glass problem indicating higher problem complexity which might partially stem from the large number of classes in the problem.

XCS's performance in heart-c1 (h-c1) is actually very similar to the performance in in heart-h (h-h) suggesting that besides the problem of missing values in heart-h, XCS tends to strongly over-fit the training data. GAssist does not suffer from this problem in these datasets.

Another interesting observation was made in the ionosphere problem (ion) in which the automatic default rule detection mechanism in GAssist is actually able to discover that the minority class results in a better problem performance. XCS tends to over-fit as indicated by the poor performance and large population size when starting too specific.

On the other hand, in the sonar problem (son) a start from the specific side is actually beneficial for XCS suggesting small special-case niches which can be separated only if the population is initialized more specific. The more generalized representation of GAssist is not advantageous in this dataset.

In the Wisconsin breast-cancer dataset (wbcd) performance of both systems is similar and the problem appears to be generally easy as indicated by the small number of rules in both systems.

Finally, wdbc is another problem in which the complexity of the problem makes it hard for GAssist to reach XCS's performance level. XCS needs a large number of classifiers to solve the problem but is able to evolve the appropriate set. Slight generalizations are possible. GAssist on the other hand learns a very general—but slightly over–general solution.

4   Summary and Conclusions

In sum, both LCS systems showed that they are suitable for data-mining applications developing very different problem solutions that nonetheless perform similarly well on the test sets. Additionally, the comparison showed that regardless of offline (GAssist) or online (XCS) learning, LCSs are suitable data-miners.

The results allowed us to infer problem properties as well as problem difficulties. We saw that the current policy of handling missing values in XCS can affect performance negatively. Also, while GAssist has the tendency to ignore additional problem complexity, XCS tends to over-fit the training data more often (dependent on the nature of the data). Additionally, GAssist has slight problems with handling many output classes as well as a huge search space suggesting the addition of special covering operators that could ensure that each individual in GAssist differentiates at least all classes in the problem at hand. On the other hand, XCS's generalization tendency needs to be revisited in the data-mining domain. Especially in smaller datasets, XCS clearly tends to over-fit the data. Due to the small size of the datasets, the natural generalization pressure due to the niche reproduction mechanism hardly applies. Thus, additional pressure towards syntactic generality becomes more important and may be reconsidered in these problem domains.

The insights gained from our study prepare the systems for a more general problem application suggesting initial testing with each learning approach for suitability and appropriate initialization. XCS may need to be improved in terms of generalization to avoid over-fitting. GAssist may be endowed with further covering mechanism to ensure that all problem classes are covered by each individual and that it is able to detect additional small but significant problem subspaces.

Acknowledgments

The authors would like to thank Professor David E. Goldberg and the whole Illi-GAL lab for their support and advise during this work. Support from the following sources is acknowledged: the Spanish Research Agency (CICYT) under grant numbers TIC2002-04160-C02-02 and TIC 2002-04036-C05-03; the Department of Universities, Research and Information Society (DURSI) of the Autonomous Government of Catalonia under grants 2002SGR 00155 and 2001FI 00514; the German research foundation (DFG) under grant DFG HO1301/4-3; the European commission contract no. FP6-511931; the Air Force Office of Scientific Research, Air Force Materiel Command, USAF, under grant F49620-03-1-0129; the Computational Science and Engineering graduate option program (CSE) at the University of Illinois at Urbana-Champaign.

References

1. J. Bacardit and J. M. Garrell. Analysis and improvements of the adaptive discretization intervals knowledge representation. In *GECCO 2004: Proceedings of the Genetic and Evolutionary Computation Conference*, pages 726–738. Springer-Verlag, LNCS 3103, 2004.
2. Jaume Bacardit. *Pittsburgh Genetics-Based Machine Learning in the Data Mining era: Representations, generalization, and run-time*. PhD thesis, Ramon Llull University, Barcelona, Catalonia, Spain, 2004.
3. Jaume Bacardit and Josep M. Garrell. Bloat control and generalization pressure using the minimum description length principle for a pittsburgh approach learning classifier system. In *Proceedings of the 6th International Workshop on Learning Classifier Systems*. (in press), LNAI, Springer, 2003.
4. Jaume Bacardit and Josep M. Garrell. Incremental learning for pittsburgh approach classifier systems. In *Proceedings of the "Segundo Congreso Espaol de Metaheurísticas, Algoritmos Evolutivos y Bioinspirados."*, pages 303–311, 2003.
5. Ester Bernadó, Xavier Llorà, and Josep M. Garrell. XCS and GALE: a comparative study of two learning classifier systems with six other learning algorithms on classification tasks. In *Fourth International Workshop on Learning Classifier Systems - IWLCS-2001*, pages 337–341, 2001.
6. C. Blake, E. Keogh, and C. Merz. UCI repository of machine learning databases, 1998. (www.ics.uci.edu/mlearn/MLRepository.html).
7. M. V. Butz. *Rule-Based Evolutionary Online Learning Systems: A Principled Approach to LCS Analysis and Design*. Studies in Fuzziness and Soft Computing. Springer-Verlag, Berlin Heidelberg, 2005.

8. Martin V. Butz, Kumara Sastry, and David E. Goldberg. Tournament selection in XCS. *Proceedings of the Fifth Genetic and Evolutionary Computation Conference (GECCO-2003)*, pages 1857–1869, 2003.

9. Kenneth A. DeJong, William M. Spears, and Diana F. Gordon. Using genetic algorithms for concept learning. *Machine Learning*, 13(2/3):161–188, 1993.

10. Xavier Llorà and Josep M. Garrell. Knowledge-independent data mining with fine-grained parallel evolutionary algorithms. In *Proceedings of the Third Genetic and Evolutionary Computation Conference*, pages 461–468. Morgan Kaufmann, 2001.

11. Stewart W. Wilson. Classifier fitness based on accuracy. *Evolutionary Computation*, 3(2):149–175, 1995.

12. Stewart W. Wilson. Get real! XCS with continuous-valued inputs. In L. Booker, Stephanie Forrest, M. Mitchell, and Rick L. Riolo, editors, *Festschrift in Honor of John H. Holland*, pages 111–121. Center for the Study of Complex Systems, 1999.

Improving the Performance of a Pittsburgh Learning Classifier System Using a Default Rule

Jaume Bacardit[1], David E. Goldberg[2], and Martin V. Butz[3]

[1] ASAP, School of Computer Science and IT, University of Nottingham, Jubilee Campus, Wollaton Road, Nottingham, NG8 1BB, UK
jqb@cs.nott.ac.uk
http://www.cs.nott.ac.uk/ jqb/
[2] Illinois Genetic Algorithms Laboratory (IlliGAL), Department of General Engineering, University of Illinois at Urbana-Champaign, 104 S. Mathews Ave, Urbana, IL 61801
deg@uiuc.edu
http://www-illigal.ge.uiuc.edu/goldberg/d-goldberg.html
[3] Department of Cognitive Psychology, University of Würzburg, 97070 Würzburg, Germany
butz@psychologie.uni-wuerzburg.de
http://www-illigal.ge.uiuc.edu/ butz/

Abstract. An interesting feature of encoding the individuals of a Pittsburgh learning classifier system as a decision list is the emergent generation of a default rule. However, performance of the system is strongly tied to the learning system choosing the correct class for this default rule. In this paper we experimentally study the use of an explicit (static) default rule. We first test simple policies for setting the class of the default rule, such as the majority/minority class of the problem. Next, we introduce some techniques to automatically determine the most suitable class.

1 Introduction

One of the ways to solve classification problems using a genetic algorithm [1,2] is called Pittsburgh approach [3] or Pittsburgh learning classifier system. The individuals of this system encode a full and variable-length rule set and the solution proposed is the best individual of the population. There are several encoding options for an individual. One of them is coding an individual as a decision list [4] (an ordered set of rules). If we apply this strategy in the evolutionary framework, often the system evolves a default rule. That is, a rule that matches any input instance.

Default rules can be very useful in combination with a decision list because the size of the rule set can be reduced significantly. For instance, for the 11-bit multiplexer we can obtain a rule set of 9 rules instead of 16 unordered ones, as represented in Figure 1. With a smaller rule set, the search space is reduced resulting in two potential advantages: (1) the learner can learn fewer rules faster (representing only the other classes of the dataset) and (2) with a smaller rule

X. Llorà et al. (Eds.): IWLCS 2003-2005, LNAI 4399, pp. 291–307, 2007.

Unordered MX-11 rule set
0 0 0 0 # # # # # # # : 0
0 0 0 1 # # # # # # # : 1
0 0 1 # 0 # # # # # # : 0
0 0 1 # 1 # # # # # # : 1
0 1 0 # # 0 # # # # # : 0
0 1 0 # # 1 # # # # # : 1
0 1 1 # # # 0 # # # # : 0
0 1 1 # # # 1 # # # # : 1
1 0 0 # # # # 0 # # # : 0
1 0 0 # # # # 1 # # # : 1
1 0 1 # # # # # 0 # # : 0
1 0 1 # # # # # 1 # # : 1
1 1 0 # # # # # # 0 # : 0
1 1 0 # # # # # # 1 # : 1
1 1 1 # # # # # # # 0 : 0
1 1 1 # # # # # # # 1 : 1

Ordered MX-11 rule set
0 0 0 0 # # # # # # # : 0
0 0 1 # 0 # # # # # # : 0
0 1 0 # # 0 # # # # # : 0
0 1 1 # # # 0 # # # # : 0
1 0 0 # # # # 0 # # # : 0
1 0 1 # # # # # 0 # # : 0
1 1 0 # # # # # # 0 # : 0
1 1 1 # # # # # # # 0 : 0
: 1

Fig. 1. Unordered and ordered rule sets for the MX-11 domain

set the system may be less sensitive to over-learning, potentially increasing the test accuracy of the system.

The objective of this paper is to investigate the potential benefits of using an explicit and static default rule in a Pitt LCS. Along those lines, the question arises which is the best default class to use. Simple strategies may use the majority class. However, our tests show that dependent on the problem, the minority class may be better as the default class choice. Thus, we develop a mechanism that is able to automatically determine the best class for the default rule.

The rest of the paper is structured as follows: Section 2 shows some related work. Next, Section 3 describes briefly the main characteristics of the system used in this paper. Later, Section 4 illustrates the motivation of using a default rule, followed by Section 5 that reports the modifications applied to the knowledge representation of the system to integrate the default rule. Next, Section 6 shows some illustrative results of the simple policies for the default rule. After the simple policies, we describe the more sophisticated ones in Section 7. Section 8 shows the experimentation results of applying the described policies. Finally, Section 9 presents conclusions and further work.

2 Related Work

We can find previous uses of a static default rule in the *LCS* field, although not in an explicit way: Classic Pitt-approach systems such as *GABIL* [3] or GIL [5], which perform concept learning (learning a concept from sets of positive/negative examples), implicitly have a default rule that covers the negative examples. The rules generated do not have an associated class because all of them cover the positive examples. However, there is no explicit policy to decide which set is the positive or negative one in order to learn better. The decision simply comes from the definition of the dataset.

Looking at the machine learning field in general we find other examples of default rules. The C4.5 rule system [6] uses an explicit default rule and, alike our system, it generates a rule set acting as a decision list. To select the class for this default rule, it uses the class that has less instances covered by the other rules in the rule set. This kind of approach seems feasible when we have induced the rule set beforehand, instead of using it during learning as our system does.

The *IREP* system [7] induces the rules in order, modeling each class of the problem (using the instances of the classes still to be learned as negative examples). The criteria of this global order is ascendant frequency of examples. Therefore, the default rule of this system uses a majority class policy.

3 Framework

GAssist [8] is a Pittsburgh genetic-based machine learning system descendant of *GABIL* [3]. The system applies a near-standard *GA* that evolves individuals that represent complete problem solutions. An individual consists of an ordered, variable-length rule set. Directly from GABIL we have taken the semantically correct crossover operator for variable-length individuals.

Dealing with variable-length individuals raises some important issues. One of the most important one is the control of the size of the evolving individuals [9]. This control is achieved in GAssist using two different operators:

1. *Rule deletion.* This operator deletes the rules of the individuals that do not match any training example. This rule deletion is done after the fitness computation and has two constraints:
   (a) The process is only activated after a predefined number of iterations (to prevent an irreversible diversity loss)
   (b) The number of rules of an individual never decreases below a threshold. This introduces some "neutral code" that can protect the individuals from the disruptive effect of the crossover operator.
2. *Minimum description length-based fitness function.* The minimum description length (*MDL*) principle [10] is a metric applied in general to a theory

(being a rule set in this paper) which balances the complexity and accuracy of the rule set. In previous work we developed a fitness function based on this principle. A detailed explanation of the fitness function can be found in [11].

The knowledge representation used for real-valued attributes is called *adaptive discretization intervals* rule representation (*ADI*) [12]. This representation uses the semantics of the *GABIL* rules (conjunctive normal form predicates), but applies non-static intervals formed by joining several neighbor discretization intervals. These intervals can evolve through the learning process splitting or merging among them potentially using several discretizers at the same time.

Parameters of the system are set as follows: Crossover probability 0.6; tournament selection; tournament size 3; population size 300; Individual-wise mutation probability 0.6; initial number of rules per individual 20; probability of "1" in initialization 0.75; Rule Deletion Operator: Iteration of activation: 5; minimum number of rules: number of active rules +3; MDL-based fitness function: Iteration of activation 25; initial theory length ratio: 0.075; weight relax factor: 0.9. ADI knowledge representation: split and merge probability: 0.05; reinitialize probability at initial iteration: 0.02; reinitialize probability at final iteration: 0; merge restriction probability: 0.5; maximum number of intervals: 5; set of uniform discretizers used: 4, 5, 6, 7, 8, 10, 15, 20 and 25 bins; iterations: maximum of 1500.

4   Motivation

In order to illustrate the benefits of the default rule, we show the results of running the system with no static default rule for the *Glass* problem from the *UCI* repository [13] in table 1. We used stratified ten-fold cross validation for the tests and a hundred random seeds for each fold (a total of 1000 runs, unlike the 15 seeds and 150 runs used in the rest of the paper).

We can see the benefits of using a default rule and, more importantly, the benefits of choosing the correct class for the default rule. The choice of the class for the default rule has a significant influence on the resulting accuracy, suggesting that a good default rule choice can improve learning performance and generality of the resulting solution.

5   Static Default Rule Mechanism

To force the usage of a default rule, few modifications are necessary: we only need to codify our individuals as decision lists, independent of the knowledge representation used. The implementation of the static default rule is very simple. Basically it affects only the matching function classifying any input instance by the default class if no rule (in the decision list) matches the instance. The pseudocode in Figure 2 clarifies this mechanism. Additionally, the default rule class is removed from the classes that can be used by the rest of the rules in the population, effectively reducing the search space. A general representation of the extended rule set is shown in Figure 3.

Table 1. How the generation of a default rule can affect the performance in the *Glass* dataset

Runs generating a default rule	736
Runs not generating a default rule	264
Accuracy of runs with a default rule	66.98±8.00
Accuracy of runs without a default rule	66.27±7.79
Average accuracy of runs using class 1 as default rule	65.45±7.39
Average accuracy of runs using class 2 as default rule	67.76±7.81
Average accuracy of runs using class 3 as default rule	59.40±5.51
Average accuracy of runs using class 4 as default rule	66.18±8.70
Average accuracy of runs using class 5 as default rule	67.66±8.58
Average accuracy of runs using class 6 as default rule	64.48±7.36

1. We determine with some criterion (in the following sections several criteria are studied) which class is the default class.
2. An individual predicts this default class when no rule matches an input instance.
3. The other rules of the individual cannot use the default class. Neither initialization nor mutation can make a regular rule of the individual point to the default class.
4. The default rule is included in the size of the rule set. This means that the rest of the system transparently sees an individual with one more rule. This affects the parts of the fitness formula that uses the size of the rule set as a variable.
5. The default rule cannot be affected by crossover, mutation nor any other recombination operator.
6. The rule deletion operator ignores the petitions to delete this rule, in the rare chance that this rule matches nothing (all problem instances are covered by other rules already).
7. The MDL-based fitness function computes a theory length for this rule supposing that the rule is totally general, that is, as if it were the emergent default rule observed before implementing this mechanism.

For the specific case of two-class domains, the classification problem is transformed into a concept learning problem and the resulting knowledge representation is quite close to the ones used in other evolutionary concept learning systems like *GABIL* [3] or *GIL* [5].

6   Simple Policies Determining the Default Rule Class

In order to answer the question of which class is suitable for being the default class we start by experimenting with two simple policies: using the most and least frequent class in the domain. In Section 8 we can see the results of these tests for several datasets. Here we show the results (in Table 2) of only two datasets (*Glass* and *Ionosphere*), also from UCI. For *Glass* the best policy is using the majority class. For *Ionosphere* the best policy is using the minority class. The point of showing these two datasets is that it is very difficult to decide

Match process
Input : *RuleSet, Instance*
Index = 0
Found = *false*
While *Index* < *RuleSet.size* and not *Found* **Do**
    If *RuleSet.rule[Index]* matches *Instance* **Then**
        Class = *RuleSet.rule[Index].class*
        Found = *true*
    Else
        Index + +
    EndIf
EndWhile
If not *Found* **Then**
    Class = *DefaultClass*
EndIf
Output : Predict class *Class* for instance *Instance*

Fig. 2. Match process using an static default rule

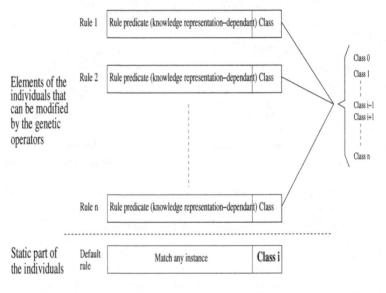

Fig. 3. Representation of the extended rule set with the static default rule

Table 2. Results using majority and minority policy for the default class in the *Glass* and *Ionosphere* datasets

Domain	Def. Class. Policy	Train accuracy	Test accuracy	Number of rules
Glass	disabled	79.9±2.6	66.4±8.1	6.4±0.7
Glass	majority	83.2±1.6	69.5±6.9	6.6±0.8
Glass	minority	80.6±2.3	66.7±8.0	7.2±0.8
Ionosphere	disabled	96.0±0.6	92.8±3.6	2.3±0.6
Ionosphere	majority	95.7±0.8	90.0±4.4	5.7±1.2
Ionosphere	minority	96.8±0.7	93.0±3.7	2.6±0.8

a priori which is the most suitable default rule class for each dataset. The values of the train accuracy and the number of rules give hints about how to combine the two policies to maximize the performance of the system. In Section 8 we show a simple combination consisting of choosing at the test stage the policy which has more train accuracy.

7 Automatically Determined Default Rule Class

Given that the majority class does not always suite best as default class, the next step is to modify the system to automatically determine the best default class. Our initial approach simply assigns a randomly chosen class as default class to each individual in the initial population. Additionally, we introduce a restricted mating mechanism to avoid crossover operations between individuals having different default classes, summarized by the code in Figure 4. Having removed the default class from the rest of the rules, crossing individuals with different default classes may create lethals with high probability. Especially in the specific case of two-class domains, the regular rules of individuals using different default classes cover completely different subsets of rules. Therefore, it is impossible to integrate the rules of these two individuals using the regular crossover operator.

```
Niched crossover algorithm
Comment To simplify the code, Parents contains only the parent individuals
Comment already selected for crossover by the probability of crossover
Input : Parents
OffspringSet = ∅
While Parents is not empty
        Parent1 = select randomly and individual from Parents
        Remove Parent1 from Parents
        Niche = default class of Parent1
        If there are individuals in Parents belonging to Niche
                Parent2 = select randomly and individual from Parents
                        belonging to Niche
                Remove Parent2 from Parents
                Offspring1, Offspring2 = apply crossover to Parent1, Parent2
                Add Offspring1, Offspring2 to OffspringSet
        Else
                Offspring = clone of Parent1
                Add Offspring to OffspringSet
        EndIf
EndWhile
Output : OffspringSet
```

Fig. 4. Code of the crossover algorithm with restricted mating

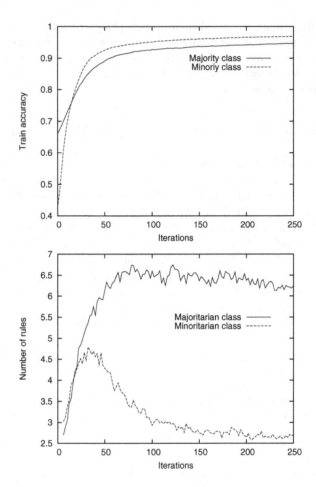

Fig. 5. Evolution of the train accuracy and the number of rules for the Ionosphere problem using majority/minority default class policies

If we run the system in this setting, we observed that usually all individuals with one default class take over the population. The question is if the system is able to choose the correct default class during the initial iterations. To answer this question, we show the evolution of the train accuracy and the number of rules for the *Ionosphere* tests described in the previous section in Figure 5. We can see that the train accuracy of the default class policy using the suitable class for this problem (that is, the minority class) is lower at the initial iterations than the accuracy of the majority class policy. Also, we can see the reason for the better test accuracy of the minority policy in the smaller (better generalized) rule set created by this policy.

Thus, it appears necessary to introduce an additional niching mechanism that preserves individuals for all default classes until the system has learned enough to decide correctly on the best default class. This niching is achieved using a modified tournament selection mechanism, inspired by [14] in which the individuals participating in each tournament are forced to belong to the same class. Also, each default class has an equal number of tournaments. This niched tournament selection is represented by the pseudocode in Figure 6. The tournament with niche preservation is used until the best individuals of each default class have similar train accuracy. After this point, the niching is disabled and the system chooses freely among the individuals. Specifically, we compute for each niche the average accuracy over the last 15 iterations of its best individual. When the standard deviation of all these averages is smaller than 0.5%, we disable the niched tournament selection, effectively enabling the superior default class to take over the whole population.

Niched tournament selection
Input : $Population, PopSize, NumNiches, TournamentSize$
$NextPopulation = \emptyset$
For $i = 1$ to $NumNiches$
    $ProportionNiche[i] = PopSize/NumNiches$
EndFor

For $i = 1$ to $PopSize$
    $Niche$ = select randomly a niche based on ProportionNiche
    $ProportionNiche[Niche] - -$
    Select $TournamentSize$ individuals from $Population$ belonging to $Niche$
    $winner=$Apply tournament
    Add $winner$ to $NextPopulation$
EndFor
Output : $NextPopulation$

Fig. 6. Pseudocode for the niched tournament selection

To summarize, the changes introduced to the default rule model by the automatic policy are the following:

1. Initialization assigns randomly to each individual a class as being the default class.
2. This class cannot be used in the regular rules of the individual.
3. Individuals having different default classes cannot exchange rules. The crossover algorithm is modified adding this mating restriction.
4. Niched tournament selection preserves an uniform proportion of individuals from all default classes in the population. This niching process is achieved reserving a quota of tournaments to each niche and only applying tournaments among individuals belonging to the same niche.
5. The niching mechanism is disabled when individuals using different default classes can compete fairly among themselves. Specifically, we compute, for each default class, the average accuracy over the last 15 iterations of its best individual. When the standard deviation of all these averages is smaller than 0.5%, the niched tournament selection is disabled and a regular tournament selection takes places until the end of the learning process.

8   Results

In this section, we show the results of comparing the three policies tested for the default class (*majority, minority, auto* to the original system (*orig*) with emergent default rule. The tests include 15 datasets used previously in [12], summarized in table 3. Each dataset has been partitioned into training/test sets using stratified ten-fold cross-validation [15], and having for each fold the tests repeated 15 times.

Table 3. Features of the datasets used in the experimentation of this paper

Domain	#Inst.	#Attr.	#Real	#Nom.	#Cla.	Dev.cla.	Maj.cla.	Min.cla.
				Dataset Properties				
bpa	345	6	6	—	2	7.97%	57.97%	42.03%
bps	1027	24	24	—	2	1.60%	51.61%	48.39%
bre	699	9	9	—	2	15.52%	65.52%	34.48%
gls	214	9	9	—	6	12.69%	35.51%	4.21%
h-s	270	13	13	—	2	5.56%	55.56%	44.44%
ion	351	34	34	—	2	14.10%	64.10%	35.90%
lrn	648	6	4	2	5	14.90%	45.83%	1.54%
mmg	216	21	21	—	2	6.01%	56.02%	43.98%
pim	768	8	8	—	2	15.10%	65.10%	34.90%
son	208	60	60	—	2	3.37%	53.37%	46.63%
thy	215	5	5	—	3	25.78%	69.77%	13.95%
veh	846	18	18	—	4	0.89%	25.77%	23.52%
wdbc	569	30	30	—	2	12.74%	62.74%	37.26%
wine	178	13	13	—	3	5.28%	39.89%	26.97%
wpbc	198	33	33	—	2	26.26%	76.26%	23.74%

Table 4 shows the results for these tests, also including a fifth configuration (*majority+minority*), in which the majority/minority policy is chosen in the test stage that obtained more training accuracy. This configuration usually chooses the correct policy (although there are some exceptions, like *bpa*). The results were analyzed using pair-wise statistical t-tests with Bonferroni correction to determine how many times each method could significantly outperform or be outperformed by the other methods. These statistical tests are summarized in table 5.

At first glance, we can see that all but two datasets (*wbcd* and *wpbc*) can benefit (by one or more of the studied default class policies) from the inclusion of a default rule. However, the achieved accuracy increase is not uniform across the datasets. Some of them, like *gls* or *son*, show a notable accuracy increase, while some others only show a small, non-significant increase. To understand these different degrees of accuracy increase we have computed the percentage of runs where the *orig* configuration was already generating a default rule emergently. Table 6 shows these results including the accuracy of the *orig* configuration as well as the accuracy of the best default class policy for each dataset (and their difference). Although it is not totally clear, we can see a correlation between the percentage of discovered default rules and the accuracy difference between using/not using the default rule. The clearest exception is the *gls* dataset. However, considering that this dataset has 6 classes, the benefits of removing the default class from the pool of classes used in the regular rules are already substantial even if the *orig* configuration was already using a default rule.

Table 4. Results of the tests comparing the studied default class policies to the original configuration using pop. size 300

Domain	Result	Default rule policy				
		Disabled	Major	Minor	Auto	Major+Minor
bpa	Train	78.6±1.6	81.4±1.3	80.1±1.6	80.8±1.4	81.4±1.3
	Test	63.8±7.4	62.9±7.8	65.2±6.5	64.0±6.9	62.9±7.8
	#rules	6.7±1.0	8.9±1.4	8.3±1.5	8.5±1.6	8.9±1.4
bps	Train	84.8±0.9	86.0±0.7	86.8±0.7	86.6±0.7	86.8±0.7
	Test	80.1±3.9	81.2±3.6	81.5±3.6	81.4±3.7	81.5±3.6
	#rules	5.1±0.4	6.1±1.1	5.7±0.9	5.6±0.8	5.7±0.9
bre	Train	97.7±0.3	98.2±0.3	98.4±0.3	98.4±0.3	98.4±0.3
	Test	95.9±2.2	95.0±2.5	95.7±2.0	95.6±2.2	95.7±2.0
	#rules	2.6±0.7	5.8±1.2	3.2±0.6	3.3±0.7	3.2±0.6
gls	Train	79.9±2.6	83.2±1.6	80.6±2.3	79.0±1.8	83.2±1.6
	Test	66.4±8.1	69.5±6.9	66.7±8.0	66.7±8.0	69.5±6.9
	#rules	6.4±0.7	6.6±0.8	7.2±0.8	6.9±0.9	6.6±0.8
h-s	Train	89.8±1.2	91.6±0.9	92.1±0.8	91.9±0.9	92.1±0.8
	Test	79.5±6.2	79.3±6.4	81.3±6.8	81.3±6.1	81.3±6.8
	#rules	6.7±0.9	7.6±1.2	7.3±1.2	7.4±1.3	7.3±1.2
ion	Train	96.0±0.6	95.7±0.8	96.8±0.7	96.8±0.7	96.8±0.7
	Test	92.8±3.6	90.0±4.4	93.0±3.7	93.1±3.9	93.0±3.7
	#rules	2.3±0.6	5.7±1.2	2.6±0.8	2.6±0.7	2.6±0.8
lrn	Train	75.2±1.9	76.8±0.8	75.4±1.4	75.4±1.0	76.8±0.8
	Test	68.5±4.7	68.9±5.7	68.9±4.5	68.6±5.6	68.9±5.7
	#rules	8.5±1.9	9.6±1.9	9.2±1.9	8.6±1.7	9.6±1.9
mmg	Train	79.7±1.8	83.2±1.3	83.1±1.3	83.0±1.4	83.2±1.3
	Test	66.2±7.8	68.9±8.3	67.8±8.4	66.8±9.0	68.9±8.3
	#rules	6.5±0.8	6.7±0.9	6.7±0.8	6.6±0.9	6.7±0.9
pim	Train	79.7±0.9	81.3±0.8	80.9±0.7	81.1±0.8	81.3±0.8
	Test	74.7±4.7	75.4±4.8	75.0±4.7	75.0±4.5	75.4±4.8
	#rules	5.2±0.4	6.2±1.0	5.6±0.8	6.1±1.0	6.2±1.0
son	Train	92.2±1.6	96.1±1.2	94.8±1.4	95.5±1.4	96.1±1.2
	Test	72.6±11.5	77.0±9.0	76.1±9.7	76.1±9.3	77.0±9.0
	#rules	6.7±1.1	7.6±1.4	7.7±1.3	7.4±1.1	7.6±1.4
thy	Train	97.4±1.0	98.4±0.7	98.4±0.7	98.1±0.8	98.4±0.7
	Test	91.9±5.6	92.8±4.8	92.3±5.3	92.3±5.6	92.8±4.8
	#rules	5.2±0.4	5.7±0.6	5.4±0.5	5.5±0.6	5.7±0.6
veh	Train	71.1±2.2	73.5±1.4	73.5±1.4	72.0±1.5	73.5±1.4
	Test	66.4±4.7	68.1±4.5	67.4±4.9	67.5±4.7	68.1±4.5
	#rules	6.6±1.2	9.3±2.0	9.9±1.6	8.0±1.8	9.3±2.0
wdbc	Train	97.2±0.8	97.8±0.6	97.8±0.6	97.8±0.6	97.8±0.6
	Test	94.1±3.0	94.2±3.1	94.0±3.0	94.3±3.1	94.2±3.1
	#rules	4.3±1.1	4.6±0.9	4.4±1.0	4.5±1.0	4.6±0.9
wine	Train	99.4±0.5	99.7±0.4	99.9±0.3	99.6±0.4	99.9±0.3
	Test	92.7±5.9	93.3±6.2	92.2±6.3	93.9±5.9	92.2±6.3
	#rules	3.8±0.7	3.6±0.6	4.1±0.5	3.8±0.6	4.1±0.5
wpbc	Train	84.3±3.0	89.4±2.0	86.4±3.4	88.7±2.3	89.4±2.0
	Test	76.0±7.3	75.8±7.4	72.6±8.5	75.2±7.5	75.8±7.4
	#rules	2.8±0.8	3.8±0.9	4.2±1.2	3.6±1.0	3.8±0.9
ave.	Train	86.9±9.0	88.8±8.4	88.3±8.8	88.3±9.0	89.0±8.5
	Test	78.8±11.4	79.5±10.7	79.3±11.0	79.5±11.3	79.8±10.9
	#rules	5.3±1.8	6.5±1.8	6.1±2.1	5.9±1.9	6.1±2.1

From the test accuracy averages and the t-test results it is clear that the *major+minor* policy is the best configuration, both in performance and robustness, because it has been never outperformed in a significant way. However, having in this configuration a run-time two times larger than in the other configurations, we have to question whether the computational cost sacrifice is worth it. Looking at the other configurations, *major* and *auto* are tied in accuracy average, but *auto* is much more robust than *major* according to the t-tests.

Table 5. Summary of the statistical t-tests applied to the experimentation results of popsize 300, with a confidence level of 0.05. Cells in table count how many times the method in the row significantly outperforms the method in the column.

Policy	Disabled	Major	Minor	Auto	Major+Minor	Total
Disabled	-	2	1	0	0	3
Major	3	-	2	1	0	6
Minor	2	2	-	0	0	4
Auto	2	1	1	-	0	4
Major+Minor	4	2	2	1	-	9
Total	11	7	6	2	0	

Table 6. Percentage of runs where *orig* configuration was already generating a default rule, accuracy difference between *orig* and the best default class policy for each dataset

Rows are sorted by the percentage of default rule generation in *orig*

Label	meaning
DRG	Percentage of runs where the default rule was generated in *orig* configuration
AccO	Accuracy of the *orig* configuration
AccDR	Accuracy of the best rule policy on the dataset
AccDif	Accuracy difference between AccO and AccDR

Dataset	DRG	AccO	AccDR	AccDif
mmg	19.33%	66.21%	68.88%	-2.67%
son	36.00%	72.58%	76.99%	-4.42%
bps	40.00%	80.10%	81.55%	-1.44%
veh	46.67%	66.43%	68.15%	-1.72%
pim	50.67%	74.65%	75.37%	-0.71%
wdbc	55.33%	94.06%	94.26%	-0.20%
h-s	57.33%	79.46%	81.31%	-1.85%
bpa	65.33%	63.79%	65.22%	-1.43%
thy	68.67%	91.92%	92.79%	-0.87%
wine	71.33%	92.74%	93.85%	-1.12%
gls	74.00%	66.37%	69.52%	-3.15%
lrn	76.00%	68.55%	68.93%	-0.39%
wpbc	82.00%	76.03%	75.78%	0.25%
ion	86.00%	92.85%	93.13%	-0.29%
bre	96.00%	95.88%	95.74%	0.14%

Nevertheless, it is important to investigate why the *auto* policy reaches a lower performance than *major+minor*. Table 7 shows the class distribution of the default rules that appear in the *auto* configuration runs. We can see that this configuration is not able to determine, which is the most suitable default class. Actually, on only 5 of the 15 datasets the chosen default class was almost or totally concentrated on a single class.

Another important issue is the number of iterations where the niched tournament selection was used. Table 8 shows these results. We can see that for some datasets, the niching process was used for quite a long time.

Table 7. Default class behavior in the auto configuration

Dataset	Major. class pos.	Minor. class pos.	Class distribution in default rule
bpa	2	1	(50.67%,49.33%)
bps	1	2	(14.67%,85.33%)
bre	1	2	(0.00%,100.00%)
gls	2	4	(14.00%,40.00%,8.67%,9.33%,14.00%,14.00%)
h-s	1	2	(32.00%,68.00%)
ion	2	1	(97.33%,2.67%)
lrn	1	5	(17.33%,35.33%,34.00%,11.33%,2.00%)
mmg	1	2	(48.00%,52.00%)
pim	1	2	(62.00%,38.00%)
son	2	1	(32.00%,68.00%)
thy	1	3	(40.67%,18.67%,40.67%)
veh	3	4	(35.33%,24.00%,13.33%,27.33%)
wdbc	2	1	(48.00%,52.00%)
wine	2	3	(4.00%,70.67%,25.33%)
wpbc	2	1	(1.33%,98.67%)

Table 8. Percentage of iterations that used the niched tournament selection in the default rule auto configuration

Dataset	Percentage of iterations
bpa	8.19%
bps	15.10%
bre	13.71%
gls	27.82%
h-s	13.33%
ion	6.72%
lrn	69.06%
mmg	10.79%
pim	9.41%
son	15.45%
thy	30.20%
veh	20.29%
wdbc	7.66%
wine	34.11%
wpbc	12.43%

It is reported in the niching literature [16] that we should increase the population size in order to guarantee that all niches can be learned properly. For this reason, a second set of tests was performed increasing the population size from 300 to 400. The results are shown in table 9. The summary of the statistical t-tests applied to these results is in table 10.

Now we can see a different picture. The increase in population size actually enables the *auto* policy to permit all niches to be learned properly. This fact is reflected by the accuracy performance of this policy, which manages to reach *major+minor*, both in accuracy and in robustness, based on the t-tests. Now that both policies are competitive, the smaller computational cost of *auto* (also

Table 9. Results of the tests comparing the studied default class policies to the original configuration using pop. size 400

Domain	Result	Default rule policy				
		Disabled	Major	Minor	Auto	Major+Minor
bpa	Train	79.3±1.7	82.0±1.4	80.7±1.4	81.0±1.6	82.0±1.4
	Test	64.0±7.5	62.6±7.5	64.4±6.9	64.5±7.3	62.6±7.5
	#rules	6.8±1.0	8.9±1.4	8.3±1.6	8.7±1.4	8.9±1.4
bps	Train	84.9±0.9	86.2±0.7	87.1±0.6	86.9±0.8	87.1±0.6
	Test	80.4±4.5	80.9±3.8	81.6±3.8	81.2±3.9	81.6±3.8
	#rules	5.1±0.4	6.1±1.1	5.9±1.0	5.8±1.0	5.9±1.0
bre	Train	97.7±0.4	98.3±0.3	98.5±0.4	98.4±0.4	98.5±0.4
	Test	95.7±2.3	95.0±2.6	95.7±1.9	95.8±1.9	95.7±1.9
	#rules	2.6±0.8	5.8±1.1	3.3±0.7	3.2±0.7	3.3±0.7
gls	Train	80.8±2.5	83.8±1.6	81.3±2.1	79.5±1.7	83.8±1.6
	Test	66.8±7.0	69.1±7.7	68.0±8.3	67.1±7.4	69.1±7.7
	#rules	6.5±0.7	6.8±0.8	7.5±0.9	6.7±0.8	6.8±0.8
h-s	Train	90.1±1.0	92.0±0.9	92.4±0.8	92.2±0.8	92.4±0.8
	Test	79.4±7.0	79.2±5.8	81.6±6.9	81.2±6.6	81.6±6.9
	#rules	6.6±0.8	7.8±1.3	7.4±1.2	7.4±1.2	7.4±1.2
ion	Train	96.1±0.6	95.9±0.8	97.1±0.7	96.9±0.7	97.1±0.7
	Test	93.5±3.5	90.4±4.3	93.4±3.5	92.8±4.0	93.4±3.5
	#rules	2.3±0.7	5.7±1.2	2.6±0.7	2.6±0.9	2.6±0.7
lrn	Train	75.7±1.7	77.2±0.8	75.8±1.4	75.7±1.0	77.2±0.8
	Test	68.0±5.0	69.1±5.4	68.7±5.2	69.1±4.9	69.1±5.4
	#rules	8.4±1.9	9.5±1.6	9.3±1.9	8.8±1.8	9.5±1.6
mmg	Train	80.3±1.7	83.4±1.3	83.4±1.3	83.5±1.1	83.4±1.3
	Test	65.9±8.3	69.0±8.0	67.3±8.9	69.7±7.7	69.0±8.0
	#rules	6.5±0.8	6.5±0.9	6.8±1.0	6.6±0.9	6.5±0.9
pim	Train	80.0±1.0	81.5±0.7	81.2±0.7	81.4±0.7	81.5±0.7
	Test	74.7±4.6	75.2±4.4	74.8±4.7	74.9±4.6	75.2±4.4
	#rules	5.3±0.6	6.3±1.1	5.8±0.9	6.1±1.0	6.3±1.1
son	Train	92.7±1.5	96.7±1.1	95.3±1.3	96.1±1.3	96.7±1.1
	Test	71.3±9.4	76.2±9.1	74.6±10.1	76.3±8.9	76.2±9.1
	#rules	6.7±1.0	7.6±1.3	7.7±1.5	7.6±1.4	7.6±1.3
thy	Train	97.6±0.9	98.6±0.7	98.6±0.7	98.3±0.8	98.6±0.7
	Test	91.5±6.2	92.0±5.2	92.4±4.8	91.4±5.6	92.4±4.8
	#rules	5.2±0.5	5.7±0.7	5.4±0.6	5.5±0.6	5.4±0.6
veh	Train	71.9±1.9	74.1±1.3	74.2±1.2	72.6±1.3	74.2±1.2
	Test	66.9±4.3	67.6±4.2	68.3±4.5	67.9±4.8	68.3±4.5
	#rules	6.5±1.3	9.4±1.8	10.0±1.8	8.4±1.8	10.0±1.8
wdbc	Train	97.2±0.8	98.0±0.5	97.9±0.6	97.8±0.6	98.0±0.5
	Test	93.9±2.9	94.4±3.1	94.4±3.2	94.4±3.1	94.4±3.1
	#rules	4.3±1.2	4.8±1.1	4.2±0.7	4.5±0.9	4.8±1.1
wine	Train	99.4±0.6	99.7±0.4	99.8±0.3	99.6±0.4	99.8±0.3
	Test	94.1±6.0	93.2±6.4	92.0±6.5	93.2±6.3	92.0±6.5
	#rules	3.8±0.7	3.7±0.6	4.2±0.5	3.8±0.7	4.2±0.5
wpbc	Train	84.9±2.8	89.9±1.8	87.1±3.3	89.0±2.1	89.9±1.8
	Test	76.6±6.7	75.3±7.0	72.4±9.1	76.3±7.1	75.3±7.0
	#rules	2.8±0.9	3.9±0.9	4.4±1.2	3.7±1.0	3.9±0.9
ave	Train	87.2±8.8	89.2±8.3	88.7±8.6	88.6±8.9	89.3±8.3
	Test	78.8±11.5	79.3±10.7	79.3±11.1	79.7±10.8	79.7±11.7
	#rules	5.3±1.7	6.6±1.7	6.2±2.1	6.0±2.0	6.2±2.2

compared to *major+minor* using a population size of 300) clearly makes it the most suitable configuration for the default class.

Moreover, we can see how the only method that degrades performance when we increase the population size is the majority class policy, suggesting that the system is sensitive to over-learning in domains where the majority class policy is not suitable. The larger average number of rules and the better training accuracy of the solutions generated by this policy confirm the over-learning problem.

Table 10. Summary of the statistical t-tests applied to the experimentation results of popsize 400, with a confidence level of 0.05. Cells in table count how many times the method in the row significantly outperforms the method in the column.

Policy	Disabled	Major	Minor	Auto	Major+Minor	Total
Disabled	-	2	1	0	0	3
Major	1	-	1	0	0	2
Minor	1	3	-	0	0	4
Auto	1	3	1	-	0	5
Major+Minor	2	3	1	0	-	6
Total	5	11	4	0	0	

9   Conclusions and Future Work

In this paper we have tested some methods that extend the rule-based and decision-list-style knowledge representations for a Pittsburgh Learning Classifier System by using a static default rule. This kind of systems tend to generate an emergent default rule, which can increase the performance of the system. By forcing the representation of a default rule, we intended to guarantee these positive effects.

Simple policies such as using the majority/minority class as the default class perform quite well compared to the original system. However, they perform poorly on certain datasets somewhat showing a lack of robustness. We can almost integrate the best results of both policies by using the simple heuristic of selecting the policy with more training accuracy. This mechanism introduces a good performance boost, but doubles the run-time.

For this reason, we have developed a mechanism that decides automatically the class for the default rule. This technique works by integrating in a single population individuals using all possible default classes and letting them compete among themselves. This approach has a problem, however, which is providing a fair competition framework, because each default rule class can yield different learning progress. In order to achieve this fairness, we use a niched tournament selection that guarantees that all niches (different default rules) survive in the population until they can compete successfully by themselves. This automatic mechanism performs best when we increase the population size, which is an usual requirement in most systems that use niching, because we have to guarantee that each niche has enough individuals to ensure sufficient diversity for building block supply and thus successful and reliable learning.

The increase in population size for the majority/minority policies, however, showed no performance increase or even some performance decrease, suggesting the amplification of the policy weaknesses This weaknesses are derived from overlearning, which is reflected in the larger training accuracy and larger average rule set sizes and also on the statistical tests.

Although the automatic policy does not outperform the major+minor policy, the accuracy difference is quite small in most datasets and the computational cost is significantly lower. Therefore, it appears that in most situations the automatic policy is the best method.

One of the main sacrifices done in the *auto* default class determination policy is the mating restriction introduced in the crossover algorithm, preventing the creation of lethals, because it is almost impossible to create competitive offspring if the parents cover different subsets of the training instances. However, it would be useful to study if there are any feasible ways to recombine successfully individuals with different default classes. If we achieve this objective, perhaps we can reduce the population size requirements of the *auto* policy.

Another alternative would be to develop more sophisticated heuristics that combine the simple default class policies. It might be possible to have a method that only requires a short run to reliably decide on the most suitable default rule class, instead of running a full test for each candidate class. To do so, it appears necessary to also investigate in general in which cases which default rule class is most appropriate. It is expected that the best default rule class does not only depend on the class distribution and class boundaries but also, mutually, on the representation of the class boundaries in the evolving rules. Future research will shine further light on this matter.

Acknowledgments

The authors acknowledge the support provided by the Spanish Research Agency (CICYT) under grant numbers TIC2002-04160-C02-02 and TIC 2002-04036-C05-03, the support provided by the Department of Universities, Research and Information Society (DURSI) of the Autonomous Government of Catalonia under grants 2002SGR 00155 and 2001FI 00514. Additional funding from the German research foundation (DFG) under grant DFG HO1301/4-3 as well as from the European commission contract no. FP6-511931 is acknowledged. Additional support from the UK Engineering and Physical Sciences Research Council (EPSRC) under grant GR/T07534/01 is acknowledged.

Also, this work was sponsored by the Air Force Office of Scientific Research, Air Force Materiel Command, USAF, under grant F49620-03-1-0129, and by the Technology Research, Education, and Commercialization Center (TRECC), at University of Illinois at Urbana-Champaign, administered by the National Center for Supercomputing Applications (NCSA) and funded by the Office of Naval Research under grant N00014-01-1-0175. The US Government is authorized to reproduce and distribute reprints for Government purposes notwithstanding any copyright notation thereon.

The views and conclusions contained herein are those of the authors and should not be interpreted as necessarily representing the official policies or endorsements, either expressed or implied, of the Air Force Office of Scientific Research, the Technology Research, Education, and Commercialization Center, the Office of Naval Research, or the U.S. Government.

References

1. Holland, J.H.: Adaptation in Natural and Artificial Systems. University of Michigan Press (1975)
2. Goldberg, D.E.: Genetic Algorithms in Search, Optimization and Machine Learning. Addison-Wesley Publishing Company, Inc. (1989)
3. DeJong, K.A., Spears, W.M., Gordon, D.F.: Using genetic algorithms for concept learning. Machine Learning **13** (1993) 161–188
4. Rivest, R.L.: Learning decision lists. Machine Learning **2** (1987) 229–246
5. Janikow, C.: Indictive Learning of Decision Rules in Attribute-Based Examples: a Knowledge-Intensive Genetic Algorithm Approach. Phd dissertation, University of North Carolina (1991)
6. Quinlan, J.R.: C4.5: Programs for Machine Learning. Morgan Kaufmann (1993)
7. Cohen, W.W.: Fast effective rule induction. In: International Conference on Machine Learning. (1995) 115–123
8. Bacardit, J.: Pittsburgh Genetics-Based Machine Learning in the Data Mining era: Representations, generalization, and run-time. PhD thesis, Ramon Llull University, Barcelona, Catalonia, Spain (2004)
9. Soule, T., Foster, J.A.: Effects of code growth and parsimony pressure on populations in genetic programming. Evolutionary Computation **6** (1998) 293–309
10. Rissanen, J.: Modeling by shortest data description. Automatica **vol. 14** (1978) 465–471
11. Bacardit, J., Garrell, J.M.: Bloat control and generalization pressure using the minimum description length principle for a pittsburgh approach learning classifier system. In: Proceedings of the 6th International Workshop on Learning Classifier Systems, (in press), LNAI, Springer (2003)
12. Bacardit, J., Garrell, J.: Analysis and improvements of the adaptive discretization intervals knowledge representation. In: GECCO 2004: Proceedings of the Genetic and Evolutionary Computation Conference, Springer (to appear) (2004)
13. Blake, C., Keogh, E., Merz, C.: UCI repository of machine learning databases (1998) (www.ics.uci.edu/mlearn/MLRepository.html).
14. Oei, C.K., Goldberg, D.E., Chang, S.J.: Tournament selection, niching, and the preservation of diversity. IlliGAL Report No. 91011, University of Illinois at Urbana-Champaign, Urbana, IL (1991)
15. Kohavi, R.: A study of cross-validation and bootstrap for accuracy estimation and model selection. In: IJCAI. (1995) 1137–1145
16. Goldberg, D.E.: Sizing populations for serial and parallel genetic algorithms. In: Proceedings of the Third International Conference on Genetic Algorithms (ICGA89), Morgan Kaufmann (1989) 70–79

Using XCS to Describe Continuous-Valued Problem Spaces

David Wyatt, Larry Bull, and Ian Parmee

University of the West of England, Faculty of Computing,
Engineering & Mathematical Sciences, Frenchay Campus, Bristol BS16 1QY
{David2.Wyatt, Larry.Bull, Ian.Parmee}@uwe.ac.uk

Abstract. Learning classifier systems have previously been shown to have some application in single-step tasks. This paper extends work in the area by applying the classifier system to progressively more complex multi-modal test environments, each with typical search space characteristics, convex/non-convex regions of high performance and complex interplay between variables. In particular, two test environments are used to investigate the effects of different degrees of feature sampling, parameter sensitivity, training set size and rule subsumption. Results show that XCSR is able to deduce the characteristics of such problem spaces to a suitable level of accuracy. This paper provides a foundation for the possible use of XCS as an exploratory tool that can provide information from conceptual design spaces enabling a designer to identify the best direction for further investigation as well as a better representation of their design problem through redefinition and reformulation of the design space.

1 Introduction

In this paper, the XCSR classifier system [35] is cast as an induction engine that is trained using a reinforcement learning approach, i.e., an external agent provides a reward for each successfully classified data instance. Once the system has completed its training, new unseen data are presented and a measure of classification accuracy made. There have been several papers published that demonstrate XCS's capabilities for data-mining through rule induction. In [3], Bernado et al. describe an experimental comparison of XCS with seven other learning schemes, including C4.5, Naive Bayes and Support Vector Machines. Fifteen UCI repository datasets [4] were used each with mixture of attribute types and differing numbers of classes and dataset sizes. The XCS system is shown to be highly competitive when compared with the other learning schemes. Wilson [36][37] has also demonstrated the capabilities of an interval based encoding when used to induce rules describing the Wisconsin Breast Cancer dataset, where XCS improved on the best known performance for that dataset.

The studies described above highlight the capabilities of XCS for data-mining through rule-induction and provide some motivation for the investigation presented here. It is hoped that this paper can provide a clear foundation for the possible use of XCS as an exploratory tool, and in particular one that can provide information from conceptual design spaces enabling a designer to identify the best direction for further

X. Llorà et al. (Eds.): IWLCS 2003-2005, LNAI 4399, pp. 308–332, 2007.

investigation [24] as well as a better representation of their design problem through redefinition and reformulation of the design space. The objective is to develop the XCS classifier system to enable the evolution of an accurate set of maximally general rules that can identify and describe the high performance regions of any given test environment, before exploring the utility of this approach within more real-world problem environments. In particular, the XCSR [35] classifier system is expected to induce rules from continuous-valued domains where some discretised classification is defined for each sample point.

Parmee [23] introduced the concept of the identification of high performance regions of complex preliminary design spaces rather than the identification of single optimal design solutions. A region of high performance is any contiguous set of points in a given design space which are considered to be exceptional solutions to a particular set of possibly conflicting design criteria. Parmee introduced a new evolutionary search method, namely the Cluster-Oriented Genetic Algorithm (COGA), which proved itself capable of rapidly discovering high performance regions of an unknown design space whilst achieving a high percentage of solutions within regions of high performance [6]. The interested reader is directed to [23], [6] and [25] for more information. We are interested in the utility of XCS for such tasks.

This paper extends work presented in [7] in which XCSR was applied to a well-known single-step task, the Boolean Multiplexor Problem. These single-step functions are traditionally defined for binary strings of length $l = k + 2^k$ under which the first k bits index into the 2^k remaining bits, returning the indexed bit. In fact, the Boolean Multiplexor Problem can be redefined such that each binary digit in the problem string is represented as a real value in the interval [0, 1] together with a fixed threshold value, usually 0.5. The threshold is used to convert a real value to its corresponding binary digit, e.g. the real-valued string 0.30, 0.7, 0.58 decodes to the binary string 011 given a threshold of 0.5. It is the task of XCSR to learn accurate general classification rules that describe the range of real values that correspond to 0's, 1's and #'s in the binary-encoded solutions, which can be simplified to the finding interval bounds for each variable. This view of finding a Boolean Multiplexor Problem solution casts the act of solving the problem as one of identifying regions of high performance - assuming some simplification of the definition given above, that is, a solution to the Multiplexor Problem is either correct or incorrect whereas a design problem will have many different levels of performance.

Initial investigations in [7] show that the XCSR system is able to identify high performance regions from a continuous multi-variable search space using a sample of training data points. The solution provided is a complete set of simple classification rules that define orthogonal regions of the solution space with attached classification labels. Investigations continued using a new simplified learning scheme with the aim of improving XCSR performance with respect to learning speed and ability to respond to changes in the underlying test environment (such as class relabelling). The new system was termed sXCSR and results showed that improvements can be made under the new learning scheme. The work presented clearly demonstrated the capability of XCSR to evolve real-valued pairs to describe interval bounds for each variable in the multi-variable problem and thereby define a set of simple classification rules for the high performance regions of an eleven variable Multiplexor-related search space.

The investigation of XCSR's capabilities for describing regions of high performance is continued in this paper by applying XCSR and sXCSR to progressively more
complex multi-modal test environments each with typical search space characteristics,
convex/non-convex regions of high performance and complex interplay between
variables. In particular, two test environments are used to investigate the effects of
different degrees of feature sampling, parameter sensitivity, training set size and rule
subsumption. These environments were previously used by Bonham [5][25] to investigate the capabilities of the COGA system. Both test environments are constructed
using a combination of functions allowing for the simple generation of training and
test points, where each sample point can be represented by a vector of continuous
values and a continuous performance measure which may be discretised as appropriate. Fixed size training datasets are used in an effort to provide some consistency in
experimentation with those design problems for which the cost of an on-line evaluation per sample point is high or for which datasets are constructed from other off-line
data sources.

The XCSR experiments presented below use an exact performance threshold, ψ,
that has been defined with prior knowledge of the test environments to allow for a
cleaner analysis of the XCSR system. The performance threshold defines a two-class
classification task where a given sample point is classified as High or Low. A similar
approach was adopted during the development of COGA utilising a similar set of test
functions [5][6]. This approach is essential in order to evaluate algorithm performance
in terms of a set of predefined criteria.

The paper is arranged as follows: the next section describes the XCSR system used
throughout; section 3 describes and presents results for a two dimensional test environment; section 4 considers the class imbalance problem; section 5 describes and
presents results for a six dimensional test environment; and finally, all findings are
discussed.

2   XCSR

In [35], Wilson presents a version of XCS [34] for problems which can be defined by
a vector of bounded continuous real-coded variables – XCSR. In that system, each
rule in the classifier system population consists of the following parameters: <condi-
tion> : <action> : prediction (p) : prediction error (ε) : fitness (F) : experience (exp) :
time-stamp (ts) : action set size (as) : numerosity (n). Given that XCSR is an accuracy-based classifier system, the three parameters p, ε and F represent how accurately
a rule predicts <action> given an input vector matched by <condition>. In [35],
Wilson defines a <condition> as consisting of interval predicates of the form
$\{\{c_1, s_1\}, \ldots \{c_n, s_n\}\}$, where c is the interval's range centre and s is the "spread" from
that centre - termed here as the Centre-Spread encoding. Each interval predicate's
upper and lower bounds are calculated as follows : $[c_n - s_n, c_n + s_n]$. If an interval
predicate goes outside the variable's defined bounds, it is truncated. In order for a rule
to match the environmental stimulus, each input vector value must sit within the interval predicate defined for that variable. The other parameters, exp, ts, as and n, are
used by the classifier system to maintain the internal dynamics of the system, such as

balancing resources across environmental niches, genetic algorithm invocation and computational issues. In [36], Wilson describes another version of XCS which could also be used for such multi-variable problems in which a vector of integer-coded interval predicates is used in the form $\{[l_1, u_1], \ldots [l_n, u_n]\}$, where l and u are the intervals' lower and upper bounds, respectively - termed here as the Interval encoding. It is clear that a real-coded version of the integer bounded interval predicates would be trivial to implement.

The form of subsumption used for both types of encoding is that a rule may subsume another if every interval predicate in the subsumee's *<condition>* can be subsumed by the corresponding predicate in the subsumer. In order to identify if a predicate may be subsumed, the subsumee's lower bound must be greater and its upper bound must be lesser than the corresponding subsuming predicate. In fact, XCS implements two different forms of subsumption, Action Set Subsumption and Genetic Algorithm Subsumption. In the first form, a single rule is defined as the most general in a given Action Set and is permitted to subsume any other rule in the Action Set providing it is sufficiently experienced and accurate enough. In [36], Wilson defines a generality measure for each rule as $\sum(u_i - l_i + 1) \ \forall i$. However, this approach was not used for the experimental results presented throughout this paper. Instead, the Action Set rule that has a lower bound lesser than and an upper bound greater than any other rule in the Action Set is defined as the most general rule. In the second form of subsumption, a newly generated offspring rule may be subsumed if either of its parents are more general than it, sufficiently experienced and accurate enough.

All other XCS processing remains as described in [8] for both the Centre-Spread and Interval encoded versions except that mutation is implemented via a random step (range $-0.1 < x < 0.1$) and cover produces rules centred on the input value with a range of s_0. It is important to note in the case of the Interval encoding, a potential problem may arise as a result of the action of the mutation operator such that it is possible for a variable predicate's upper bound to become smaller than its lower bound. There are two ways to deal with this problem, termed here as Ordered Interval and Unordered Interval [27]. The first way uses a repair operator to enforce an ordering restriction on the predicates by swapping the offending values to ensure that all interval predicates in the *<condition>* remain feasible, i.e., in the form $\{[l_1, u_1], \ldots [l_n, u_n]\}$. The second way lifts the ordering restriction such that an interval $[l_n, u_n]$ is equivalent to $[u_n \ l_n]$. The reader is referred to [27] for a discussion of the issues related to the differences between interval encodings.

The results for each parameter setting of the XCSR system are averaged over five independent runs and presented together with a standard deviation for that sample. Any conclusions made in this paper are based on the application of Mann-Whitney Rank Sum Test which makes no assumptions about the distribution of population from which the runs where sampled. It should also be made clear that those figures showing classification accuracy for a given parameter setting represent the performance of the system on the test dataset during the entire learning phase. The apparent improvement in performance during the first 10-20% of each figure should only be taken as an indicator of successful learning due to cover.

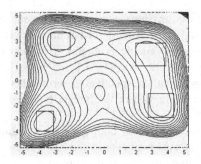

Fig. 1. The Modified Himmelblau Function Contour with Four High Performance Regions

3   A Two Dimensional Test Environment

The two dimensional test environment used in this paper is the multi-modal modified Himmelblau function [2]. The equation for the modified Himmelblau function, which is used to evaluate each sample point, is given in the Appendix. There are four optima of approximately equal magnitude. This function is used to define a two-class classification task to investigate the effects of different degrees of feature sampling, parameter sensitivity, training set size and rule subsumption on the performance of the XCSR and sXCSR classifier systems. In particular, an exact threshold value of $\psi =$ 184, where $\psi \in$ [-1986, 200], is used to define High/Low class decision boundaries. Figure 1 shows a contour plot of the function, clearly indicating the four regions of high performance as defined by the threshold value given above.

The XCSR system was trained using two different training datasets and tested using a single test dataset generated from a uniform random distribution which has been manipulated in such a way as to provide an equal number of test points per classification as shown in Figure 2. In particular, n points are sampled from a uniform random distribution and evaluated according to the current environment. The sample points are sorted in descending order of performance and the top $2m$ points are used to define the test dataset, where m equals the total number of High points generated. All three datasets have two defined classes, High and Low. The two training datasets, Figure 2(a) and Figure 2(b), were both generated from a Halton Sequence Leaped (HSL) sequence [14] with 500 and 2000 sample points, respectively. The HSL is a quasi-random sequence which provides a set of real numbers whose degree of uniformity is high.   The test dataset used for this section of the paper was generated from a uniform random distribution with 2116 sample points, where 1073 points are defined as High - shown as faint dots in Figure 2 (c) below. In particular, the test dataset was manipulated to include sample points from both classifications near to the classification decision boundaries. It is hoped that results using this dataset will give clear evidence of the classifier system's capability to evolve rules that define those boundaries.

Experiments were conducted to discover if there were any underlying trends in the parameter space for this problem. This involved altering subsumption parameters, training dataset size, rule encodings and using the new update technique for payoff prediction [7]. However, an approach is required to enable comparisons of performance to be made between different parameter sets. One common way to define the performance of a classification system is to use a confusion matrix [15] of size $L \times L$, where L is the number of different classifications. The matrix contains information about the actual and predicted classifications resulting from the classification task and provides a simple format to record and analyse a system's performance. Figure 3 gives an example of a 2 x 2 confusion matrix with the two possible classifications used in this section of the paper.

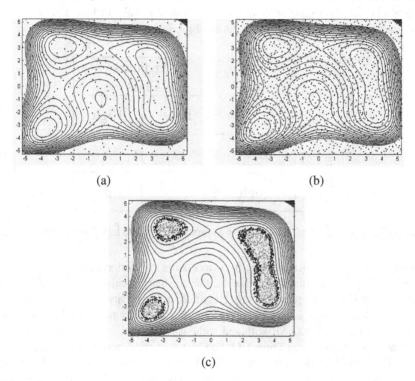

(a)   (b)

(c)

Fig. 2. Two (a & b) Training Datasets and One Test Dataset (c) : (a) 500 HSL-generated Points, (b) 2000 HSL-generated Points, (c) Uniform Random (equal points per classification)

There are a number of possible measures of classification accuracy based on confusion matrices, which include the Lewis and Gale's F-measure [21], the geometric mean as defined by Kubat et al in [18], using ROC graphs to examine classifier performance [28] and Kononenko and Bratko's information-based evaluation criterion [16]. In fact, all of the above measures were developed to overcome problems associated with analysis where the number of examples in each classification is significantly different. Given that all the test datasets used in this paper have been manipulated such that the number of examples per classification are nearly equal, a simple accuracy measure will suffice for basic analysis. For this two-class classification problem, the accuracy measure is defined as the number of examples correctly classified as

High plus the number of examples correctly classified as Low divided by the total number of examples classified, that is, $(a + d) / (a + b + c + d)$ using the cells shown in Figure 3. However, it should be made clear that this measure may fail to provide usable analytical data when the differences between the percentage correct for each classification are too large. The percentage of High and Low points correctly classified are traditionally known as the sensitivity and specificity, respectively. These terms frequently appear in the medical literature and are mainly used to describe the result of medical trials for disease prevention, but have come to be used in many non-medical classification tasks including information retrieval. In fact, a similar set of performance metrics were introduced for the EpiCS [11] system.

		Predicted Class	
		Low	High
Actual Class	Low	a	b
	High	c	d

Fig. 3. An example 2 x 2 Confusion Matrix with High and Low Classifications

Table 1 shows three performance measures using the Uniform Random test dataset for each parameter combination, that is, the accuracy measure defined above, the percentage of High examples correctly classified, or sensitivity, and the percentage of Low examples correctly classified, or specificity. The parameter combinations used include running XCSR/sXCSR with and without Action Set Subsumption using three different encodings - Centre-Spread, Ordered Interval and Unordered Interval.

It is clear from Table 1 that the system performed well on the two dimensional Himmelblau test problem in terms of correct classification of unseen data, between 62.3% and 91.1% depending on rule encoding, subsumption type and training sample size. The system also performed at a very high level, around 98%, when presented with "probe points" randomly distributed across the entire search space (not shown), providing empirical evidence of the ability of XCS to accurately describe the entire search space. The performance gain for XCSR when Action Set Subsumption is turned off is remarkably clear in Table 1. In fact, the difference between XCSR with and without Action Set Subsumption is statistically significant (>99%) for all other parameter settings shown. Figure 4 shows the performance gain for the Ordered Interval XCSR encoding using a 2000 HSL-generated training dataset. This improvement may be a result of permitting initially weaker rules enough time to show their true potential by reducing the early domination of more numerate rules in a given Action Set as was suggested in [7].

Table 1. Classification Accuracy for XCSR Using a Uniform Random Test Dataset based on the modified Himmelblau Function, where $Trials = 200000$, $N = 8000$, $\beta = 0.2$, $\alpha = 0.1$, $\varepsilon_0 = 10$, $v = 5$, $\theta_{GA} = 12$, $\chi = 0.8$, $\mu = 0.04$, $\theta_{del} = 20$, $\delta = 0.1$, $p_I = 10$, $\varepsilon_I = 0$, $F_I = 0.01$, $\theta_{mn\ a} = 2$, $\theta_{sub} = 20$, $m = \pm10\%$ and $s_0 = 2\%$. Table 1 presents results for each parameter combination in the format : % Accuracy$_{(1\ s.d.)}$ over Sensitivity & Specificity.

		Centre-Spread Encoding		Ordered Interval Encoding		Unordered Interval Encoding	
		500 HSL	2000 HSL	500 HSL	2000 HSL	500 HSL	2000 HSL
With AS-Subsumption	XCSR	**64.4**$_{(3.5)}$	**75.0**$_{(5.9)}$	**62.3**$_{(4.8)}$	**70.1**$_{(6.6)}$	**64.5**$_{(8.0)}$	**68.5**$_{(3.9)}$
		34.0 95.7	53.2 97.5	35.9 89.5	42.1 98.8	40.1 89.6	38.5 99.4
	sXCSR	**63.0**$_{(4.5)}$	**66.1**$_{(2.6)}$	**63.2**$_{(3.2)}$	**64.5**$_{(3.7)}$	**65.7**$_{(4.0)}$	**66.9**$_{(4.9)}$
		35.0 91.8	34.3 98.7	32.9 94.2	30.8 99.2	42.4 89.7	36.2 98.5
Without AS-Subsumption	XCSR	**76.3**$_{(2.2)}$	**90.0**$_{(1.3)}$	**79.7**$_{(2.0)}$	**91.1**$_{(0.6)}$	**79.9**$_{(2.1)}$	**90.0**$_{(0.5)}$
		75.1 77.6	89.4 90.6	82.4 76.9	91.2 91.1	80.1 79.7	88.9 91.2
	sXCSR	**76.5**$_{(2.3)}$	**89.6**$_{(0.7)}$	**80.0**$_{(2.0)}$	**89.9**$_{(0.7)}$	**80.3**$_{(1.5)}$	**89.7**$_{(1.7)}$
		73.2 80.0	89.6 89.5	82.9 77.0	89.1 90.8	82.6 77.9	88.3 91.1

Regarding the different rule encodings, Table 1 shows that the Ordered and Unordered Interval encodings out-perform the Centre-Spread encoding on smaller training datasets with Action Set Subsumption turned off. Although this difference in performance is statistically significant (>95%), there does not appear to be any significant performance differences between the Ordered and Unordered Interval encodings. Figure 5 shows the learning speed and system error for the Ordered Interval XCSR and Ordered Interval sXCSR encoded systems using a 2000 HSL-generated training dataset with No Action Set Subsumption. It is clear from Table 1 and Figure 5 that there are no statistically significant differences in final performance between the sXCSR and XCSR versions of the system or in learning speed between the two systems but that there is a small improvement in system error for the sXCSR system.

According to Table 1 and Figure 6, as the number of training points increases the performance of the system improves, suggesting that the system can gain more information about the dimensionality of the high performance regions when more points are sampled in that area. However, this statement is only statistically significant (>99%) for those XCSR parameters settings with Action Set Subsumption turned off. Figure 7 shows that there is little difference in the learning speed and system error for the number of training points used here given that the classifier system is able to describe the training set perfectly within a few thousand trials. It should be made clear that the classifier system is expected to learn accurate general classification rules that describe regions of the search space rather than single training data points and so simply stopping as soon as the training data has been learned is not enough to complete the region discovery aspects of learning – this should follow as the classifier system evolves general rules that cover regions of the search space.

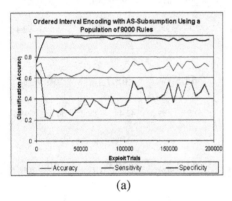 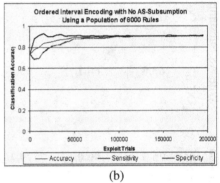

(a)   (b)

Fig. 4. Comparison of Classification Accuracy for Different Subsumption Settings (a) with Action Set Subsumption and (b) without Action Set Subsumption, where $Trials = 200000$, $N = 8000$, $\beta = 0.2$, $\alpha = 0.1$, $\varepsilon_0 = 10$, $\nu = 5$, $\theta_{GA} = 12$, $\chi = 0.8$, $\mu = 0.04$, $\theta_{del} = 20$, $\delta = 0.1$, $p_I = 10$, $\varepsilon_I = 0$, $F_I = 0.01$, $\theta_{mn\,a} = 2$, $\theta_{sub} = 20$, $m = \pm10\%$ and $s_0 = 2\%$

There is an important issue regarding the imbalanced nature of the training sets used for the experiments discussed above. Given that only a small fraction, around 5%, of the training examples are from the high performance region and that the classifier system is expected to form accurate descriptions of these regions, it is clear that some degree of re-sampling may be necessary in more complex environments. The next section provides a definition of this problem and details a small number of the published attempts to overcome it.

4   The Class Imbalance Problem

The class imbalance problem [13] can be defined as a problem encountered by any inductive learning system in domains for which one class is under-represented and which assume a balanced class distribution in the training data. For a two-class problem, the class defined by the smaller set of examples is referred to as the minority class while the other class is referred to as the majority class. However, much of the following discussion also applies to multi-class problems. A few papers detailing attempts to deal with this problem are discussed below.

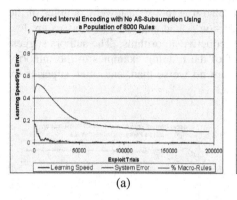
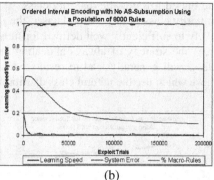

(a)   (b)

Fig. 5. Comparison of Learning Speed and System Error for (a) XCSR and (b) sXCSR , where $Trials = 200000$, $N = 8000$, $\beta = 0.2$, $\alpha = 0.1$, $\varepsilon_0 = 10$, $v = 5$, $\theta_{GA} = 12$, $\chi = 0.8$, $\mu = 0.04$, $\theta_{del} = 20$, $\delta = 0.1$, $p_I = 10$, $\varepsilon_I = 0$, $F_I = 0.01$, $\theta_{mn\,a} = 2$, $\theta_{sub} = 20$, $m = \pm 10\%$ and $s_0 = 2\%$

In [13], Japkowicz and Stephen discuss several methods for dealing with the class imbalance problem including minority over-sampling, majority under-sampling and Elkan's Cost Sensitive Learning method [9]. Minority over-sampling refers to the repeated sampling of, or duplicating of, examples from the minority class with re-placement until the number of examples in the minority class equals some pre-defined fraction of the majority class size. Majority under-sampling refers to the elimination of examples from the majority class until its' size equals some pre-defined multiple of the minority class. Both of these approaches have known drawbacks. For instance, majority under-sampling may cause potentially useful information to be lost and, depending upon the degree of minority over-sampling, there may be an increased chance of overfitting. Elkan's Cost Sensitive Learning method modifies the relative cost associated with misclassifying an example to compensate for the imbalance and provides a theorem that shows "how to change the proportion of minority examples in a training set in order to make optimal cost-sensitive classification decisions using a classifier learned by a standard non-cost-sensitive learning method".

In [31], Weiss and Provost provide an interesting empirical study of the effect of class distribution on classifier performance for twenty-five published datasets. A majority under-sampling approach was used to form a two class distribution, that is, stratified random sampling without replacement of the majority examples with no replication of any minority example. The authors suggest that the reasons why classi-fier systems perform differently on the minority class versus the majority class can be justified by two observations. The first observation is that "classification rules that predict the minority class tend to have a much higher error rate than those that predict the majority class". They reason that higher error rate for minority classification is in part because of conventional wisdom which suggests that a test-set should match the natural class distribution of the underlying domain and also that rules classifying the minority class are generally formed from fewer training examples than their majority counterparts. The second observation is that "test examples belonging to the minority class are misclassified more often than test examples belonging to the majority class". One reason for this is that the marginal probabilities of the natural class distribution are biased towards the majority class. It is also true that with fewer examples of the minority class in any given dataset, the decision boundaries between classes are less likely to sufficiently well defined for most inductive algorithms. The authors suggest that "the strategy of always allocating half of the training examples to the minority class...will generally lead to results which are no worse than, and often superior to, those which use the natural class distribution."

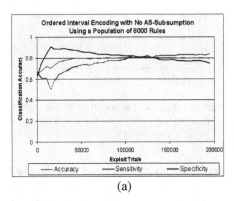   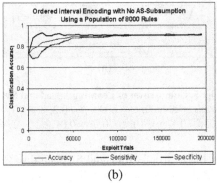

(a)   (b)

Fig. 6. Comparison of Classification Accuracy for Different Training Set Sizes (a) 500 HSL-generated data-points and (b) 2000 HSL-generated data-points , where $Trials = 200000$, $N = 8000$, $\beta = 0.2$, $\alpha = 0.1$, $\varepsilon_0 = 10$, $v = 5$, $\theta_{GA} = 12$, $\chi = 0.8$, $\mu = 0.04$, $\theta_{del} = 20$, $\delta = 0.1$, $p_I = 10$, $\varepsilon_I = 0$, $F_I = 0.01$, $\theta_{mn\,a} = 2$, $\theta_{sub} = 20$, $m = \pm10\%$ and $s_0 = 2\%$

Kubat and Matwin [17] addressed the problem by using an "intelligent" under-sampling method which attempts to remove majority examples according to their membership of the following four groups : examples suffering from *class-label noise*, *borderline* examples, *redundant* examples or *safe* examples. An attempt to reduce the *redundant* majority examples is defined as that of constructing a subset C of the main training set S by taking every minority example from S and one randomly chosen majority example. Examples remaining in S are classified by subset C using the 1-Nearest Neighbour (using a Euclidean distance measure) with any misclassified examples added to C as suggested by Hart in [10][33]. The method also makes use of the concept of Tomek links [29] to identify *noisy* or *borderline* examples, that is, choose a pair of examples (x, y), one from each class; the distance between examples is denoted $\delta(x, y)$; the pair (x, y) are called a Tomek link if no example z exists such that $\delta(x, z) < \delta(x, y)$ or $\delta(y, z) < \delta(y, x)$; remove any majority class example from the subset C that participates in a Tomek link. The remaining examples represent the new training set T. A counter-example of this technique is discussed in [30]. In [19], Laurikkala develops another method of "intelligent" under-sampling which is based on the one-sided selection principle developed by Kubat and Matwin [17]. The new method was designed to work with multi-class problems and considers the quality of data to be removed from the majority subset of the training dataset more rigorously. In particular, the new method uses a *neighbouthood cleaning rule* that is less sensitive to noise. The new method utilizes Wilson's edited nearest neighbour rule [33], to identify noisy data, by removing examples whose classification differs from the majority class of the k nearest neighbours (where k normally equals 3). In fact, the method will "clean" neighbourhoods by removing data that misclassify examples of the class of interest, i.e. the minority class. In addition, a heterogeneous value difference metric [32] is used which treats nominal attributes more appropriately. Laurikkala [19] was able to show that the *neighbourhood cleaning rule* is significantly better than Kubat and Matwin's method using ten real-world datasets.

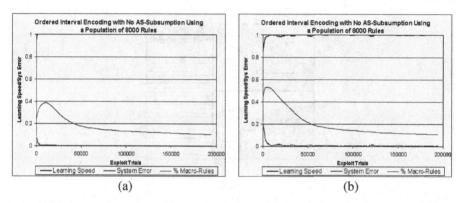

Fig. 7. Comparison of Learning Speed and System Error for Different Training Set Sizes (a) 500 HSL-generated data-points and (b) 2000 HSL-generated data-points , where *Trials* = 200000, $N = 8000$, $\beta = 0.2$, $\alpha = 0.1$, $\varepsilon_0 = 10$, $v = 5$, $\theta_{GA} = 12$, $\chi = 0.8, \mu = 0.04$, $\theta_{del} = 20$, $\delta = 0.1$, $p_I = 10, \varepsilon_I = 0$, $F_I = 0.01$, $\theta_{mn\,a} = 2$, $\theta_{sub} = 20$, $m = \pm 10\%$ and $s_0 = 2\%$

In [20], Lee describes an oversampling method to help overcome the Class Imbalance Problem which produces "noisy replicates of the rare cases while keeping the dominant class cases unchanged", that is, new examples of the minority class are created such that there will be m_I replicates of the minority examples with added noise. Each noisy example is generated by adding a small noise term, ε, to each dimension of its corresponding minority example. The noise term, ε, can be defined as $N(0, \sigma^2_{noise})$ where the noisy replicates become just m_I exact copies of each minority example as the variance of ε tends toward zero. The original majority and minority examples are added to the noisy replicates to form a new training set T^*. In particular, Lee suggests that when the new data is used with some form of bootstrapping technique, a regularisation or "smoothing" effect is produced on the minority examples to help avoid overfitting and improve generalisation. This effect has been studied extensively in the Neural Network domain for which [26] provides a good starting point. In [1], An suggests that the addition of noise to inputs in neural network training significantly reduces the generalisation error of the network while Holmström and Koistinen [12] show that the method of interpreting the addition of noise to the network inputs for generating additional training data is asymptotically consistent, that is, as the size of training dataset approaches infinity and variance of added noise

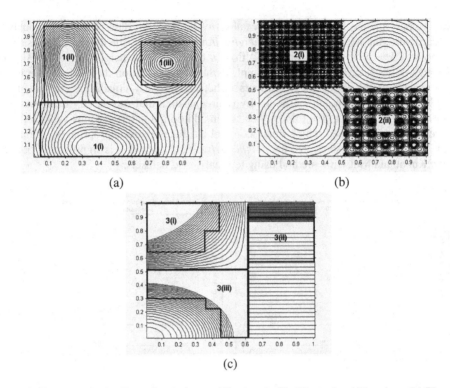

(a)   (b)

(c)

Fig. 8. Three local, two dimensional planes of Bonham's Six Dimensional Function : (a) Plane 1 with three high performance regions, (b) Plane 2 with two high performance regions, and (c) Plane 3 with three high performance regions

approaches zero, the method is equivalent to minimising the true error function. It is not clear whether the same effects would be seen for the classifier system, but it is certainly an approach worth investigating further.

Ling and Li [22] define the class imbalance problem in terms of a direct marketing domain in which class distribution can be extremely imbalanced. The authors motivate the need for data mining and machine learning in direct marketing and discuss the need for a finer distinction among buyers and non-buyers to allow flexibility in decisions about the means of promotion used. The exact details of their approach can be found in [22]. An aspect of their work which is of interest here, is a solution suggested for the class imbalance problem which uses both minority oversampling and majority undersampling. In particular, the minority class is oversampled with replacement until some pre-defined multiple, n, of the original sample size is achieved, that is, by setting n equal to two, twice as many minority class examples are used during training as those occurring in the original dataset. The majority class is undersampled without replacement until a number of examples equal to those sampled from the minority class have been defined. In addition, to the balancing of examples volumes per class, which has already been shown to be advantageous, no information about the search space is discarded unnecessarily.

Ling and Li's solution [22] to the Class Imbalance Problem is used in the following section to demonstrate XCSR's capability for accurately describing the high perform- ance regions of a six dimensional multi-modal environment. The approach provides a straightforward solution to a major data sampling problem common to many inductive learning systems. Weiss and Provost [31] suggest that one characteristic of a good solution to the imbalance problem is to ensure half of the training examples are allo- cated to the minority class while Japkowicz and Stephen [13] suggest several strate- gies to achieve this re-balancing of class examples through minority oversampling or majority undersampling. Ling and Li's solution combines the minority oversampling and majority undersampling strategies in an attempt provide an equality to the volume of examples per class as well as addressing the issues of a potential loss of useful information implicit in the majority examples. Other approaches should also be in- vestigated for their effects on performance of XCSR in the six dimensional test envi- ronment detailed below as well as other test environments which suffer from the class imbalance problem.

5   A Six Dimensional Test Environment

The six dimensional test environment used in this paper is a multi-modal function developed by Bonham and Parmee, [5] and [6], and is described in the Appendix at the end of this paper. It is defined by the additive effect of three different two dimen- sional planes as shown in contour plot form in Figure 8. Each plane has an associated "local" fitness value and the "global" fitness value of the six dimensional function is defined by adding each of these "local" fitness values together, that is, $fitness_{global} = fitness_{plane1} + fitness_{plane2} + fitness_{plane3}$. Each sample point is defined by a six dimen- sional vector of the form $\{a, b, c, d, e, f\}$, where $a...f \in [0, 1]$. In this two class prob- lem, a sample point is classified as either High or Low. It is classified as High only when each "local" fitness value is greater than the exact threshold value $\psi = 0.35$, where $\psi \in [0, 0.5]$, and the "global" fitness value is greater than exact threshold value $\psi_G = 1.20$, where $\psi_G \in [0, 1.5]$, otherwise the point is classified as Low. By combin- ing local regions of high performance, an environment of eighteen unique regions of globally high performance are defined, that is, three local high performance regions in Plane 1, two in Plane 2 and three in Plane 3. An important advantage of using this environment is that visualisation and subsequent interpretation of rules produced by XCSR is made less problematic by being able to identify rules that accurately cover each of the eighteen regions. This would be much more difficult with a non- decomposable six dimensional function.

As before, the XCSR system was trained using two different training datasets and tested using a single test dataset generated from a uniform random distribution. An attempt was made to replicate the decision boundary manipulation used for the two dimensional environment above, that is, the top $2m$ sorted sample points taken from a uniform random distribution are used to form the test dataset with 1693 sample points, where 813 points are defined as High. The two training datasets (not shown) were generated using a HSL sequence with 6000 and 12000 sample points, respectively.

Table 2. Classification Accuracy for XCSR Using a Uniform Random Test Dataset based on the Six Dimensional Test Function, where Trials = 500000, $N = 8000$, $\beta = 0.2$, $\alpha = 0.1$, $\varepsilon_0 = 10$, $v = 5$, $\theta_{GA} = 12$, $\chi = 0.8$, $\mu = 0.04$, $\theta_{del} = 20$, $\delta = 0.1$, $p_I = 10$, $\varepsilon_I = 0$, $F_I = 0.01$, $\theta_{mna} = 2$, $\theta_{sub} = 20$, $m = \pm10\%$ and $s_0 = 25\%$. Table 2 presents results for each parameter combination in the format : % $\text{Accuracy}_{(1\ s.d.)}$ over Sensitivity & Specificity.

		Centre-Spread Encoding		Ordered Interval Encoding		Unordered Interval Encoding	
		6000 HSL	12000 HSL	6000 HSL	12000 HSL	6000 HSL	12000 HSL
With AS-Subsumption	XCSR	$68.8_{(1.6)}$	$75.6_{(2.1)}$	$67.3_{(3.2)}$	$75.4_{(1.1)}$	$66.9_{(3.5)}$	$74.9_{(2.7)}$
		55.7 80.8	76.2 75.0	53.8 79.8	74.0 76.6	55.9 77.0	75.4 74.5
	sXCSR	$69.2_{(2.4)}$	$76.9_{(1.6)}$	$68.0_{(1.2)}$	$74.4_{(1.4)}$	$67.9_{(0.6)}$	$74.7_{(1.2)}$
		56.3 81.2	77.8 76.0	55.0 80.0	75.8 73.0	55.0 79.8	76.9 72.8
Without AS-Subsumption	XCSR	$69.1_{(2.3)}$	$77.3_{(0.9)}$	$69.0_{(0.9)}$	$78.6_{(1.4)}$	$71.4_{(1.5)}$	$77.5_{(1.2)}$
		54.8 82.4	73.8 80.5	56.5 80.5	79.4 77.9	61.1 81.2	75.3 79.5
	sXCSR	$70.4_{(1.4)}$	$78.5_{(2.0)}$	$71.0_{(2.2)}$	$80.2_{(1.9)}$	$72.2_{(1.4)}$	$79.1_{(1.6)}$
		57.2 82.5	77.6 79.3	61.5 80.1	82.6 78.0	62.6 81.0	80.3 78.0

Initial experiments for the six dimensional test problem, using the same approach as for the two dimensional Himmelblau test problem, highlighted the concerns expressed in the previous section over the class imbalance problem. In fact, a HSL-generated sample of the search space provides as little as 1.6% high performance region cover, which is clearly too small for the system to learn anything other than an accurate description of the majority class, that is, the low performance regions of this space. XCSR achieves this by generating a completely general rule for the environment, {[0, 1], [0, 1], [0, 1], [0, 1], [0, 1], [0, 1]}, which is correct for 98.4% of the training data points and approximately 50% of the test data points (not shown). In fact, this problem was evident in [35] in which Wilson applied XCSR to a problem where a small number of input patterns were much less prevalent than that of the majority of input patterns. An explanation for this problem was provided in [27], where Stone and Bull showed that by oversampling data points from the minority class, the effects of the imbalance were mitigated.

Table 3. Classification Accuracy for XCSR Using a Uniform Random Test Dataset based on the Six Dimensional Test Function, where Trials = 500000, $N = 2000$, $\beta = 0.2$, $\alpha = 0.1$, $\varepsilon_0 = 10$, $v = 5$, $\theta_{GA} = 12$, $\chi = 0.8$, $\mu = 0.04$, $\theta_{del} = 20$, $\delta = 0.1$, $p_I = 10$, $\varepsilon_I = 0$, $F_I = 0.01$, $\theta_{mna} = 2$, $\theta_{sub} = 20$, $m = \pm 10\%$ and $s_0 = 25\%$. Table 3 presents results for each parameter combination in the format : % Accuracy$_{(1\ s.d.)}$ over Sensitivity & Specificity.

		Centre-Spread Encoding		Ordered Interval Encoding		Unordered Interval Encoding	
		6000 HSL	12000 HSL	6000 HSL	12000 HSL	6000 HSL	12000 HSL
With AS-Subsumption	XCSR	$77.5_{(2.5)}$ 75.2 79.7	$82.0_{(1.6)}$ 90.5 74.1	$76.9_{(2.8)}$ 76.4 77.3	$80.9_{(1.0)}$ 88.7 73.6	$78.1_{(1.9)}$ 77.6 78.6	$81.6_{(0.9)}$ 86.2 77.4
With AS-Subsumption	sXCSR	$77.4_{(2.7)}$ 78.0 76.9	$79.7_{(2.6)}$ 89.4 70.7	$74.6_{(1.8)}$ 73.2 76.2	$79.6_{(2.8)}$ 87.8 72.0	$76.7_{(1.2)}$ 77.9 75.6	$79.5_{(1.0)}$ 87.3 72.3
Without AS-Subsumption	XCSR	$76.2_{(2.8)}$ 71.4 81.1	$82.4_{(2.7)}$ 88.0 77.3	$76.5_{(2.9)}$ 73.5 79.7	$81.4_{(0.6)}$ 88.8 74.8	$76.1_{(2.9)}$ 72.6 80.3	$81.9_{(1.0)}$ 90.7 73.8
Without AS-Subsumption	sXCSR	$77.1_{(1.3)}$ 73.9 80.0	$82.5_{(3.9)}$ 89.6 76.0	$77.5_{(2.1)}$ 75.0 79.9	$82.6_{(0.1)}$ 90.6 75.2	$77.2_{(1.8)}$ 75.5 79.0	$82.4_{(0.5)}$ 89.7 75.7

In order to overcome the severe level of class imbalance exhibited by the six dimensional test environment, the approach suggested by Ling and Li [22] is used. In particular, the minority class is oversampled with replacement for a predefined multiple of $n = 32$, and the majority class undersampled without replacement until a number of examples equal to those sampled from the minority class have been defined. This re-balancing of sample points per classification in the original HSL-generated datasets creates two new datasets with 5824 and 11968 sample points, that is, within

3% of their original size. In particular, experimentation and analysis for the six dimensional environment concentrates on two different population sizes, $N = 2000$ and $N = 8000$, in order to identify and explain any similarities or differences between them. Using the same format as Table 1, Table 2 and Table 3 show the accuracy measure, the percentage of High examples correctly classified and the percentage of Low examples correctly classified using the Uniform Random test dataset defined above for the population sizes, $N = 8000$ and $N = 2000$, respectively. The parameter combinations used are the same as used in the two dimensional environment.

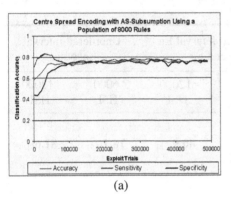 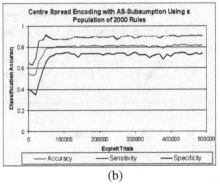

(a)   (b)

Fig. 9. Comparison of Classification Accuracy for Different Population Sizes (a) $N = 8000$ and (b) $N = 2000$, where $Trials = 500000$, $\beta = 0.2$, $\alpha = 0.1$, $\varepsilon_0 = 10$, $\nu = 5$, $\theta_{GA} = 12$, $\chi = 0.8$, $\mu = 0.04$, $\theta_{del} = 20$, $\delta = 0.1$, $p_I = 10$, $\varepsilon_I = 0$, $F_I = 0.01$, $\theta_{mna} = 2$, $\theta_{sub} = 20$, $m = \pm 10\%$ and $s_0 = 25\%$

It is clear from Table 2 and Table 3 and also from Figure 9 that when $N = 2000$ the system performs significantly better than when $N = 8000$, that is, there is between 2.3% and 11.2% improvement for different parameter combinations. These improvements are statistically significant at a level of >95%. A closer investigation of this result shows that the improvement is due to the system's inability to accurately classify the High test dataset points represented by the sensitivity line when $N = 8000$. Both population sizes produce the same level of accuracy for Low test data-points suggesting that a system with too large a population may lead to overfitting the over-sampled High training data-points thereby reducing the performance of the system on previously unseen test data. Figure 10 shows that a system with $N = 8000$ actually performs at a higher level for the training dataset than a system with $N = 2000$, implying perhaps that the larger system has learnt the training dataset at the expense of performance on the test dataset.

The results shown in Table 2 and Table 3 indicate that there is no clear advantage in performance for the smaller population on the test dataset to be gained when Action Set Subsumption is turned off, but that there is a statistically significant improvement (>95%) for the Ordered and Unordered Interval encoded versions of the $N = 8000$ system. There was a pronounced effect in the two dimensional problem. The disruptive effects of Action Set Subsumption may be weakened as a result of the difference in complexity of the problem. The number of training sample points is almost an order higher in the more complex six dimensional problem and so it is less likely for a highly numerate rule to dominate an Action Set early in an experimental run. Although this effect may provide some implicit protection for the weaker rules, an explicit form of protection may be afforded to the weaker rules by increasing the Subsumption Experience threshold (as suggested in [7]), that is, those rules that have the potential to subsume other members of an Action Set are expected to have taken part in many more Action Sets before they are permitted to subsume. Clearly, the same form of protection could be used for less complex environments while gaining some of the generalisation advantages inherent in the Action Set Subsumption technique.

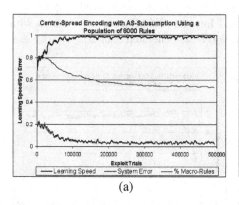 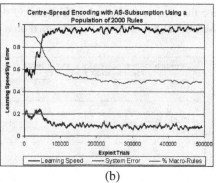

(a)   (b)

Fig. 10. Comparison of Learning Speed and System Error for Different Population Sizes (a) N = 8000 and (b) N = 2000 where $Trials$ = 500000, β = 0.2, α = 0.1, ε_0 = 10, v = 5, θ_{GA} = 12, χ = 0.8, μ = 0.04, θ_{del} = 20, δ = 0.1, p_I = 10, ε_I = 0, F_I = 0.01, θ_{mna} = 2, θ_{sub} = 20, m = ±10% and s_0 = 25%

It is also true that as the number of training points increases the test dataset performance of the system improves, regardless of training dataset size, suggesting that the system can gain more information about the dimensionality of the high performance regions when more points are sampled in that area. In fact, the statistical significance of the difference in performance for the larger population is >99% while for the smaller population the difference is significant at the level of >95%. Figure 11 clearly shows the improvement gain for the 12000 HSL-generated training dataset over the 6000 HSL and, in particular, shows the sensitivity measure to be the main difference between them, that is, the system performs better when there are more sample points from the high performance regions in the training dataset. In fact, Figure 12 suggests that this improvement in classification accuracy is gained at no extra expense in terms of learning speed.

Given that this paper represents a preliminary investigation into the use of XCS as an exploratory tool, it makes sense to discuss some of the issues related to computational load. Experimental timings have shown that the system takes four times longer to complete a 500,000 exploit trial experiment for the N = 8000 population sized system than the N = 2000 population sized system, that is, there is a linear decrease in computational time as the population size drops from 8000 to 2000. In fact, the type of encoding used for each rule may also have an effect on the computational time. A comparison of timing by encoding for the larger populated system shows that it takes longer to complete the experiment using Centre-Spread than for the other encodings (not shown). However, there are no statistically significant differences between encodings for this six dimensional environment and so this timing issue means very little. It was expected that the sXCSR system would provide some speed-up in learning as was demonstrated in [7], but for both the two and six dimensional environments used in this paper there does not appear to be any significant difference in performance or learning time. Figure 13 shows the learning speed and system error for the Ordered Interval XCSR and Ordered Interval sXCSR encoded systems with Action Set Subsumption and trained on a 12000 HSL-generated dataset.

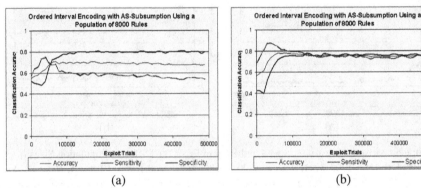

(a)   (b)

Fig. 11. Comparison of Classification Accuracy for Different Training Set Sizes (a) 6000 HSL-generated data-points and (b) 12000 HSL-generated data-points, where *Trials* = 500000, N = 8000, $\beta = 0.2$, $\alpha = 0.1$, $\varepsilon_0 = 10$, $\nu = 5$, $\theta_{GA} = 12$, $\chi = 0.8$, $\mu = 0.04$, $\theta_{del} = 20$, $\delta = 0.1$, $p_I = 10$, $\varepsilon_I = 0$, $F_I = 0.01$, $\theta_{mna} = 2$, $\theta_{sub} = 20$, $m = \pm10\%$ and $s_0 = 25\%$

6 Conclusion

The motivation for this work was to investigate how accurately XCS is able to describe high performance regions in a design-oriented environment given its previously demonstrated capabilities in the main-stream field of data-mining. The objective is to develop the XCS classifier system to evolve a complete and accurate set of maximally general rules that identify and describe the high performance regions of real-world problem environments with particular emphasis on an interactive process of design evolution. In this role, XCS would act as an information-gathering tool that is capable of providing aid to the designer in decision making relevant to a better representation of their design problem through the reformulation and redefinition of the design problem. It is hoped that XCS can provide compact understandable rules that describe the regions of interest to the designer. This aspect of the investigation provides a great deal of potential for further research given the apparent mismatch between the use of orthogonal rules and the complex regions of design space. It is also clear that

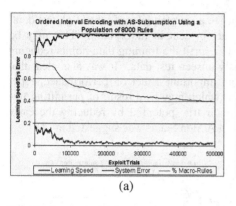
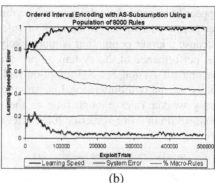

(a)   (b)

Fig. 12. Comparison of Learning Speed and System Error for Different Training Set Sizes (a) 6000 HSL-generated data-points and (b) 12000 HSL-generated data-points, where *Trials* = 500000, $N = 8000$, $\beta = 0.2$, $\alpha = 0.1$, $\varepsilon_0 = 10$, $v = 5$, $\theta_{GA} = 12$, $\chi = 0.8$, $\mu = 0.04$, $\theta_{del} = 20$, $\delta = 0.1$, $p_I = 10$, $\varepsilon_I = 0$, $F_I = 0.01$, $\theta_{mna} = 2$, $\theta_{sub} = 20$, $m = \pm 10\%$ and $s_0 = 25\%$

some further work will be required to fully understand XCS's capabilities in design-oriented problem environments.

An analysis of the many experiments performed on the six dimensional environment show that performance on the test dataset levels off at around 150,000 exploit trials, regardless of population size. The same is also true of the two dimensional test environment, where performance on the test dataset levels off at around 100,000 exploit trials. In fact, this leveling off coincides with a marked reduction in cover operator usage, that is, shortly after the system's ruleset has a rule for every training dataset point, it also achieves its best classification of the test dataset points. It is likely that further trials of the system will only result in a compaction of the ruleset, that is, most of the learning takes place during the first 150,000 trials. However, this suggests that an initial period of covering is required by the system before an accurate description of the high performance regions is possible which also implies any extraction of simple meaningful and accurate rules that describe the regions of high performance will have a minimum temporal overhead of around 150,000 trials.

It is clear from comparing results of the sXCSR system with those of the original XCSR system that the sXCSR attains the same level of performance as XCSR. Although, the learning speed-up apparent in [7] does not appear to have been matched in this set of experiments, the results in that paper were based on an eleven variable problem. There is some evidence that improvements under the new learning scheme are related to the complexity of the environment, given that results for a six variable multiplexor problem showed no learning speed-up. However, it is unclear at this point whether any real improvement in learning speed will be seen in sXCSR for a more complex eleven variable design problem with a finite training set.

Results for both XCSR and sXCSR showed a statistically significant improvement in performance, between 5% and 10%, when the population size was reduced from 8000 to 2000 rules in the six dimensional test environment. In fact, the improvements appear to be a result of the system's inability to accurately classify the High test data-set points when a larger number of rules are present in the system. This may be because a larger population is more likely to overfit the training data-points reducing the performance of the system on previously unseen test data. It was also clear that turning off Action Set Subsumption had an important effect on performance, especially for the two dimensional environment, which may be a result of permitting initially weaker rules enough time to show their true potential by reducing the early domination of more numerate rules in a given Action Set as was suggested in [7].

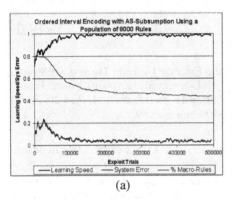 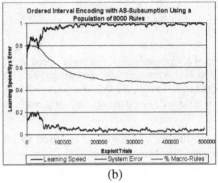

(a)   (b)

Fig. 13. Comparison of Learning Speed and System Error for (a) XCSR and (b) sXCSR, where $Trials = 500000$, $N = 8000$, $\beta = 0.2$, $\alpha = 0.1$, $\varepsilon_0 = 10$, $v = 5$, $\theta_{GA} = 12$, $\chi = 0.8$, $\mu = 0.04$, $\theta_{del} = 20$, $\delta = 0.1$, $p_I = 10$, $\varepsilon_I = 0$, $F_I = 0.01$, $\theta_{mna} = 2$, $\theta_{sub} = 20$, $m = \pm10\%$ and $s_0 = 25\%$

The investigation presented in this paper clearly shows XCSR/sXCSR's capabilities for describing regions of high performance for two complex multi-modal test environments both of which embody typical search space characteristics. In particular, issues concerning data sampling and performance measures were raised as well as an attempt to provide empirical evidence of the efficacy of the new update mechanism introduced in [7]. We also compared three different real-coded interval encodings that may be used with XCS (see Stone and Bull [27] for a discussion of the issues related to the differences between interval encodings). Results showed that there was little or no difference between the different encodings for the two test environments used.

References

1. An, G. (1996), The Effects of Adding Noise During Backpropagation Training on a Generalization Performance, *Neural Computation* 8:643-674
2. Beasley, D., Bull, D. & Martin, R. (1993), A Sequential Niche Technique for Multimodal Function Optimisation, *Evolutionary Computation*, 1(2):101-125
3. Bernadó, E., Llorà, X. & Garrell, J. (2001) XCS and GALE: a Comparative Study of Two Learning Classifier Systems with Six Other Learning Algorithms on Classification Tasks, In Lanzi, P. L., Stolzmann, W., and S. W. Wilson (Eds), *Advances in Learning Classifier Systems.4th International Workshop (IWLCS-2001)*, LNAI-2321. Springer-Verlag, pp. 115-133

4. Blake, C. & Merz, C. (1998) UCI Repository of Machine Learning Databases, *University of California, Irvine.* Available at htttp://www.ics.uci.edu/~mlearn/MLRepository.html
5. Bonham, C. (2000), Evolutionary Decomposition of Complex Design Spaces, *PhD Thesis,* University of Plymouth
6. Bonham, C. and Parmee, I. (1999), An Investigation of Exploration and Exploitation Within Cluster-Oriented Genetic Algorithms (COGAs). In W. Banzhaf, J. Daida, A. Eiben, M. Garzon, V. Honavar, M. Jakiela and R. Smith (Eds), *Proceedings of the Genetic and Evolutionary Computation Conference 1999,* Morgan Kaufmann, pp. 1491-1497.
7. Bull, L., Wyatt, D. & Parmee, I. (2002), Initial Modifications to XCS for use in Interactive Evolutionary Design. In J. Merelo, P. Adamidis, H-G. Beyer, J-L. Fernandez-Villacanas & H-P. Schwefel (eds) *Parallel Problem Solving From Nature – PPSN VII,* Springer Verlag, pp. 568-577
8. Butz, M. and Wilson, S. (2001) An algorithmic description of XCS. In Lanzi, P. L., Stolzmann, W., and S. W. Wilson (Eds.), *Advances in Learning Classifier Systems. 3rd International Workshop (IWLCS-2000),* LNAI-1996. Springer-Verlag, pp. 253-272
9. Elkan, C. (2001) The Foundations of Cost-Sensitive Learning, *Proceedings of the 17th International Joint Conference on Artificial Intelligence,* pp. 973-978
10. Hart, P. (1968) The Condensed Nearest Neighbor Rule, *IEEE Transactions on Information Theory,* IT-14:515-516
11. Holmes, J. (1996), A Genetics-Based Machine Learning Approach to Knowledge Discovery in Clinical Data, *Journal of the American Medical Informatics Association Supplement 883*
12. Holmström, L. & Koistinen, P. (1992), Using Additive Noise in Back-Propagation Training, *IEEE Transactions on Neural Networks* 3:24-38
13. Japkowicz, N. & Stephen, S. (2002) The Class Imbalance Problem: A Systematic Study, *Intelligent Data Analysis,* 6(5):429-450
14. Kocis, L. & Whiten, W. J., Computational Investigations in Low Discrepancy Sequences, *ACM Transactions on Mathematical Software,*23(2):266-294
15. Kohavi, R. & Provost, F. (1998) Glossary of Terms, *Machine Learning 30,* pp. 271-274.
16. Kononenko, I. & Bratko, I. (1991) Information-Based Evaluation Criterion for Classifier's Performance, *Machine Learning* 6:67-80
17. Kubat, M. & Matwin, S. (1997) Addressing the Curse of Imbalanced Data Sets: One-Sided Sampling. In D. Fisher (Ed), *Proceedings of the 14th International Conference on Machine Learning,* Morgan Kaufmann, pp. 179-186
18. Kubat, M., Holte, R. & Matwin, S. (1997) Learning when Negative Examples Abound, *Proceedings of the 9th European Conference on Machine Learning,* LNCS 1224, Springer-Verlag, pp. 146-153
19. Laurikkala, J. (2001) Improving Identification of Difficult Small Classes by Balancing Class Distribution, *Proceedings of the 8th Conference on Artificial Intelligence in Medicine in Europe,* LNCS 2101, Springer-Verlag, pp. 63-66
20. Lee, S. (2000) Noisy Replication in Skewed Binary Classification, *Computational Statistics and Data Analysis,* Vol. 34:165-191
21. Lewis, D. & Gale, W. (1994) A Sequential Algorithm for Training Text Classifiers, *Proceedings of SIGIR-94, 17th ACM International Conference on Research and Development in Information Retrieval,* ACM/Springer, pp. 3-12
22. Ling, C. & Li, C. (1998) Data Mining for Direct Marketing: Problems and Solutions, *Proceedings of ACM SIGKDD International Conference on Knowledge Discovery and Data Mining (KDD-98),* AAAI, pp. 73-79

23. Parmee, I. (1996), The Maintenance of Search Diversity for Effective Design Space Decomposition using Cluster-Oriented Genetic Algorithms (COGAs) and Multi-Agent Strategies (GAANT), *Proceedings of 2nd International Conference on Adaptive Computing in Engineering Design and Control*, PEDC, University of Plymouth, pp. 128-138.

24. Parmee, I. (2002), Improving Problem Definition through Interactive Evolutionary Computation, *Journal of Artificial Intelligence in Engineering Design, Analysis and Manufacture*, 16 (3)

25. Parmee, I. & Bonham C. (2001) Improving Cluster-Oriented Genetic Algorithms for High-Performance Region Identification, *Proceedings US United Engineering Foundation's 'Optimisation in Industry' Conference*, Tuscany, Italy 2001, Springer Verlag.

26. Raviv, Y. & Intrator, N. (1995) Bootstrapping with Noise: An Effective Regularisation Technique, *Connection Science, Special issue on Combining Estimators*, 8:356-372

27. Stone, C. & Bull, L. (2003), For Real! XCS with Continuous-Valued Inputs, *Evolutionary Computation* 11(3):299-336.

28. Swets, J. (1988) Measuring the Accuracy of Diagnostic Systems, *Science 240:*1285-1293

29. Tomek, I. (1976) Two Modifications to CNN, *IEEE Transactions on Systems, Man and Communications*, SMC-6:769-772

30. Toussaint, G. (1994) A Counter-Example to Tomek's Consistency Theorem for a Condensed Nearest Neighbor Decision Rule, *Pattern Recognition Letters*, 15:797-801

31. Weiss, G. & Provost, F. (2001) The Effect of Class Distribution on Classifier Learning: An Empirical Study, *Technical Report ML-TR-44*, Rutgers University

32. Wilson, D. & Martinez, T. (1997) Improved Heterogeneous Distance Functions, *Journal of Artificial Intelligence Research 6:*1-34

33. Wilson, D. & Martinez, T. (1998) Reduction Techniques for Exemplar-Based Learning Algorithms, *Machine Learning 38(3):* 257-286

34. Wilson, S. (1995) Classifier fitness based on accuracy, *Evolutionary Computation 3(2):*149-175

35. Wilson, S. (2000) Get real! XCS with Continuous-valued inputs. Lanzi, P. L., Stolzmann, W., and Wilson, S. W., eds. *Learning Classifier Systems. From Foundations to Applications*, LNAI-1813, Springer-Verlag, pp. 209-222

36. Wilson, S. (2001) Compact Rulesets for XCSI, In Lanzi, P. L., Stolzmann, W., and S. W. Wilson (Eds.), *Advances in Learning Classifier Systems. 4th International Workshop (IWLCS-2001)*, LNAI-2321. Springer-Verlag, pp. 197-210

37. Wilson, S. (2001) Mining Oblique Data with XCS, In Lanzi, P. L., Stolzmann, W., and S. W. Wilson (Eds.), *Advances in Learning Classifier Systems. 3rd International Workshop (IWLCS-2000)*, LNAI-1996. Springer-Verlag, pp. 158-177

Appendix: Test Environment Descriptions

Modified Himmelblau Function

$$f(x_1, x_2) = 200 - (x_1^2 + x_2 - 1 \ )^2 - (x_2^2 + x_1 - 7)^2$$

Chris Bonham's 6D Function

$$a, b, c, d, e, f \in [0,1]$$

PLANE 1

$$z_1 = \frac{0.41}{\dfrac{(0.8-a)^2}{0.04} + \dfrac{(0.7-b)^2}{0.04} + 1}$$

$$z_2 = \frac{.04}{\dfrac{(0.2-a)^2 5}{0.0225} + \dfrac{(0.7-b)^2}{.90} + 1}$$

$$z_3 = \frac{.04}{\dfrac{(0.4-a)^2 5}{0.2025} + \dfrac{(0.0-b)^2}{.90} + 1}$$

$$fitness_{plane1} = \min \begin{cases} 0.5 \\ z_1 + z_2 + z_3 \end{cases}$$

PLANE 2

$$z_1 = 0.5 - \frac{0.5}{\dfrac{(0.25-c)^2}{0.09} + \dfrac{(0.75-d)^2}{0.09} + 1}$$

$$z_2 = 0.5 - e^{(\cos(48\pi \times c) + \cos(48\pi \times d))/14.8}$$

$$z_3 = 0.5 - \frac{0.5}{\dfrac{(0.75-c)^2}{0.09} + \dfrac{(0.25-d)^2}{0.09} + 1}$$

$$z_4 = 0.5 - e^{(\cos(24\pi \times c) + \cos(24\pi \times d))/14.8}$$

$$z_5 = \max \begin{cases} 0.0 \\ z_2 - z_1 \end{cases}$$

$$z_6 = \max \begin{cases} 0.0 \\ z_4 - z_3 \end{cases}$$

$$z_7 = \begin{cases} z_5 & if\,(c < 0.5)\,AND\,(d > 0.5) \\ 0.0 & otherwise \end{cases}$$

$$z_8 = \begin{cases} 0.0 & if(c < 0.5)\,OR\,(d < 0.5) \\ \dfrac{0.35}{\dfrac{(0.75-c)^2}{0.09} + \dfrac{(0.75-d)^2}{0.09} + 1} & otherwise \end{cases}$$

$$z_9 < \begin{cases} 0.0 & \text{if}(c = 0.5)\text{OR}(d = 0.5) \\ \dfrac{0.35}{\dfrac{(0.25 - c)^2}{0.09} + \dfrac{(0.25 - d)^2}{0.09} + 1} & \text{otherwise} \end{cases}$$

$$z_{10} = \begin{cases} z_6 & \text{if}(c > 0.5)AND(d < 0.5) \\ 0.0 & \text{otherwise} \end{cases}$$

$$fitness_{plane2} = z_7 + z_8 + z_9 + z_{10}$$

PLANE 3

$$z_1 = \min \begin{cases} 0.5 \\ 1.8e^2 + 3f^2 \end{cases}$$

$$z_2 = \begin{cases} 0.0 & \text{if}(e > 0.6)OR(f > 0.5) \\ z_1 & \text{otherwise} \end{cases}$$

$$z_3 > \begin{cases} 0.625f & \text{if}(f = 0.8) \\ 0.5 & \text{if}(f - 0.8)AND(f = 0.85) \\ \dfrac{10(1 - f)}{3} & \text{if}(f \geq 0.85) \end{cases}$$

$$z_4 = \begin{cases} z_3 & \text{if}(e > 0.6) \\ 0.0 & \text{otherwise} \end{cases}$$

$$z_5 = \min \begin{cases} 0.5 \\ z_4 \end{cases}$$

$$z_6 = \min \begin{cases} 0.5 \\ 1.5(0.6 \times e) \times 2.5(f \times 0.5) \end{cases}$$

$$z_7 = \begin{cases} z_6 & \text{if}(e - 0.6)AND(f > 0.5) \\ 0.0 & \text{otherwise} \end{cases}$$

$$fitness_{plane3} = z_2 + z_5 + z_7$$

The EpiXCS Workbench:
A Tool for Experimentation and Visualization

John H. Holmes and Jennifer A. Sager

Center for Clinical Epidemiology and Biostatistics
University of Pennsylvania School of Medicine
Philadelphia, PA 19104 USA
jholmes@cceb.med.upenn.edu
Department of Computer Science
University of New Mexico
Albuquerque, NM 87131, USA
sagerj@cs.unm.edu

1 Introduction

The EpiXCS Workbench is a knowledge discovery tool that provides the user with the capability for knowledge discovery and visualization in medical data. The foundation for the workbench is the XCS paradigm [1]. The workbench is designed to benefit both expert learning classifier systems (LCS) researchers and inexperienced end-users in a variety of domains, especially clinical, epidemiologic, and public health researchers. It was implemented in Microsoft Visual C++, Version 6.0, using the GNU Scientific Library, using the XCSlib class library developed by Lanzi [2]. EpiXCS is designed to run on Intel Pentium processor environments at 1.0GHz and higher. No special graphics or other co-processors are required.

1.1 Demonstration Data

The Wisconsin Breast Cancer dataset [3] is used here to illustrate the various features of EpiXCS. The dataset consists of nine predictor features (shown in Section 2.1) and one class feature (malignant/non-malignant).

2 Architecture

While using the XCS class library implemented by Lanzi, EpiXCS implements several additional features that tailor the XCS paradigm to the demands of epidemiologic data and users who are not familiar with learning classifier systems. These features include specialized data encoding, evaluation metrics, reinforcement, missing values handling, classifier ranking, and risk assessment. Finally, a *workbench-style interface* is used for visualization and parameterization.

2.1 Environment Data Encoding

Training and testing data can be encoded in variety of ways for use with EpiXCS. Binary, categorical, ordinal, and real formats are all acceptable, even in the same

X. Llorà et al. (Eds.): IWLCS 2003-2005, LNAI 4399, pp. 333–344, 2007.

dataset. Data files are constructed using a modified Attribute-Relation File Format (ARFF), developed at the University of Waikoto [4]. ARFF files include a header and data in a single file, as shown below. In the header, features are declared with the ATTRIBUTE token, and enumerated from 0. Any missing values are declared using the WILD token, followed by the specific values enclosed in double quotes. The type of the feature is declared next. Finally, the name of the feature is declared; this name will be used as the feature reference in visualizing the evolved rule sets in natural language format after training the system. The action feature declaration completes the header; actions are limited to binary decisions ($\in \{0,1\}$) in this version of EpiXCS. The data immediately follow the header, using space-delimited format.

```
ATTRIBUTE 0 <WILD "99"><REAL><STRING "Clump Thickness">
ATTRIBUTE 1 <WILD "99"><REAL><STRING "Uniformity of Cell Size">
ATTRIBUTE 2 <WILD "99"><REAL><STRING "Uniformity of Cell Shape">
ATTRIBUTE 3 <WILD "99"><REAL><STRING "Marginal Adhesion">
ATTRIBUTE 4 <WILD "99"><REAL><STRING "Single Epithelial Cell Size">
ATTRIBUTE 5 <WILD "99"><REAL><STRING "Bare Nuclei">
ATTRIBUTE 6 <WILD "99"><REAL><STRING "Bland Chromatin">
ATTRIBUTE 7 <WILD "99"><REAL><STRING "Normal Nucleoli">
ATTRIBUTE 8 <WILD "99"><REAL><STRING "Mitoses">
ACTION   9                      <STRING "Malignant">
 5 1 1 1 2 1 3 1 1 0
 6 8 8 1 3 4 3 7 1 0
 ...
```

2.2 Classifier Encoding

Classifiers are represented in EpiXCS using the "center-spread" approach described by Wilson [5]. This representation uses two genes for each feature, one for the "center" value and the other for the "spread," which corresponds to an interval within which matches to input data from the environment are made. Initialization of the full classifier population is automatically performed at start-up, facilitated by a routine that scans the training set for minimum and maximum values for each feature. This routine selects a hypothetical "center" value for each feature, based on its mean, median, or randomly selected value within its empirically observed range. The spread for each feature is determined by its range in the training data.

2.3 Evaluation Metrics

EpiXCS uses the same set of evaluation metrics as its predecessor, EpiCS [6]. These include crude accuracy, sensitivity, specificity, positive and negative predictive values, and area under the receiver operating characteristic curve, or *AUC*. These are provided in Equations 1 through 5, respectively. The calculation of AUC is described in detail in [7].

As classified by EpiXCS	Gold standard (class value in data)	
	Positive	Negative
Positive	A	B
Negative	C	D

Fig. 1. 2x2 confusion matrix for a two-choice decision problem. The columns represent the "gold standard," or the classifications as they exist in the data. The rows represent the classification. A=True positives; B=False positives; C=False negatives; D=True negatives.

$$\text{Crude Accuracy} = \frac{A+D}{A+B+C+D} \tag{1}$$

$$\text{Sensitivity} = \frac{A}{A+C} \tag{2}$$

$$\text{Specificity} = \frac{D}{B+D} \tag{3}$$

$$\text{Positive predictive value} = \frac{A}{A+B} \tag{4}$$

$$\text{Negative predictive value} = \frac{D}{C+D} \tag{5}$$

These metrics are calculated at every 100th iteration during training and then on the testing set. These metrics are displayed in text and graphical formats during training and in text format at testing. In addition, the predictive values are used to rank the macroclassifiers obtained after training. Specifically, classifiers advocating positive-class decisions are ranked by their positive predictive value; those advocating negative decisions are ranked by the negative predictive value. This helps the user to focus on the most highly accurate rules in the knowledge discovery process.

In addition to these classical metrics, the convergence rate, λ, is calculated during learning:

$$\lambda = \left(\frac{AUC_{Shoulder}}{Shoulder} \right) 1000 \tag{6}$$

Shoulder is the iteration at which 95% of the maximum AUC obtained during training is first attained, and $AUC_{Shoulder}$ is the AUC obtained at the shoulder. Thus, the higher the value of λ, the faster the system reaches convergence on the training data. As the first AUC is not measured until the 100th iteration, and the maximum AUC measurable is 1.0, the maximum value of λ is 10.0. The minimum λ is 0.0.

2.4 Missing Value Handling

Missing values in input data are dealt with during creation of the match sets by means of simple pass-through. During covering, one of four approaches may be used to represent missing values present in the non-matching input case. These include:

- Wild-to-Wild: missing values are covered as #s, equivalent to "don't care"
- Random within range: a covering value is selected randomly from the range for the feature with the missing value
- Population average: the population average for the feature with the missing value is used for the covering value
- Population standard deviation: a random value within the standard deviation for the feature is selected as the covering value

3 The EpiXCS Workbench Interface

The basic EpiXCS Workbench interface is shown in Figure 2:

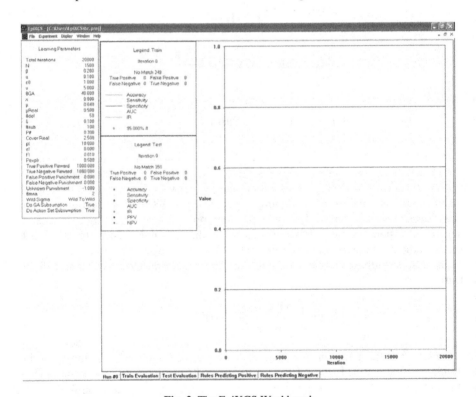

Fig. 2. The EpiXCS Workbench

The left pane is an interface to the learning parameters which are adjustable by the user prior to runtime. These parameters correspond to those discussed in Butz and Wilson [8], and include additional parameters for modifying the reward given to classifiers based on the accuracy of the decision they advocate. The middle pane contains the results of the classification metrics, which are obtained at each 100th iteration during training (top half) and at testing (bottom half). This pane also includes the legend to the graph in the right pane, which displays classification performance as plots of the classification metrics in real time during training. The tabs at the bottom of the figure are described in Section 3.1.

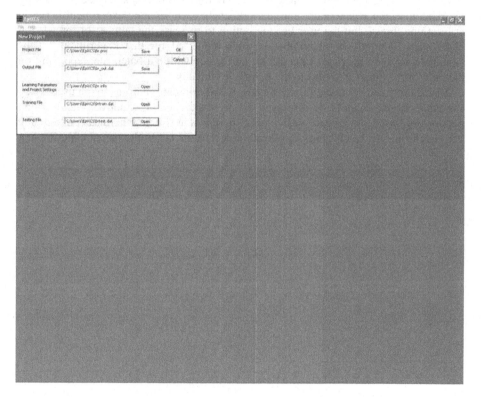

Fig. 3. Selection of project, parameter, output, and data files

3.1 Experimentation

EpiXCS is intended to support the needs of LCS researchers as they experiment with such problems as parameterization. For the purposes of this paper, an *experiment* is a set of one or more runs that use the same learning parameters and input and output. A *run* is defined as a single excursion through a training-testing cycle. A *batch run* is defined as a group of multiple similar runs contained within an experiment. *Multitasking* enables the user to perform multiple simultaneous experiments as well as multiple simultaneous batch runs for a particular experiment.

Multitasking for multiple similar experiments enables the user to run several experiments with different learning parameters, data input files, and output files. For example, this feature could be used to compare different learning parameter settings or to run different datasets at the same time. Multitasking in a batch run allows the user some choice in the way in which the processor's resources are used by the program. In a batch run the user is given the option to choose the number of threads to perform the requested number of runs in order to take advantage of any efficiency gains from multithreading. The order and start times of the runs do not significantly affect the experiment results because the variation in results in a batch run, which have the same data set and learning parameters for each run, is solely determined by the random number generator. In the workbench, each separate experiment is run in at least one thread, but the user has the option to increase the number of threads per experiment up to the total number of runs requested for that experiment. Thus simultaneous and sequential runs can be performed within a batch run depending on the ratio of the number of runs to the number of threads.

Visual clues convey the difference between experiments and runs to the user. Each simultaneous experiment is displayed in a separate child window. The learning parameters can be set for each individual experiment at run-time by means of a dialog box. Each run in a batch run is displayed on a separate tab in the parent experiment's

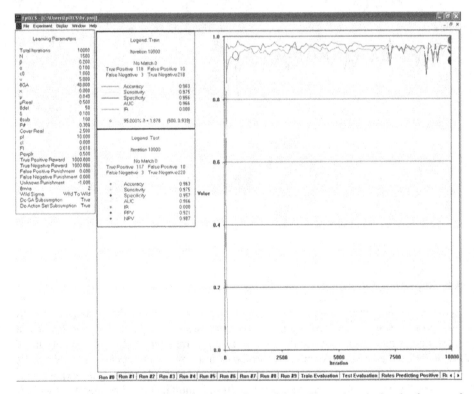

Fig. 4. Plot of classification metrics obtained during training. Shown here is the plot for one of the 10 runs on the Wisconsin Breast Cancer data.

window. Each *run tab* displays the classification metrics in both graphical and textual forms. Since the overall results of a batch run may be of interest to the user, additional tabs are provided for the training and testing evaluations. These tables include the mean and standard deviation of the aforementioned metrics calculated for all runs in the experiment. In addition to these metrics, detailed information such as the final population set is dumped to a permanent text file for further use.

Figure 4 illustrates the training epoch of a batch run consisting of 10 individual runs on the Wisconsin Breast Cancer data. The user can select the plots obtained during training for each run by selecting the appropriate tab.

Figure 5 illustrates the summary statistics obtained at testing over the 10 runs; similar results are also available for the results obtained during training by selecting the appropriate tab.

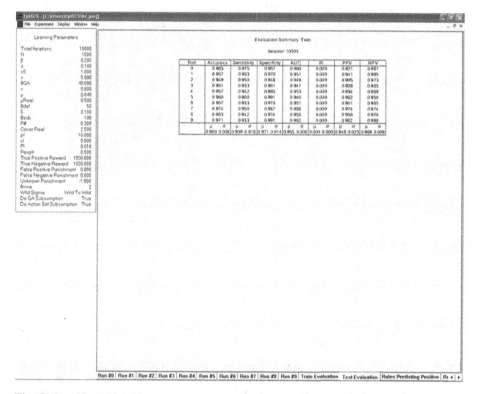

Fig. 5. Classification metrics obtained on the testing set, summarized for the 10 runs on the Wisconsin Breast Cancer data

3.2 Rule Visualization

Often, obscure coding is used for taxons and actions. As a result, translating the classifiers in final populations and understanding their significance is usually very difficult, especially without understanding the coding and how classifier fitness, prediction, prediction error, numerosity, and accuracy interact.   The EpiXCS workbench attempts to help alleviate this problem by providing two additional tabs on the interface that allow the user to examine the classifiers, expressed as condition-action rules in natural language, that predict positive and negative, respectively. Each of the two tabs summarizes the best-predicting classifiers from all runs. Classifiers are chosen to be displayed on these tabs by their predictive value, which indicates their accuracy, once in hand, as a posterior probability. The classifiers are represented by "human readable strings", if-then statements which use strings to represent the knowledge held in each attribute position. For numerically-encoded data, a graph is used to show the center and spread within the minimum and maximum possible range. This feature is illustrated in Figure 6:

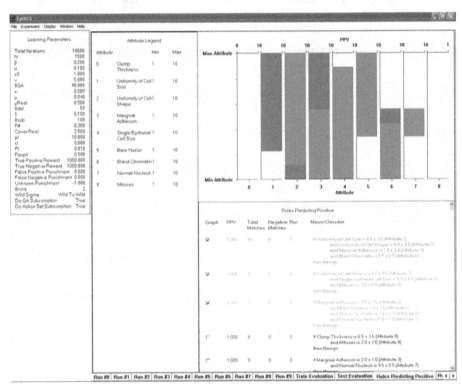

Fig. 6. Rule visualization tool in EpiXCS. Rules are presented in natural language in the bottom pane. Clicking on the checkbox next to a rule causes the values to be plotted in the graphic in the upper pane. Multiple rules can be plotted for comparison and for visual identification of patterns that tend to recur from rule to rule.

4   Evaluation of EpiXCS

4.1   Source of Data

Classification and rule discovery performance of EpiXCS were evaluated using five well-known datasets from the Machine Learning Repository at the University of

California at Irvine [9]. In addition to the Wisconsin Breast Cancer dataset, the Echocardiogram [10], Cleveland Heart [11], Hepatitis [12], and Pima Indians Diabetes [13] datasets were used as the test suite. These datasets were selected because of their popularity as benchmarks in the machine learning and knowledge discovery (KDD) literature, and their application to a variety of medical domains. Thus, even though they are rather small in size, especially for a KDD exercise, they afford a degree of comparability with other approaches. In addition, they offer a breadth of degree of difficulty in terms of contradictions, conflicts, and imbalanced class distributions. There were some additional constraints that influenced the selection of this particular suite of datasets. These included a requirement for dichotomous class attributes (EpiXCS does not support polytomous or real-encoded classes at this time) and a sufficient number of number of records for creation of training and testing sets. The datasets selected for this investigation are described in Table 1.

Table 1. Datasets used for evaluating EpiXCS

Dataset	Positive class	Negative Class	Total
Breast cancer	241	458	699
Echo	43	89	132
Heart	164	139	303
Hepatitis	32	123	155
Pima	500	268	768

The datasets were prepared for analysis by standardizing class attribute values to 0 (Negative) and 1 (Positive). Ordinarily, this meant that "positive" cases were those with a disease, but in the case of the Echo dataset, the positive class represented living cases.

4.2 Comparison Method

See5 [14] a well-known decision tree and rule inducer, was selected as the comparison method. See5 has been used extensively in KDD and it provides an ensemble of rule-based classifiers that are similar in format to those provided in EpiXCS.

4.3 Experimental Procedure

This study focused on a supervised learning problem, and therefore required the creation of training-testing set pairs. However, even with random selection, there is the possibility that the pairs may not represent the dataset as a whole. This can be addressed by various methods, including "leave-one-out" sampling or n-fold cross-validation. Although See5 supports cross-validation internally, EpiXCS does not yet do so. In order to ensure that the two systems were exposed to the same training and testing data, 10 training-testing set pairs were created for each dataset, using different random number seeds. The distributions of each predictor and class attribute were found to be comparable within each training-testing set pair and across all training and testing sets. EpiXCS and See5 were trained using a given training set and then immediately tested using the testing set in the pair. To ensure that EpiXCS arrived at a stable state of learning, and to account for perturbations in the random distribution of the genetic operators, 20 runs were completed on each dataset. The system was parameterized specifically for each dataset in order to obtain optimal results on classification. In general, the default parameters described in Butz and Wilson [8] provided satisfactory classification performance; only the population size and number of iterations required alteration of the defaults for two of the datasets, Pima and Hepatitis, described below. The same training-testing set pairs were used for the See5 runs. Sensitivity, specificity, and area under the receiver operating characteristic curve were obtained on the testing set, and for the EpiXCS runs, averaged over the 20 runs. Then, the results were averaged over the 10 training-testing set pairs separately for EpiXCS and See5 runs.

4.4 Classification Performance

The classification performance of EpiXCS and See5 on the testing sets is shown in Table 2. EpiXCS demonstrates at least equal accuracy in classifying novel cases. Of some surprise is EpiXCS' significantly better performance on the Heart and Hepatitis datasets. It should be noted that these data as well as the Pima Indians dataset required more iterations (50,000, compared to 10,000 for he other datasets), and in the case of Pima, a larger population size (3,000, compared to 500 for the other sets). This parameterization reflects the comparable complexity among the datasets: Pima and Hepatitis contain numerous conflicts in the data, mandating a larger exploration space, expressed in longer training times needed to reach convergence and larger classifier populations.

Table 2. Classification performance on testing sets (as AUC) of EpiXCS and See5. Numbers in parentheses are one standard deviation of the mean of 10 training-testing set pairs.

	EpiXCS	See5
Breast cancer	0.96 (0.02)	0.91 (0.01)
Echo	0.85 (0.07)	0.84 (0.05)
Heart	0.78 (0.02)	0.64 (0.05)
Hepatitis	0.98 (0.01)	0.77 (0.04)
Pima Indians	0.82 (0.04)	0.81 (0.03)

4.5 Rule Discovery

A sample of highly positive-predictive rules discovered by EpiXCS in the five datasets are shown below. Qualitatively, these compare very favorably with those discovered by See5 on the same dataset. However, there were some differences between them. For example, the EpiXCS rulesets were as much as 20% larger, probably due to pruning in See5. In addition, rules in EpiXCS demonstrated more complex structure, in that they contained more conjuncts. However, these were not ordinarily superfluous, and they tended to provide increased coverage when applied to test data.

Breast: IF Uniformity of Cell Shape is 6.0 ± 4.0
                AND Single Epithelial Cell Size is 2 ± 1
                AND Mitoses is 3 ± 0
            THEN Malignant

Echo: IF Months-surviving is 30±265.
                AND Age-At-MI is 53±1.5
            THEN Died

Heart: IF Pain is Severe
                AND Exercise-Angina is Present
            THEN Died

Hepatitis: IF Ascites is Present
                AND Age>30
            THEN Hepatitis-Positive

Pima: IF Glucose is 170±27.0
                AND BMI is 39±16
                AND Age is 53±19
            THEN Diabetes-Positive

5 Future Work

Several key areas of future work remain in developing EpiXCS. First, it is currently restricted to two-class problems. While these are perhaps most common in medical data, there are situations where polytomous classification is used, such as in cancer staging, where there may be three, four, or five classes. Second, the performance of EpiXCS degrades with larger problems. For example, in testing EpiXCS on 56,000 records with 35 attributes in a database of fatal motor vehicle accidents occurring in the United States, the training time was too slow to make its use practical. Future work will focus on incorporating partitioning and boosting algorithms to improve training performance. Finally, the interface will continue to be a focus of research by testing the described version of EpiXCS on actual users to learn about more about these users' needs.

Acknowledgement

This project was funded by a University of Pennsylvania Research Foundation Grant.

References

1. Wilson, S.W.: Classifier fitness based on accuracy. Evolutionary Computation (1995) 3(2):149-175.
2. (http://xcslib.sourceforge.net/)

3. Mangasarian O.L. and Wolberg H.:   Cancer diagnosis via linear programming. SIAM News (1990) 23(5):1, 18.
4. (http://www.cs.waikato.ac.nz/~ml/weka/arff.html)
5. Wilson, S.W.: Get real! XCS with continuous-valued inputs. In: Lanzi, P. L., Stolzmann, W., and Wilson, S. W. (eds.): Learning Classifier Systems. From Foundations to Applications. Lecture Notes in Artificial Intelligence (LNAI-1813), Berlin: Springer-Verlag (2000), 209-222.
6. Holmes, JH, Lanzi, PL, Stolzmann W, and Wilson SW: Learning classifier systems: new models, successful applications. Information Processing Letters (2002) 82(1):23-30.
7. Hanley, JA; McNeil, BJ: The meaning and use of the area under a receiver operating characteristic (ROC) curve. Radiology (1982) 143:29-36.
8. Butz MV and Wilson SW: An algorithmic description of XCS.   In: Lanzi, P. L., Stolzmann, W., and Wilson, S. W. (eds.):, Advances in Learning Classifier Systems. Third International Workshop, Lecture Notes in Artificial Intelligence (LNAI-1996). Berlin: Springer-Verlag (2001), 211-230.
9. www.ics.uci.edu/~mlearn/MLRepository.html
10. Steven Salzberg, Johns Hopkins University, Baltimore, MD.
11. Detrano R. V.A. Medical Center, Long Beach and Cleveland Clinic Foundation.
12. Cestnik B: Jozef Stefan Institute, Ljubljana, Slovenia.
13. National Institute of Diabetes and Digestive and Kidney Diseases.
14. Rulequest Systems, www.rulequest.com

Author Index

Bacardit, Jaume   59, 282, 291
Baronti, Flavio   80
Bernadó-Mansilla, Ester   161
Bilker, Warren B.   181
Booker, Lashon B.   219
Bull, Larry   25, 308
Butz, Martin V.   104, 282, 291

Dabrowski, Grzegorz   17
Davis, Lawrence   258

Gao, Yang   93
Garrell, Josep Maria   59
Gérard, Pierre   144
Goldberg, David E.   40, 104, 291
Gu, Da-qian   93

Hamzeh, Ali   115
Holmes, John H.   181, 270, 333
Huang, Joshua Zhexue   93

Katai, Osamu   1

Landau, Samuel   144
Lanzi, Pier Luca   104
Llorà, Xavier   40

Marín-Blázquez, Javier G.   193

O'Hara, Toby   25
Orriols-Puig, Albert   161

Parmee, Ian   308
Passaro, Alessandro   80
Picault, Sébastien   144

Rahmani, Adel   115
Rong, Hongqiang   93

Sager, Jennifer A.   181, 333
Sastry, Kumara   40
Schulenburg, Sonia   193
Shimohara, Katsunori   1, 128
Sigaud, Olivier   144
Starita, Antonina   80

Takadama, Keiki   1, 128

Unold, Olgierd   17

Wada, Atsushi   1, 128
Wilson, Stewart W.   239
Wyatt, David   308

Vol. 4200: I.F.C. Smith (Ed.), Intelligent Computing in Engineering and Architecture. XIII, 692 pages. 2006.

Vol. 4198: O. Nasraoui, O. Zaïane, M. Spiliopoulou, B. Mobasher, B. Masand, P.S. Yu (Eds.), Advances in Web Mining and Web Usage Analysis. IX, 177 pages. 2006.

Vol. 4196: K. Fischer, I.J. Timm, E. André, N. Zhong (Eds.), Multiagent System Technologies. X, 185 pages. 2006.

Vol. 4188: P. Sojka, I. Kopeček, K. Pala (Eds.), Text, Speech and Dialogue. XV, 721 pages. 2006.

Vol. 4183: J. Euzenat, J. Domingue (Eds.), Artificial Intelligence: Methodology, Systems, and Applications. XIII, 291 pages. 2006.

Vol. 4180: M. Kohlhase, OMDoc – An Open Markup Format for Mathematical Documents [version 1.2]. XIX, 428 pages. 2006.

Vol. 4177: R. Marín, E. Onaindía, A. Bugarín, J. Santos (Eds.), Current Topics in Artificial Intelligence. XV, 482 pages. 2006.

Vol. 4160: M. Fisher, W. van der Hoek, B. Konev, A. Lisitsa (Eds.), Logics in Artificial Intelligence. XII, 516 pages. 2006.

Vol. 4155: O. Stock, M. Schaerf (Eds.), Reasoning, Action and Interaction in AI Theories and Systems. XVIII, 343 pages. 2006.

Vol. 4149: M. Klusch, M. Rovatsos, T.R. Payne (Eds.), Cooperative Information Agents X. XII, 477 pages. 2006.

Vol. 4140: J.S. Sichman, H. Coelho, S.O. Rezende (Eds.), Advances in Artificial Intelligence - IBERAMIA-SBIA 2006. XXIII, 635 pages. 2006.

Vol. 4139: T. Salakoski, F. Ginter, S. Pyysalo, T. Pahikkala (Eds.), Advances in Natural Language Processing. XVI, 771 pages. 2006.

Vol. 4133: J. Gratch, M. Young, R. Aylett, D. Ballin, P. Olivier (Eds.), Intelligent Virtual Agents. XIV, 472 pages. 2006.

Vol. 4130: U. Furbach, N. Shankar (Eds.), Automated Reasoning. XV, 680 pages. 2006.

Vol. 4120: J. Calmet, T. Ida, D. Wang (Eds.), Artificial Intelligence and Symbolic Computation. XIII, 269 pages. 2006.

Vol. 4118: Z. Despotovic, S. Joseph, C. Sartori (Eds.), Agents and Peer-to-Peer Computing. XIV, 173 pages. 2006.

Vol. 4114: D.-S. Huang, K. Li, G.W. Irwin (Eds.), Computational Intelligence, Part II. XXVII, 1337 pages. 2006.

Vol. 4108: J.M. Borwein, W.M. Farmer (Eds.), Mathematical Knowledge Management. VIII, 295 pages. 2006.

Vol. 4106: T.R. Roth-Berghofer, M.H. Göker, H.A. Güvenir (Eds.), Advances in Case-Based Reasoning. XIV, 566 pages. 2006.

Vol. 4099: Q. Yang, G. Webb (Eds.), PRICAI 2006: Trends in Artificial Intelligence. XXVIII, 1263 pages. 2006.

Vol. 4095: S. Nolfi, G. Baldassarre, R. Calabretta, J.C.T. Hallam, D. Marocco, J.-A. Meyer, O. Miglino, D. Parisi (Eds.), From Animals to Animats 9. XV, 869 pages. 2006.

Vol. 4093: X. Li, O.R. Zaïane, Z. Li (Eds.), Advanced Data Mining and Applications. XXI, 1110 pages. 2006.

Vol. 4092: J. Lang, F. Lin, J. Wang (Eds.), Knowledge Science, Engineering and Management. XV, 664 pages. 2006.

Vol. 4088: Z.-Z. Shi, R. Sadananda (Eds.), Agent Computing and Multi-Agent Systems. XVII, 827 pages. 2006.

Vol. 4087: F. Schwenker, S. Marinai (Eds.), Artificial Neural Networks in Pattern Recognition. IX, 299 pages. 2006.

Vol. 4068: H. Schärfe, P. Hitzler, P. Øhrstrøm (Eds.), Conceptual Structures: Inspiration and Application. XI, 455 pages. 2006.

Vol. 4065: P. Perner (Ed.), Advances in Data Mining. XI, 592 pages. 2006.

Vol. 4062: G.-Y. Wang, J.F. Peters, A. Skowron, Y. Yao (Eds.), Rough Sets and Knowledge Technology. XX, 810 pages. 2006.

Vol. 4049: S. Parsons, N. Maudet, P. Moraitis, I. Rahwan (Eds.), Argumentation in Multi-Agent Systems. XIV, 313 pages. 2006.

Vol. 4048: L. Goble, J.-J.C.. Meyer (Eds.), Deontic Logic and Artificial Normative Systems. X, 273 pages. 2006.

Vol. 4045: D. Barker-Plummer, R. Cox, N. Swoboda (Eds.), Diagrammatic Representation and Inference. XII, 301 pages. 2006.

Vol. 4031: M. Ali, R. Dapoigny (Eds.), Advances in Applied Artificial Intelligence. XXIII, 1353 pages. 2006.

Vol. 4029: L. Rutkowski, R. Tadeusiewicz, L.A. Zadeh, J.M. Zurada (Eds.), Artificial Intelligence and Soft Computing – ICAISC 2006. XXI, 1235 pages. 2006.

Vol. 4027: H.L. Larsen, G. Pasi, D. Ortiz-Arroyo, T. Andreasen, H. Christiansen (Eds.), Flexible Query Answering Systems. XVIII, 714 pages. 2006.

Vol. 4021: E. André, L. Dybkjær, W. Minker, H. Neumann, M. Weber (Eds.), Perception and Interactive Technologies. XI, 217 pages. 2006.

Vol. 4020: A. Bredenfeld, A. Jacoff, I. Noda, Y. Takahashi (Eds.), RoboCup 2005: Robot Soccer World Cup IX. XVII, 727 pages. 2006.

Vol. 4013: L. Lamontagne, M. Marchand (Eds.), Advances in Artificial Intelligence. XIII, 564 pages. 2006.

Vol. 4012: T. Washio, A. Sakurai, K. Nakajima, H. Takeda, S. Tojo, M. Yokoo (Eds.), New Frontiers in Artificial Intelligence. XIII, 484 pages. 2006.

Vol. 4008: J.C. Augusto, C.D. Nugent (Eds.), Designing Smart Homes. XI, 183 pages. 2006.

Vol. 4005: G. Lugosi, H.U. Simon (Eds.), Learning Theory. XI, 656 pages. 2006.

Vol. 4002: A. Yli-Jyrä, L. Karttunen, J. Karhumäki (Eds.), Finite-State Methods and Natural Language Processing. XIV, 312 pages. 2006.

Vol. 3978: B. Hnich, M. Carlsson, F. Fages, F. Rossi (Eds.), Recent Advances in Constraints. VIII, 179 pages. 2006.

Vol. 4200: I.F.C. Smith (Ed.), Intelligent Computing in Engineering and Architecture. XIII, 692 pages. 2006.

Vol. 4198: O. Nasraoui, O. Zaïane, M. Spiliopoulou, B. Mobasher, B. Masand, P.S. Yu (Eds.), Advances in Web Mining and Web Usage Analysis. IX, 177 pages. 2006.

Vol. 4196: K. Fischer, I.J. Timm, E. André, N. Zhong (Eds.), Multiagent System Technologies. X, 185 pages. 2006.

Vol. 4188: P. Sojka, I. Kopeček, K. Pala (Eds.), Text, Speech and Dialogue. XV, 721 pages. 2006.

Vol. 4183: J. Euzenat, J. Domingue (Eds.), Artificial Intelligence: Methodology, Systems, and Applications. XIII, 291 pages. 2006.

Vol. 4180: M. Kohlhase, OMDoc – An Open Markup Format for Mathematical Documents [version 1.2]. XIX, 428 pages. 2006.

Vol. 4177: R. Marín, E. Onaindía, A. Bugarín, J. Santos (Eds.), Current Topics in Artificial Intelligence. XV, 482 pages. 2006.

Vol. 4160: M. Fisher, W. van der Hoek, B. Konev, A. Lisitsa (Eds.), Logics in Artificial Intelligence. XII, 516 pages. 2006.

Vol. 4155: O. Stock, M. Schaerf (Eds.), Reasoning, Action and Interaction in AI Theories and Systems. XVIII, 343 pages. 2006.

Vol. 4149: M. Klusch, M. Rovatsos, T.R. Payne (Eds.), Cooperative Information Agents X. XII, 477 pages. 2006.

Vol. 4140: J.S. Sichman, H. Coelho, S.O. Rezende (Eds.), Advances in Artificial Intelligence - IBERAMIA-SBIA 2006. XXIII, 635 pages. 2006.

Vol. 4139: T. Salakoski, F. Ginter, S. Pyysalo, T. Pahikkala (Eds.), Advances in Natural Language Processing. XVI, 771 pages. 2006.

Vol. 4133: J. Gratch, M. Young, R. Aylett, D. Ballin, P. Olivier (Eds.), Intelligent Virtual Agents. XIV, 472 pages. 2006.

Vol. 4130: U. Furbach, N. Shankar (Eds.), Automated Reasoning. XV, 680 pages. 2006.

Vol. 4120: J. Calmet, T. Ida, D. Wang (Eds.), Artificial Intelligence and Symbolic Computation. XIII, 269 pages. 2006.

Vol. 4118: Z. Despotovic, S. Joseph, C. Sartori (Eds.), Agents and Peer-to-Peer Computing. XIV, 173 pages. 2006.

Vol. 4114: D.-S. Huang, K. Li, G.W. Irwin (Eds.), Computational Intelligence, Part II. XXVII, 1337 pages. 2006.

Vol. 4108: J.M. Borwein, W.M. Farmer (Eds.), Mathematical Knowledge Management. VIII, 295 pages. 2006.

Vol. 4106: T.R. Roth-Berghofer, M.H. Göker, H.A. Güvenir (Eds.), Advances in Case-Based Reasoning. XIV, 566 pages. 2006.

Vol. 4099: Q. Yang, G. Webb (Eds.), PRICAI 2006: Trends in Artificial Intelligence. XXVIII, 1263 pages. 2006.

Vol. 4095: S. Nolfi, G. Baldassarre, R. Calabretta, J.C.T. Hallam, D. Marocco, J.-A. Meyer, O. Miglino, D. Parisi (Eds.), From Animals to Animats 9. XV, 869 pages. 2006.

Vol. 4093: X. Li, O.R. Zaïane, Z. Li (Eds.), Advanced Data Mining and Applications. XXI, 1110 pages. 2006.

Vol. 4092: J. Lang, F. Lin, J. Wang (Eds.), Knowledge Science, Engineering and Management. XV, 664 pages. 2006.

Vol. 4088: Z.-Z. Shi, R. Sadananda (Eds.), Agent Computing and Multi-Agent Systems. XVII, 827 pages. 2006.

Vol. 4087: F. Schwenker, S. Marinai (Eds.), Artificial Neural Networks in Pattern Recognition. IX, 299 pages. 2006.

Vol. 4068: H. Schärfe, P. Hitzler, P. Øhrstrøm (Eds.), Conceptual Structures: Inspiration and Application. XI, 455 pages. 2006.

Vol. 4065: P. Perner (Ed.), Advances in Data Mining. XI, 592 pages. 2006.

Vol. 4062: G.-Y. Wang, J.F. Peters, A. Skowron, Y. Yao (Eds.), Rough Sets and Knowledge Technology. XX, 810 pages. 2006.

Vol. 4049: S. Parsons, N. Maudet, P. Moraitis, I. Rahwan (Eds.), Argumentation in Multi-Agent Systems. XIV, 313 pages. 2006.

Vol. 4048: L. Goble, J.-J.C.. Meyer (Eds.), Deontic Logic and Artificial Normative Systems. X, 273 pages. 2006.

Vol. 4045: D. Barker-Plummer, R. Cox, N. Swoboda (Eds.), Diagrammatic Representation and Inference. XII, 301 pages. 2006.

Vol. 4031: M. Ali, R. Dapoigny (Eds.), Advances in Applied Artificial Intelligence. XXIII, 1353 pages. 2006.

Vol. 4029: L. Rutkowski, R. Tadeusiewicz, L.A. Zadeh, J.M. Zurada (Eds.), Artificial Intelligence and Soft Computing – ICAISC 2006. XXI, 1235 pages. 2006.

Vol. 4027: H.L. Larsen, G. Pasi, D. Ortiz-Arroyo, T. Andreasen, H. Christiansen (Eds.), Flexible Query Answering Systems. XVIII, 714 pages. 2006.

Vol. 4021: E. André, L. Dybkjær, W. Minker, H. Neumann, M. Weber (Eds.), Perception and Interactive Technologies. XI, 217 pages. 2006.

Vol. 4020: A. Bredenfeld, A. Jacoff, I. Noda, Y. Takahashi (Eds.), RoboCup 2005: Robot Soccer World Cup IX. XVII, 727 pages. 2006.

Vol. 4013: L. Lamontagne, M. Marchand (Eds.), Advances in Artificial Intelligence. XIII, 564 pages. 2006.

Vol. 4012: T. Washio, A. Sakurai, K. Nakajima, H. Takeda, S. Tojo, M. Yokoo (Eds.), New Frontiers in Artificial Intelligence. XIII, 484 pages. 2006.

Vol. 4008: J.C. Augusto, C.D. Nugent (Eds.), Designing Smart Homes. XI, 183 pages. 2006.

Vol. 4005: G. Lugosi, H.U. Simon (Eds.), Learning Theory. XI, 656 pages. 2006.

Vol. 4002: A. Yli-Jyrä, L. Karttunen, J. Karhumäki (Eds.), Finite-State Methods and Natural Language Processing. XIV, 312 pages. 2006.

Vol. 3978: B. Hnich, M. Carlsson, F. Fages, F. Rossi (Eds.), Recent Advances in Constraints. VIII, 179 pages. 2006.